TRADITIONAL FISHING BOATS OF BRITAIN & IRELAND

Steam drifters crowding the quayside at Lowestoft during the Autumnal Herring Fishery.

TRADITIONAL FISHING BOATS OF BRITAIN & IRELAND

Design, History and Evolution

MIKE SMYLIE

AMBERLEY

To Angus Martin

Front cover: The fifie *Reaper*, owned by the Scottish Fisheries Museum at Anstruther, under full sail. This boat is the sole survivor of this size of vessel and is crewed by volunteers, who cope easily with its massive rig. (*Scottish Fisheries Museum*)

Back cover: The Yorkshire coble *Gratitude*, built by Hector Handyside in 1976. (*Dave Wharton*)

First published 1999 by Waterline Books

This softback edition 2011

Amberley Publishing
The Hill, Stroud
Gloucestershire GL5 4EP

www.amberleybooks.com

All photos are credited to the person known to hold copyright.
All others from the Mike Smylie Collection.
All illustrations by the author except where stated.

British Library Cataloguing in Publication Data.
A catalogue record for this book is available from the British Library.

ISBN 978-1-4456-0252-3

Typeset in 10pt on 12pt Sabon.
Typesetting and Origination by Amberley Publishing.
Printed in the UK.

Contents

Acknowledgements

As always, I presume, with a book like this, there are so many people to whom the author is indebted for their assistance. Robert Simper not only proof-read some chapters, but also produced a foreword on the basis of those few chapters. May the west wind blow once more so that you can smell my kippers again!

Thanks to the following for information, help with gathering it, or for hospitality during my sojourns around Britain: The Scottish Fishing Museum; Helen Prescott; Len Wilson; Dennis Davidson; Nathan Wilson; George Bergius; Nick Miller, who also proof-read; Nigel Dalziel; Len Lloyd; Tom Smith; Mike Clark; Robin Board; Edgar Readman and Brian Smith. Thanks also to Ernest Bristowe, now living in Australia, for sending me his memories.

My gratitude to those who supplied photographs without reproduction charge: Rudiger Bahr, who gave me access to his huge personal library of negatives; Michael Craine; Lancaster Museum; Earle Bloomfield; Len Wilson; The Swan Trust; The Excelsior Trust; The Scottish Fishing Museum; The Shetland Museum; Don Windley; Robert & Johnathan Simper; Angus Martin; Len Lloyd; Dennis Davidson; Donal MacPholin; The Guernsey Maritime Museum; The Tenby Museum & Art Gallery; Struan Cooper; Mike Clark; John Lord; Judy Brickhill; Pauline Oliver; Ernest Bristowe; Bryan Roberts; and various others who have lent me photos over the last ten years or so.

Thanks also to all those members of the 40+ Fishing Boat Association – the only UK association that concerns itself with *all* fishing boats of historical interest or importance – who have passed on information, to all those fishermen who have spent time talking to me about their memories of their vessels and to all those who visited my 'Herring Exhibition' over the last two summers, and parted with even the tiniest fragment of information. This, to me, is the 'information highway'! Thanks also to the various fishery offices, record offices and museums which have had to put up with my questions or investigations.

Some of what is in this book was first published in the *Boatman* magazine by Pete Greenfield, to whom I am grateful for the initial impetus to print. At their demise this role was continued by Nic Compton of *Classic Boat*, and even later, by *Watercraft*. Thanks to them.

And finally thanks to my family, Maria and Christoffer, who have endured endless hours of waiting in all corners of the country while I have 'agone researching'!

Mike Smylie

A Collection at Sea

To me
my fishing boat
is strength extreme.

Oak;
the power
that retains the sea,
its massive self in frames it be:
while diesel fumes and fish
that smell
linger in the grainy planking.

And well-worn deck, trod
by past fishermen's feet;
Carlins, too, are smoothed by ice,
by stormy seas, by those same feet; are moulded
into soft shape, kissed by yesterday's sun,
so that each time I descend the hatchway ladder,
they ooze the shadows of those before.

Winches lie idle, that before
did beat the drum
as nets hauled aboard with the silver darlings.

Wheelhouse, all small and warm at sea,
musty spokes on the Wheel,
damp charts linger under varnished shelf,
sliding panes of glass open to fresh air
as the heat from the J4 rises from below,
its throb a constant moan. Pumps
that throw water around,
as each moment the sea fights back against the
hard little ship.

Crew space is tight,
but tonight the coal stove is lit, adding its homely
glow
to these cell-like surroundings,
as orange flickers dance across the solid beams.

It's quiet tonight, the moon has dropped,
the proud skiff is silent
as the current rushes past,
rasping those eager planks, chewing at weed.
But soon, with anchor raised,
we'll away up the loch:

we'll away, boys, to sea.

Introduction

This book is primarily about fishing boats: it is not meant as a social history of Britain's fishing communities. Too many good books have been written, each concentrating on its particular area of the coast. Although one can quite rightly argue that the fishermen themselves were just as responsible for the development of their working craft as were the other important factors – type of fishing, geographical limits, wind and tide, nature of coastline, etc. – the reader must turn to the bibliography at the end to delve deeper into this aspect of 'the fishing'.

Nostalgia plays an important part in capturing people's desire to investigate what has gone before. I know of many who are repulsed by this, and believe that there is no place for this particular emotion, especially in fishing, where they feel it an insult to the memory of those who fought, day in, day out, at sea, under some of the worst conditions humans can expect to work in. Yet I disagree, as I find that in these modern days this is the only way to attract interest into these 'ways of old'. The same critics will, no doubt, say that there is no need for people to venture into the past. To them, I say, that we can only look forward by seeing back. Considering today's problems within the industry, for that is now what it is, it's a shame we didn't learn a bit more from our forebears, especially concerning greed.

However, it must never be forgotten that this nostalgia is only evoked upon the backs of those that suffered, and that it is usually those who fail to understand all the implications of the past that do the most evoking. There's not much sense of nostalgia among those within the industry to whom fishing is purely a way of life, and one with many opportune moments to reflect. Having said that, there are many fishermen that I meet and talk to, especially among the older generation, who longingly recall their times at sea, who speak beautifully of their own vessels that kept them safely alive for years, and who now, with a particular sadness in their eyes, speak of an industry whose degeneration is all too apparent these days. The natural hunter skills have been replaced with screens to stare at, and the vessels, once loved members of the family, are mere tools at the catcher's grasp. The spirit of fishing is disappearing fast. And, of course, the critics are pleased.

Quaysides lie idle to the hum of distant city life; fishermen join the dole queues as huge technologically superb boats sweep up all the fish. Amusement arcades sit where once barrels of herring were stacked ten high. Communities lying at the extremities are torn apart as the earners move out, and their houses are sought by the weekender, whose only contribution to the community is to arrive occasionally in his Volvo and complain about the insularness of the locals. The harbours contain their mass-produced fibreglass yachts, toys that seldom feel the joy of the ocean beneath their keels.

And what remains of this once proud industry? Turmoil as it fights with multi-nationals, European quotas, bureaucracy, throwing back good fish, decommissioning and other arguments that mostly stem from faceless people discussing the issues of which they do not really have any comprehension. We watch good wooden boats being chopped up and smashed in what can only be described as acts of legalised vandalism, sponsored by the state in the form of cash payments; and we wonder how it can be happening. So how does the industry react? It splinters into various groups, all negotiating with each other and hating each other at the same time. Meanwhile, the authorities pick and choose who they talk to, and the situation plummets to new lows.

Fishing boats were once as varied as the very shores they worked off; reflections, in fact, of the communities they used to serve. If this book has one purpose, it is to highlight their past importance among these coastal settlements so that we, mankind, may be brought to face our stupidity, and realise how, through progression, we may have forgotten how to live.

This page and overleaf: The Hastings fleet remains one of the few that still work off the sloping beach. The designs of the boats have not altered much since the days of sail, although the hulls are fuller since motors became adopted. Tractors are used to pull and push the mostly wooden craft up and down the beach, and wooden bearers are used as runners under the keels. Yet how long will these pretty fleets be able to survive under increasing European legislation that constantly threatens fishermen's livelihoods?

RX87
RX133

Foreword by Robert Simper

The regional work boats of the British Isles are a fascinating study. Each boat has slowly evolved to suit a different set of working conditions. There are no set rules as to how they evolved, but most work boats can be traced back to the eighteenth century and it is difficult to go much further. Most however did actually have their distant origins in the clinker double-ended boat from northern Europe. It is probable that the clinker-boat building traditions of England and Scotland originated from the early Anglo-Saxon boats, rather than the later Viking ships, although the daring Viking voyages sparked off northern European quests to sail to the furthest corners of the world.

The west and southern parts of the British Isles had more contact with the Latin countries and embraced carvel hulls far earlier than the work boats of the eastern coast of England and Scotland and the Northern Isles, but there are no set rules. Each boat type has to be considered against its own background.

The smacks, bawleys and sailing punts of the Suffolk and Essex coasts were the sailing fishing boats of the Victorian and Edwardian era. They reached their peak of development around 1890 and by 1914 engines were being fitted and no more new boats were being built. Many of these former sailing craft were fitted with engines and went on being used into the 1950s and 1960s, by which time people began to buy them and restore them back to sail. All the sailing work boats on the east coast have been saved by individual people making considerable sacrifices in their leisure time and income to keep them going

The Essex smacks are recognised by their low hulls with elegant counter sterns and straight stems. They have long bowsprits and a gaff cutter rig with a generous sail area and the Colchester and Maldon registration numbers of CK and MN. The prime use for the smaller smacks was oyster dredging in the Rivers Blackwater and Colne but the larger CK smacks over 40 feet long were used to fish all over the Thames Estuary. They trawled for flat fish and shrimp in the summer and went 'stowboating' for sprat in the winter.

The bawley was seldom used for oyster dredging and was primarily used for shrimping from Harwich and Leigh. They have the same straight stem, but the hull is very beamy and ends in a wide transom stern giving plenty of deck space. Some of the bawleys from Leigh had LO for London while others had the HH for Harwich.

The Suffolk beach punt was an open clinker boat, but had two masts and set a dipping lug rig. They worked off the beaches between Aldeburgh and Lowestoft and had the IH for Ipswich or LT for Lowestoft numbers. Their main 'harvest' was sprat in the autumn, but they fished all year round off the beaches.

Mike Smylie is just the person to analyse the evolution of the everyday work boats. Sailor, relentless maritime researcher and traveller, he is probably best known to the general public as the man smoking herrings at every maritime event from Penzance to the Western Isles. With almost missionary zeal Mike has campaigned to keep alive the art of smoking herring combined with the dual purpose of gathering information and perpetuating interest in the local fishing boats. He is the co-founder of the 40+ Fishing Boat Association and was the man that started the campaign to try and end the British Government's interpretation of the EEC decommissioning of fishing boats. Two British Governments have taken the view that boats withdrawn from fishing must be cut up and scrapped. Since the fishermen were paid far more than their elderly boats were worth on the open market, many historical regional craft have been destroyed needlessly.

The campaign Mike started led to some five historic craft being saved by 1998, so at least some good has come of it, but another ninety fishing boats have been cut up. This makes this type of book more important because they are preserving knowledge of our maritime past after the actual boats have gone. Perhaps we shall follow the French example and rebuild our maritime heritage with new boats.

Preface to the Second Edition

It is some fifteen years since I wrote the main text to this book and things have changed. A monumental change, in fact, that has seen the emergence of a strong voice of maritime matters of a historical flavour in both the UK and Europe.

Across the Continent the European Maritime Heritage body has been speaking out in defence of legislation from Brussels designed mostly to thwart heritage. Nationally in the UK it has been left to National Historic Ships under the chairmanship of Robert Prescott to attract government interest, although, as I write, their very existence is at risk because of 'the cuts'. Heritage Afloat has just formed a Welsh wing while the 40+ Fishing Boat Association quietly continues its work. Various local groups fight for recognition for vessels in their particular area. At last the establishment is waking up to the fact that much of our maritime heritage has disappeared, and if we are not careful, it will all go the same way.

But these are all official or semi-official bodies and far more important are the men and women who, through sheer determination, expense and passion, discover rotting vessels in situ in mud berths, corners of boatyards or redundant quays. These they nurture back to life and sail. Others might not quite achieve rebuilds but nevertheless restore, refit and revitalise aged vessels. Over these last fifteen years dozens more vessels from various walks of working boat life have joined the growing fleet. Many are fishing boats. However, although I've added a few in places, it hasn't been my intention to include all those restored vessels that fit into the compartments within the book. I make no apology for this. My desire has only been to correct the text from the various slight errors.

Although the acknowledgments of the first edition are current, I'd like to add a few names. Thanks are also due to Simon and Ann Cooper, Stephen Perham and Lynn Craine for their photographs, Robert Prescott for his faith back in St Andrews in 2002, and those folk who have, over the last decade, added to the accuracy of the original edition. Their notes and criticisms have been heeded in most cases though the drawings have not been altered. Finally I must thank Campbell McCutcheon and the team at Amberley for taking on this edition and the forthcoming follow-up to this book, 'Traditional Fishing Boats of Europe', that will follow in several months.

Mike Smylie, 21 October 2010

CHAPTER 1

The History of the Herring – *Clupea harengus*

Of all the fish in the sea, the herring is the king.

As a role-player in Britain's history, there is no doubt that 'king herring' surpasses all other fish, even 'king cod'. In fact it could be said that the herring has influenced political decisions within all countries that border the North Sea.

The herring belongs to the pelagic group of fish that includes the mackerel, tuna, anchovy, pilchard and sprat. Sardines and whitebait are often included as being pelagic, although the former is an immature pilchard and the latter usually the herring or sprat fry. Herring are found all over the world, the Pacific species being caught off Japan, China and North America, while the Atlantic herring is taken off the coast of North America between Labrador and Long Island, off Greenland, Iceland, the White Sea, the Baltic, Biscay as well as the North Sea and western coasts of Britain, Ireland and France. Together, they represent well over half of the fish caught within North European waters. They swim in huge shoals in the deep water of the ocean within these boundaries, only to swim into shallower waters to spawn. To do this, the female swims on her side depositing her eggs on the sand or gravel bottom and the male swims behind, fertilising as he goes. It is during this time, or when the herring feed off the plankton near the surface, that the fishermen have their greatest chance of prize catches, and this is always at night. The shoals are massive, sometimes 2 miles long, and they consist of millions of fish so tightly packed together that the porpoises swimming alongside can only peck at the outer edges rather than cause havoc amid the shoals. Only mankind, with his increasingly more effective ways of netting the shoals, has managed to have any impact on their numbers, and as a consequence these numbers are now well into decline.

The herring fishery of Great Britain has undoubtedly existed for centuries. Evidence of excavations shows that prehistoric men ate herring, as did the Romans. However, the first actual fishery, as against subsistence fishing, originated off the shores of East Anglia. The Annals of the monastery of Barking in Essex, founded in AD 670, inform us that a tax was levied upon herrings, and that this was known as the 'herring silver'. In the monastery of Evesham, founded in AD 709, there are references to the

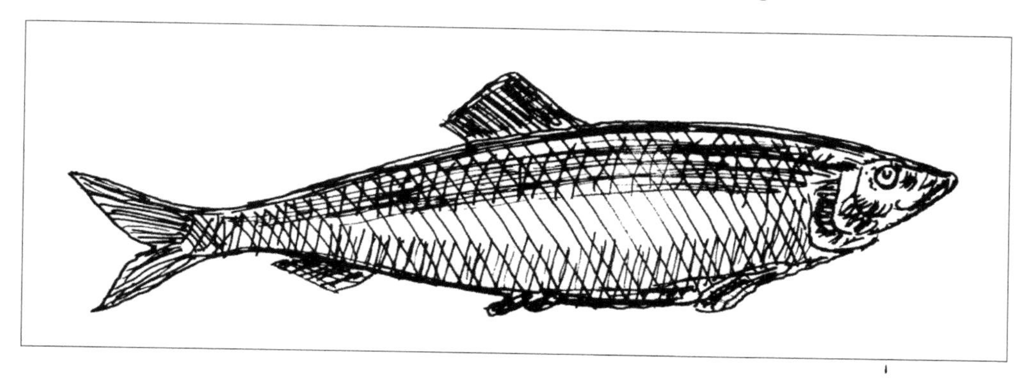

herring fishery on the west coast of Britain around that time. In the eleventh century 30,000 herrings were paid as an annual tribute to the Abbey of St Edmond. This was increased to 60,000 by William the Conqueror.

There is no doubt that the herring fishing was widely prosecuted in those days, and that herring was considered a good, nutritious source of food wherever it appeared. The plentiful herring in the Baltic led early in the twelfth century to the founding of a fishery on the southern Swedish coast, which then belonged to Denmark. Around this fishery grew the powerful Hanseatic League, controlled by the North German merchants. At its peak, around the year 1400, there were over 7,500 small boats fishing, and the merchants carefully controlled the catch, its subsequent curing and its distribution. Unfortunately for them, though, the herring mysteriously disappeared in about 1425, due to a change in rainfall and exceptionally strong tides, according to the Swedish oceanographer Otto Pettersson. It has been further suggested that these extremely high tides were caused by planetary conditions that only occur once every eighteen centuries. However, the power base of the merchants disappeared as quickly as the fish. The change in migration proved an advantage to the Dutch, English and Scottish because the herring simply moved into the wider waters of the North Sea.

In 1138, a charter was granted to the Abbey of Holyrood which included the right 'to fish herrings at Renfrew'. In 1163 large quantities of herring were caught in the River Meuse, and in 1199, Dunwich was created a free burgh on payment of 2,400 herrings and other items. In 1240, the Scottish government for the first time realised the huge scope for revenue-raising provided by herring and forthwith imposed a tax of fourpence on a last of herring (3,200 fish), a further penny per 1,000 on cured herring and another penny allowing them to be sold at market.

Herrings were a favourite food of all classes of society in the thirteenth and fourteenth centuries. In 1242, King Henry III purchased from the Yarmouth fishermen £50 worth of herring. In 1272, a ferryman near the Priory of St Olave in Suffolk was paid with bread and herring. In 1300, some 18,500 herrings were sent to the castle of Stirling and forty lasts of herring were brought from Yarmouth for the King and his household. In 1302, the same household received thirty-two lasts of herring at four marks per last, and the next year fifty lasts at the same price. In 1304, forty lasts were bought for 46*s* 8*d* per last, and the freight to London was 2*s* 9*d* per last.

Herring lassies at the farlane, gutting and packing fish on Scarborough quay. (*National Maritime Museum*)

Large landings of herring were made at Boston in 1284. In 1344, there were as many as 250 fishing boats in Great Yarmouth. In 1362, the burgesses of Yarmouth made an annual payment of 'one last of red-herrings, dried and well cleansed' to the free chapel of St George of Windsor, yet whether this was a gift as a sign of good fishing or as a fine imposed for the murder of a Cinque Port bailiff has never been determined. In 1383, large amounts of herring were caught off the French coast near Dieppe. In 1394, huge shoals were seen off Whitby.

The curing of herrings with salt was patented by a Yarmouth merchant, Peter Chivalier, in the fourteenth century. Later on in that century, in 1383, the Dutchman William Van Beukels found a superior method to preserve the fish. This method of curing them in barrels in salt, closely packed, remained the traditional method of curing into the twentieth century, and was largely the reason for Holland's dominance of the North Sea fishery.

Other fishermen from the Continent, as well as the Dutch, were fishing for herring around the coast of Shetland and the Outer Hebrides in the fourteenth and fifteenth centuries. At one time there were 2,200 Dutch busses in Bressay Sound. Queen Elizabeth I granted a petition to allow a number of Dutch fishing families to settle in England, provided they taught the Boston fishermen the Dutch method of catching, preparing and packing herrings. The Scots fleets received a boost when, in 1702, the French sank 400 Dutch fishing boats at Bressay, simultaneously dealing the Dutch fishery a death blow.

In 1707 massive shoals of herring appeared off the east coast of Scotland and, supposedly, an enormous amount of fish was chased ashore at Cromarty by 'whales, cods, dogfishes and other enemies', so that the beach was covered with them to a considerable depth, and 'salt and casks to cure and preserve them failed, and the residue was carted away for manure'.

Bounties were introduced in 1750 when each buss received 30*s* a ton to fish. To encourage smaller boats, this was reduced to 20*s* in 1785, and bounties of 2*s* 8*d* paid per barrel of white herring, 1*s* 9*d* of full red herring and 1*s* per barrel of empty herring. This system survived until 1830.

Communication by sea flourished in the eighteenth century, and cured herring was exported to many parts of the world. Whole armies were fed on it and the slaves in the West Indies survived mainly on it.

Off the Irish coast herring was plentiful though it wasn't until the late nineteenth century that this fishery developed into anything beyond subsistence except in parts of the north-west. Off the Donegal coast in the eighteenth century huge amounts were being taken and exported throughout Europe, though it does appear that the fishery was controlled by English merchants.

The West Indies market collapsed with the abolition of slavery in 1833. The Irish market suffered with the famine in the 1840s. Markets opened up in Russia, Germany and Scandinavia. Trawling commenced in earnest in the latter half of the nineteenth century. In 1880, there were 14,500 fishing boats in Scotland alone.

Although the advent of steam capstans allowed far greater catches to be handled, and overfishing became a genuine threat, the fishery continued to grow. The market then peaked in 1913, when some half a million tons of herring were landed in Britain. Yet the following year, disaster occurred in the shape of war, which saw the demise of the hugely important Russian and German markets, and which ultimately signalled the final decline of this huge herring fishery. Herring, however, is still fished today albeit in much less regal numbers that are controlled by quotas and the whims of Brussels. Fishing was banned totally for several years in the late 1970s/early 1980s to let stocks recover, but, in 1996, the same decline in stocks meant that the year's quota had to be cut in half. For those who have invested in bigger boats, the future looks grim. In the first decade of the twenty-first century stocks appear to have levelled off as fishing continues, though most of the catch comes from a small fleet of large pelagic trawlers. The days of fleets of small boats sailing out to earn their keep, and that of the crews, has long gone. Sadly much of the catch goes straight to processing factories to have its oil extracted and the protein mixed with grain for animal and fish feeds. Good-quality fresh herring can indeed be difficult to buy throughout the year and only a handful of localised coastal fisheries have survived, such as that each autumn at the Devon village of Clovelly.

CHAPTER 2

Early Fishing Boats

Herring – our goldmine in the North Sea.
—An old Dutch saying

HERRING BUSSES

Although the herring busses are the first documented type of fishing boat, the term appears more endemic to the Dutch fleets than the British. The term 'buss' originates from the time during the Viking era, when Harald Hardrada came north after seeing the ships of ancient Byzantium. He arrived in a buza-ship which had been built at Eyrar, and this vessel appears to have been the forerunner of what is now regarded as a buss – that sixteenth-century herring boat with which the Dutch commanded the North Sea fishery.

The fishermen of Holland were the first to exploit the lucrative North Sea herring fishing after the collapse of the herring in the Baltic after 1425. Although the Hanseatic towns had blossomed to create a potent stronghold through the strength alone of the herring, it disappeared as quickly as the herring did at their decline. The Dutch were keen to create a similar power base from any economic advantages the sea offered them. They were helped along in this desire by the invention of the drift-net in Hoorn in the fifteenth century, and later by Van Beukels, who in 1483 found a superior method by which the herring could be preserved using salt. The preservation of the catch was particularly important for the herring fishery as herring is a fish that deteriorates quickly once it has been landed, and the state of its freshness is always obvious by the brightness, or dullness, of its eyes.

In 1633 there were reportedly 3,000 Dutch busses working in the North Sea, and the herring were cured aboard each boat. Alongside the busses were the 'jagers', fast sloops that loaded up the cured herring and sailed it back home. Yet by 1779 there were just 162 of these busses left.

The Dutch busses were large, bluff-bowed vessels with a great amount of tumble-home, symbolising Dutch designs of the time. A typical buss of the seventeenth century was a square-sailed three-masted craft about 50 feet in overall length, 16 feet

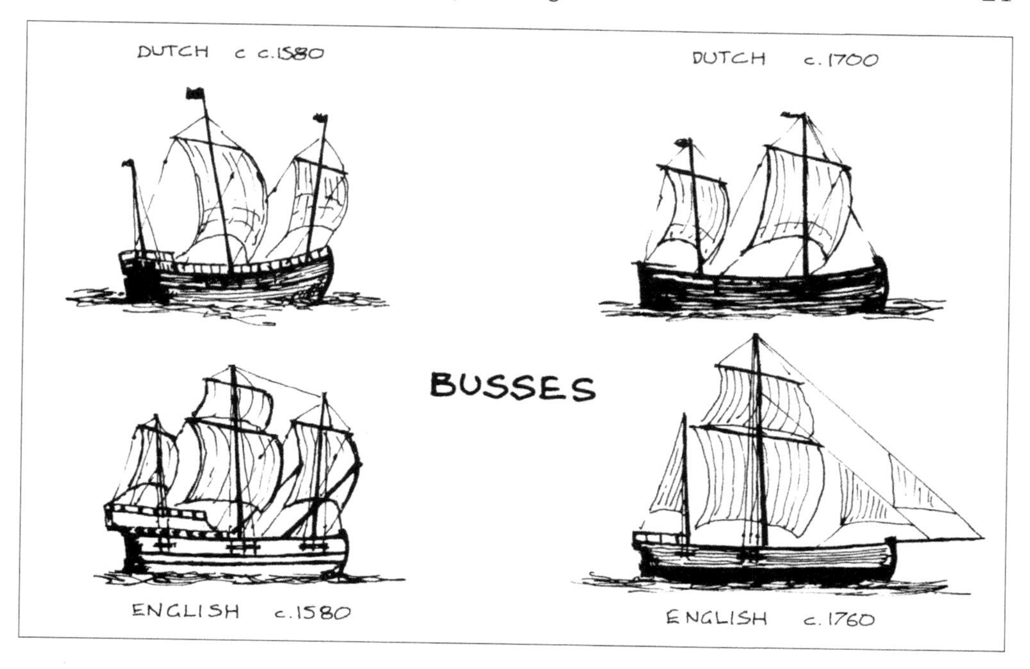

on the beam, and cost about £300 complete. They were heavily built to enable them to lie to their nets in the treacherous conditions of the North Sea. Good quantities of boatbuilding timber were not in short supply in those days!

At the beginning of the eighteenth century Scotland thought it, too, would like to obtain some of the benefits of this North Sea fishery, especially as the Dutch were fishing close to the Scottish shore. King James I tried to levy a tax of one barrel of herring upon each Dutch buss working in Scottish waters, but this was impractical as no one was quite sure what the limits were. The Dutch treated it with contempt and continued as before. Similarly the English set a 14-mile limit, and warships were sent out to fire at the Dutch, who then agreed to pay a 10 per cent commission to the Crown, although it seems not a penny was ever paid. The only remaining course both for the English and the Scots, and the most sensible it seems, was to catch the herring themselves. So two busses were built on the Thames, *Pelham* and *Carteret*, and launched in 1750. These two vessels were 70 tons burthen each, cost about £500, and were crewed by seventeen men, nine of whom were curing the fish. It appears that most of the crews were Dutch. Three years later there were thirty-eight English busses fishing from Bressay Sound, alongside the Scots.

These English busses were entirely different from the Dutch craft. Although the Dutch fishery was much larger than any other, the men of East Anglia had been at it much longer – it is said that East Anglia is home to the oldest known herring fishery in the world at more than 1,200 years. The English busses, therefore, were East Anglian in design and experience, and were developed from the Humber keels, the great square-sailed vessels that operated in the confined canals and rivers, and along much of the coast of that area. Consequently the English busses were more barge-like and not as heavy as their Dutch counterparts.

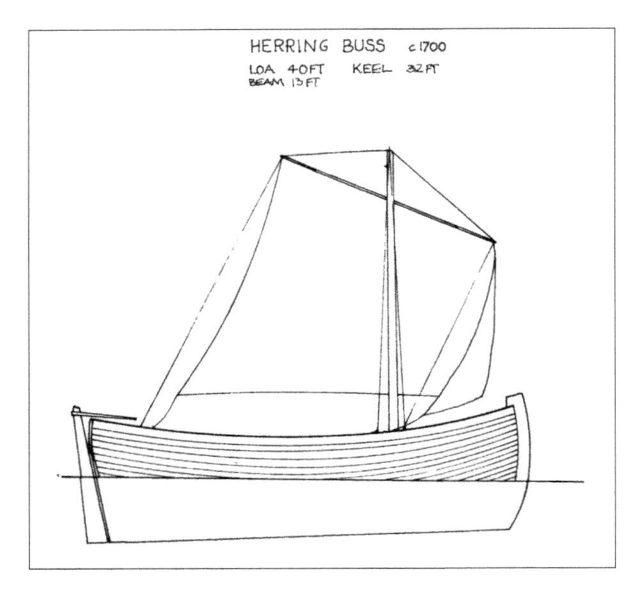

The Scots began to build busses within a decade of the English ones, and these were about 50 tons displacement. Many of these craft were built at Campbeltown on the peninsula of Kintyre on the west coast. They were crewed by local fishermen, sailing north to fish for several months at a time off the Outer Hebrides or the north-east coast. Wages were £1 5*s*–£1 16*s* per month, although sometimes the boats returned only half full at the end of the season. In 1785 there were fifty such vessels – total 3,005 tons – working from Campbeltown, and they employed 675 men. An average of 7,500 barrels were landed per year. Campbeltown was regarded as having 'one of the best harbours in the world as their pride and boast'. Other than Campbeltown, Greenock was the only other west coast Scottish port where, with the introduction of a bounty of 30*s* per ton of boat fishing in 1750, payments were made. In 1791, no fewer than 129 similar busses worked from Greenock, 53,488.5 barrels being landed that year.

The bounty system was introduced specifically to encourage the buss fishery. As well as a tonnage bounty, a bounty of 2*s* 8*d* was paid on each barrel of herring landed. Because of this, many people in Campbeltown had 'on speculation built many expensive houses of stone, lime and slate'. When the bounties ceased in 1829, these properties fell in value, a sign of the importance of the herring fishery to the local economy at that time.

Wick, in 1790, had thirty-two vessels 'on the bounty', measuring 1,610 tons in total. A year later there were forty-four such boats. These busses were also three-masted, and were typically 58 feet overall, 47 feet on the keel, 15 feet on the beam and 7 feet 6 inches depth of hold. Like the Dutch, the fishermen lowered the foremast and mainmast to rest in a crutch while they were riding their nets. They carried a crew of twelve men.

The English busses landed most of the herring at Great Yarmouth alongside Dutch and French busses. In 1751 there were 250 Dutch boats, 120 French, 120 smaller Dutch boats and ninety-five Yorkshire boats working from there, with only sixty-nine local busses.

THE THREE-MASTED LUGGERS

During the latter part of the eighteenth century, luggers were developed from the older square-sailed vessels. This change to fore and aft rig coincided with the Napoleonic

A Yarmouth three-masted lugger YH181, *c.* 1830. (*Great Yarmouth Museum*)

Wars, which completely wiped out the Dutch herring fishery. Unlike the British fishery (which also suffered), the Dutch never recovered. Although the busses continued to work offshore, the luggers, to begin with, worked inshore, gradually sailing further in their search for the shoals. The early luggers were similar in shape to the busses, but over the next century their shape changed drastically to become more akin to that better known these days.

The three-masted luggers from Yorkshire were called 'Five-man' boats, referring to the five joint owners that worked them, always employing another three paid hands. They were less bluff-looking than the busses, were about 56 tons and 60 feet overall, and cost about £600 by the turn of the century. Running costs for nets and general fitting out over a year were said to be about £100. They were all built in Yorkshire, and were

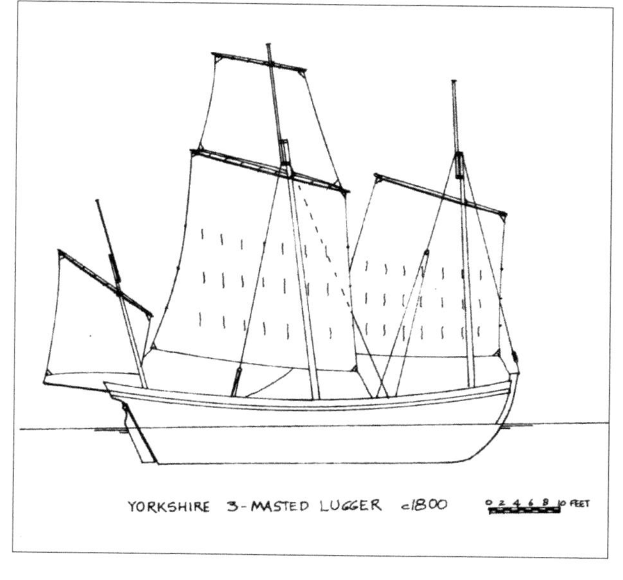

clinker-planked and fully decked. Mostly they drift-netted for herring, only occasionally using a trawl.

Similar three-masted luggers were to be found in East Anglia and all along the south coast of England. The influence for these vessels seems to have originated from the French *Chasse-Marée* – the great three-masted *lougres* (luggers) that sailed across the English Channel, often causing havoc among the inshore fleets as they were supposedly crewed by 'pirates and vagabonds'. The Cornishmen especially were so impressed with these vessels that they copied them and built themselves lug-rigged boats. At this time, around the end of the eighteenth century, the deepwater fishing fleets were travelling great distances in search of fish. The east coast boats sailed down to Devon and Cornwall in the New Year for the mackerel season, which could last until June; the Irish and Welsh waters provided rich pickings during the summer after the mackerel; east coast boats then returned home for the herring between June and November. In this way, the three-masted boats gained preference on the east coast. Furthermore, the first documented Lowestoft boat sailed to Scotland in 1776, though the first Scottish boat did not come to Lowestoft until 1863. But when the Scots did eventually arrive, they came in numbers, so that, by the end of the century, Scottish boats outnumbered East Anglian boats by three to one.

The Great Yarmouth luggers were clinker-built, fully decked and lay low in the water, whereas the Cornish ones were carvel-built. They carried three lugsails, topsails, mizzen, foresail and jib, and cost nearly £1,000. They usually left their mainmast ashore in winter and hundreds of these vessels were based in the harbour there,

A print drawn by T. Dighton showing a three-masted vessel at Eastbourne, 1829.

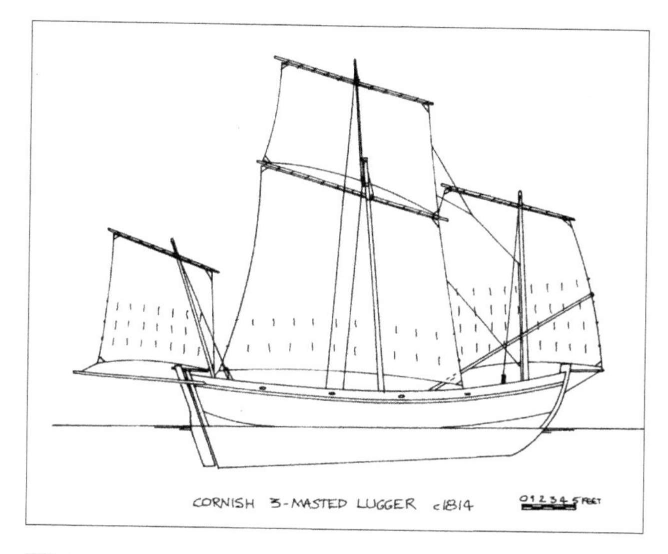

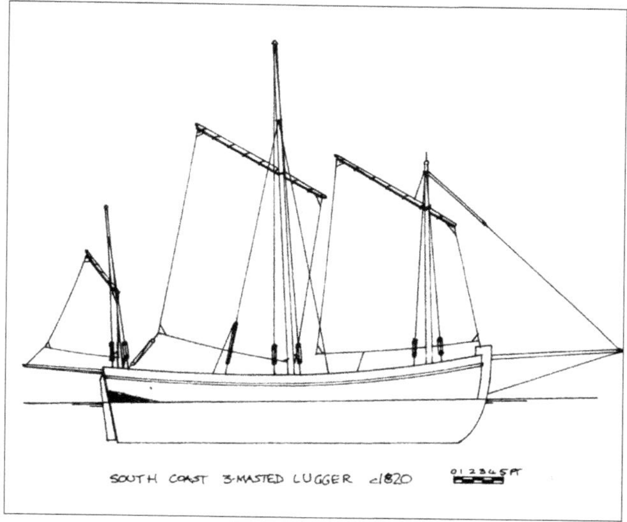

and at Lowestoft after 1830, for the great autumn herring fishery. Unlike the Yorkshire Five-man boats, but similar to the French and the Cornish craft, these luggers used a bowsprit. A typical lugger was 61 feet overall, 44 feet on the keel, 19 feet 6 inches on the beam and displaced some 90 tons. Their shape became finer and less upright – possibly a French influence, or from the fact that many fishermen crewed the powerful, less clumsy naval ships during the frequent wars. The square-headed lugsails, similar in shape to the earlier square sails, became higher-peaked at the same time.

During the 1830s, the three-masted boats all lost the mainmast as the crews soon learnt to leave this ashore, it being superfluous to their needs while fishing. Its removal created more deck space, a point they soon appreciated.

By 1840 the three-masted luggers had all but disappeared, this development amounting to the final stage of the evolution of sailing fishing boats prior to the arrival of steam, and subsequently the internal combustion engine. Competition among the fishing fleets grew, and efficiency became paramount, and, consequently, the new two-masted luggers probably were the most effective of all the sailing fishing boats.

THE VIKING INFLUENCE

On the northern and western coasts of Britain, and most of the Irish coast, the story was different. Here the bond formed during the great Viking era from AD 800 to 1100 continued to influence life, especially among the more outlying parts of these inaccessible coasts. The lines of communication remained poor, or even non-existent,

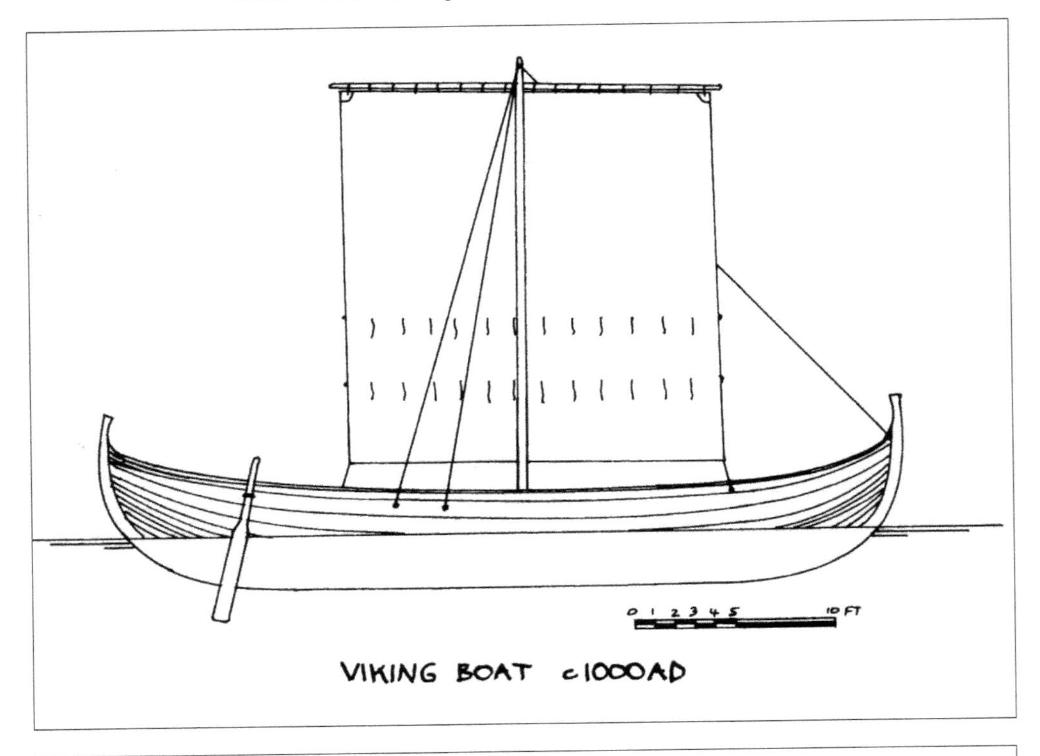
0 1 2 3 4 5 10 FT
VIKING BOAT c1000AD

IMPRESSION OF EARLY 18TH CENTURY YAWL
EAST COAST SCOTLAND
0 1 2 3 4 5 FEET

until well into the nineteenth century. Nearly all the trade was done by sea, and many trading links continued to flourish between these areas and Scandinavia. This produced boats that were reminiscent of those Viking boats of many centuries ago, often the boats themselves coming direct rather than in the form of the ideas. Boatbuilding timber was also imported into these areas direct from Scandinavia.

This then forms the basis of fishing boat design from Devon right round the coast clockwise to Norfolk and the Thames, as well as to parts of Sussex. Along with the other two distinct influences – that of the Dutch and the French – these together form the roots of all British fishing boat design. The boats themselves have evolved through generations of use into working tools fit for the local conditions they were originally built to work and fish in. Today, most small craft have origins in the fishing boats of yesteryear.

THE WASHINGTON REPORT OF 1849

The gale that struck the north-east coast of Scotland on the night of 19 August 1848, although disastrous in its consequences upon the fleets, was fortuitous for historians one hundred and fifty years later in that it now gives a good insight into the boats of that time. Wick, situated high up on the Caithness coast, was the herring capital of Europe where some 800 boats employed 3,500 fishermen in the height of the season.

As was usual, the Wick fleet sailed out at dusk on that particular night, as did the fleets from the chain of harbours and beach landings strung all along the east coast. The gale that hit in the early hours of that morning was more than typically ferocious and, fortunately, some boats had recognised the signs and had already hauled in their nets and run for shelter into the harbour at high water just after midnight. Those left out who failed to see the signs of the approaching storm, or those that chose to ignore them, were stuck out in the bay when the full force of the wind hit them, although they sailed in towards the lee of the land. But, on a dropping tide, only five feet of water covered the harbour entrance, not enough to enter in such vicious conditions. Some did try only to be swamped or smashed against the quay, and their families could only stare from the shore as forty-one boats were destroyed and twenty-five fishermen drowned. Twelve more were lost in deeper water when they were submerged by the tremendous waves that were crashing and foaming all around the bay. The story was the same along the coast. In all, 100 lives were lost and 124 boats either wrecked or severely damaged.

Captain John Washington, RN, was ordered by the House of Commons to proceed directly north and to undertake a thorough investigation of the causes of this horror and to make specific suggestions as to possible legislation to prevent a similar occurrence.

Washington took statements from many witnesses to that night – fishermen, fishery officers, harbourmasters and others involved in the fishery. From Wick, where he began, he travelled south along the coast listening to evidence everywhere he went. In his report, as well as making a number of criticisms concerning the lack of proper harbours with access at all states of the tide, the general lack of lighting, the practice of engaging hired crews from among the landsfolk and the practice of the fish curers paying the fishermen in whisky, his main criticism was directed at the designs of the boats themselves.

He summed this up by saying that the boats would have had a much better chance of staying out at sea in those conditions if they were fully decked, that it was obviously not the case with the open boats and that most of the fatalities occurred for this reason. He suggested that all boats should be decked, either wholly or partly.

In his report he produced plans of many boats from around Britain, and it is these that give us an excellent picture of the type of vessel favoured by fishermen in these areas. He also produced plans of two proposed fishing boats drawn up by the renowned naval architect James Peake from the Royal Dockyard at Woolwich.

The boats that he studied were as follows:

Penzance boat
St Ives boat
Isle of Man Nickie
Kinsale Hooker
Galway Hooker
Loch Fyne or Fairlie boat*
Loch Fyne or Rothesay boat*
Wick herring boat
Buckie herring boat
Peterhead boat
Fraserburgh boat
Aberdeen boat
Newhaven boat
Dunbar herring boat*
Scarborough Five-man boat*
Scarborough yawl*
Yarmouth lugger
Yarmouth punt
Lowestoft lugger
Deal lugger
Hastings lugger
Proposed boat 1
Proposed boat 2
Proposed boat 3*
Proposed boat 4*

*Not produced in plan form, although comments are made as to their suitability.

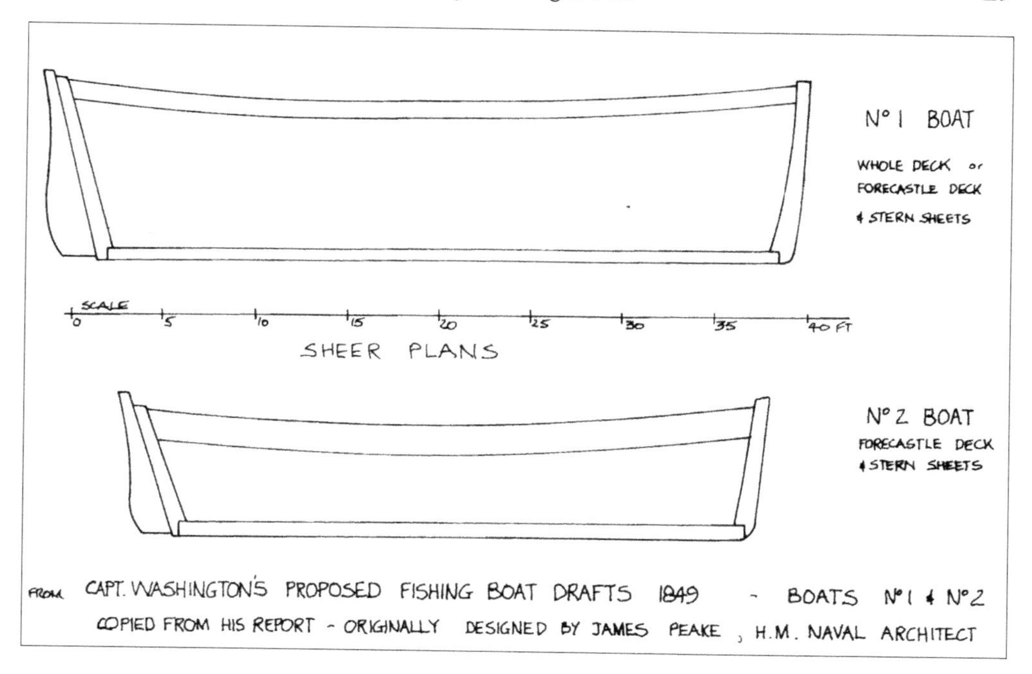
N° 1 BOAT
WHOLE DECK or FORECASTLE DECK & STERN SHEETS
SCALE
0 5 10 15 20 25 30 35 40 FT
SHEER PLANS
N° 2 BOAT
FORECASTLE DECK & STERN SHEETS
FROM CAPT. WASHINGTON'S PROPOSED FISHING BOAT DRAFTS 1849 - BOATS N°1 & N°2
COPIED FROM HIS REPORT - ORIGINALLY DESIGNED BY JAMES PEAKE, H.M. NAVAL ARCHITECT

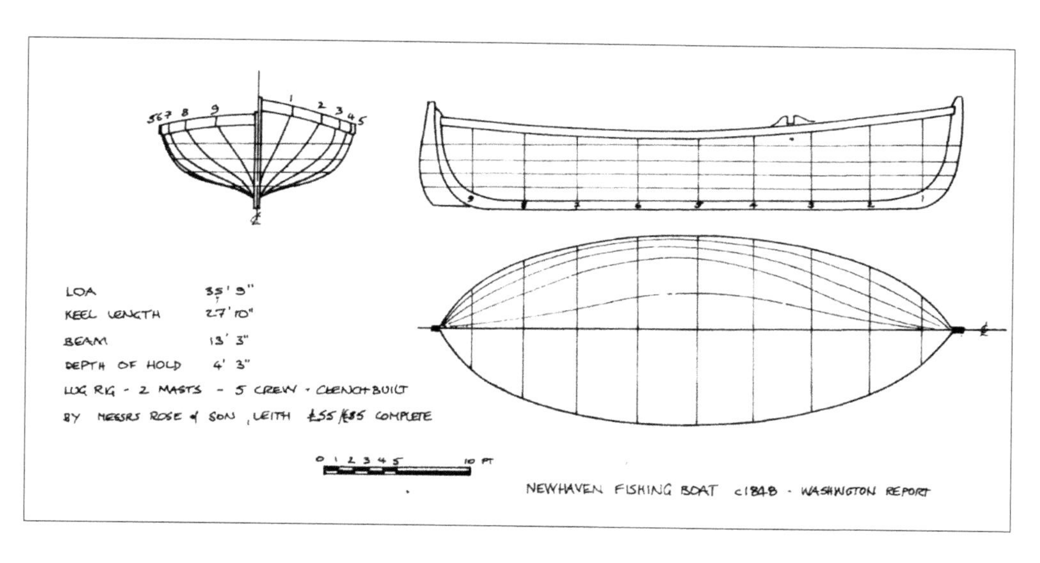
LOA 35' 9"
KEEL LENGTH 27' 10"
BEAM 13' 3"
DEPTH OF HOLD 4' 3"
LUG RIG - 2 MASTS - 5 CREW - CLENCH BUILT
BY MESSRS ROSE & SON, LEITH £55/£85 COMPLETE
0 1 2 3 4 5 10 FT
NEWHAVEN FISHING BOAT c1848 - WASHINGTON REPORT

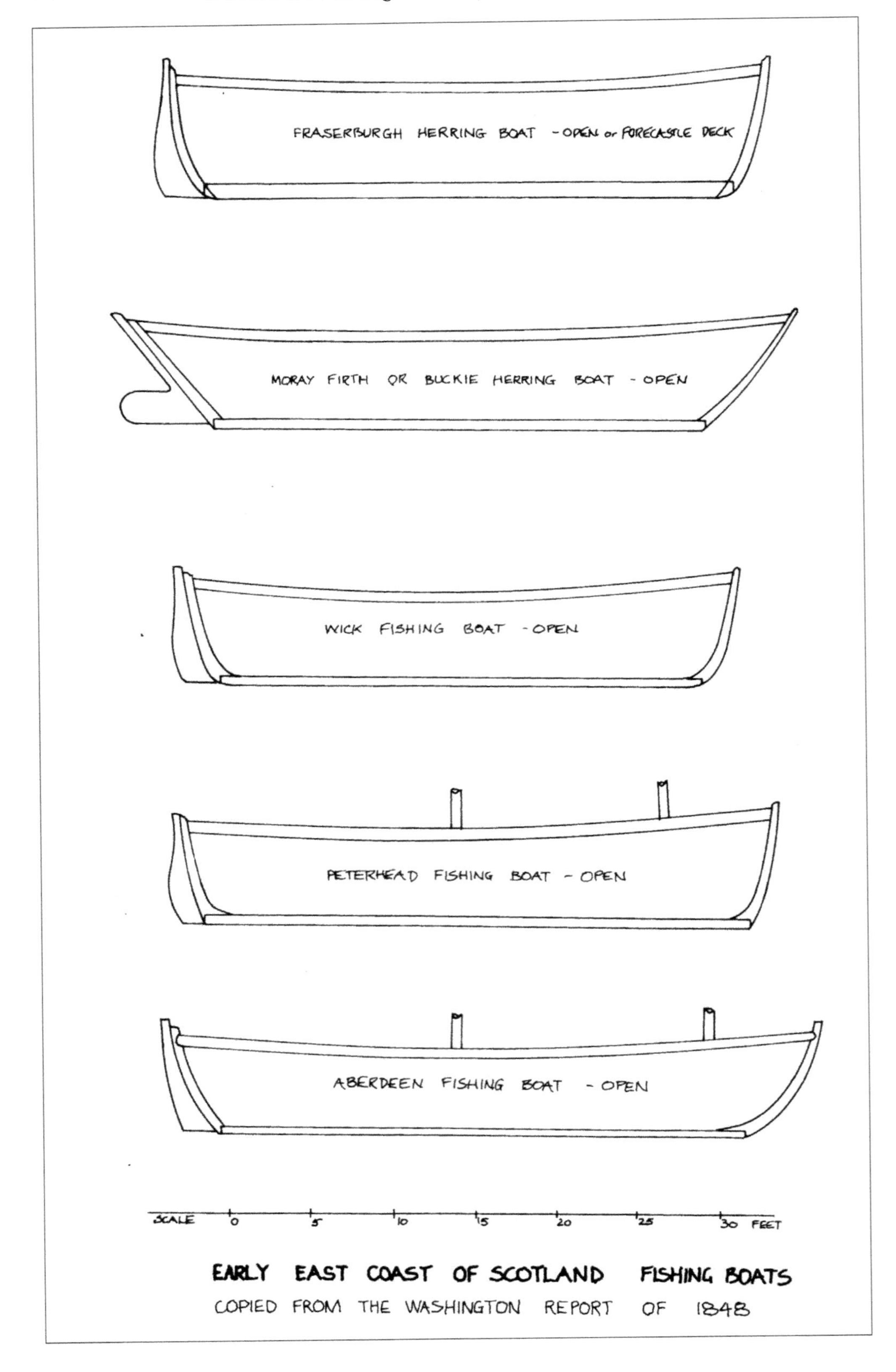

EARLY EAST COAST OF SCOTLAND FISHING BOATS
COPIED FROM THE WASHINGTON REPORT OF 1848

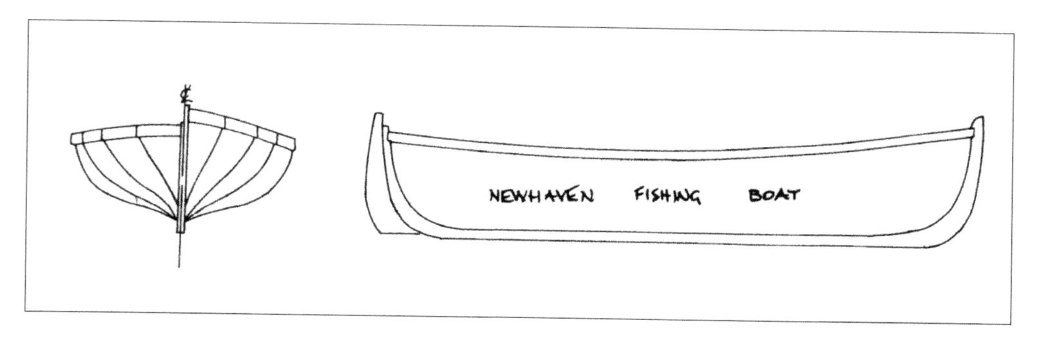

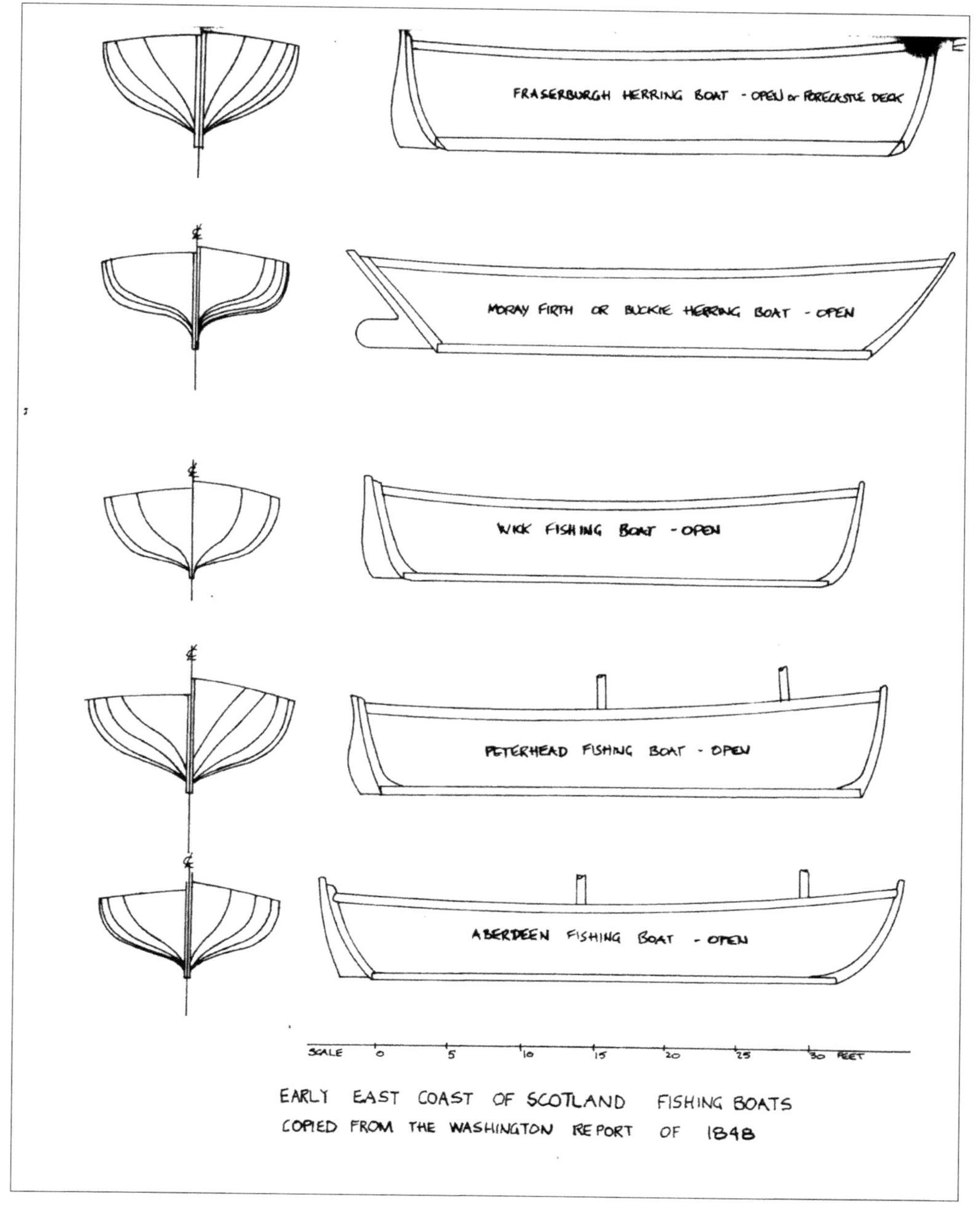

EARLY EAST COAST OF SCOTLAND FISHING BOATS
COPIED FROM THE WASHINGTON REPORT OF 1848

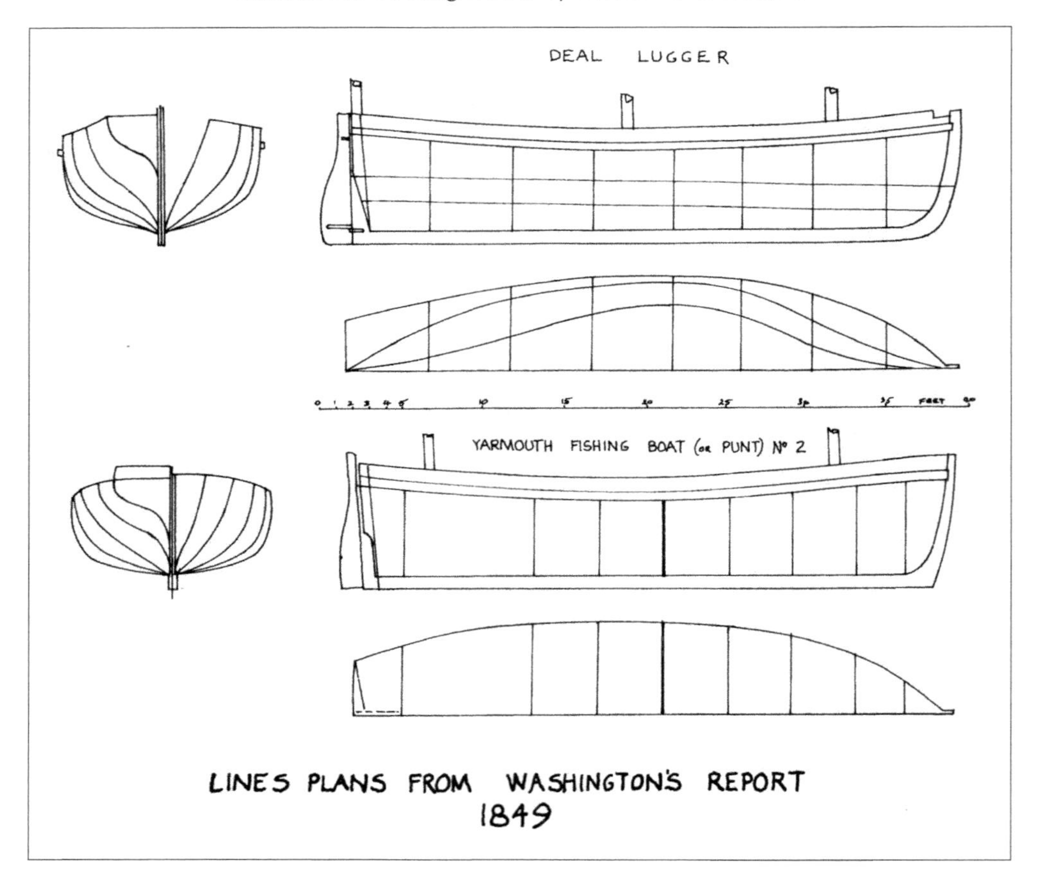
DEAL LUGGER
0 1 2 3 4 5 10 15 20 25 30 35 FEET 40
YARMOUTH FISHING BOAT (or PUNT) No 2
LINES PLANS FROM WASHINGTON'S REPORT
1849

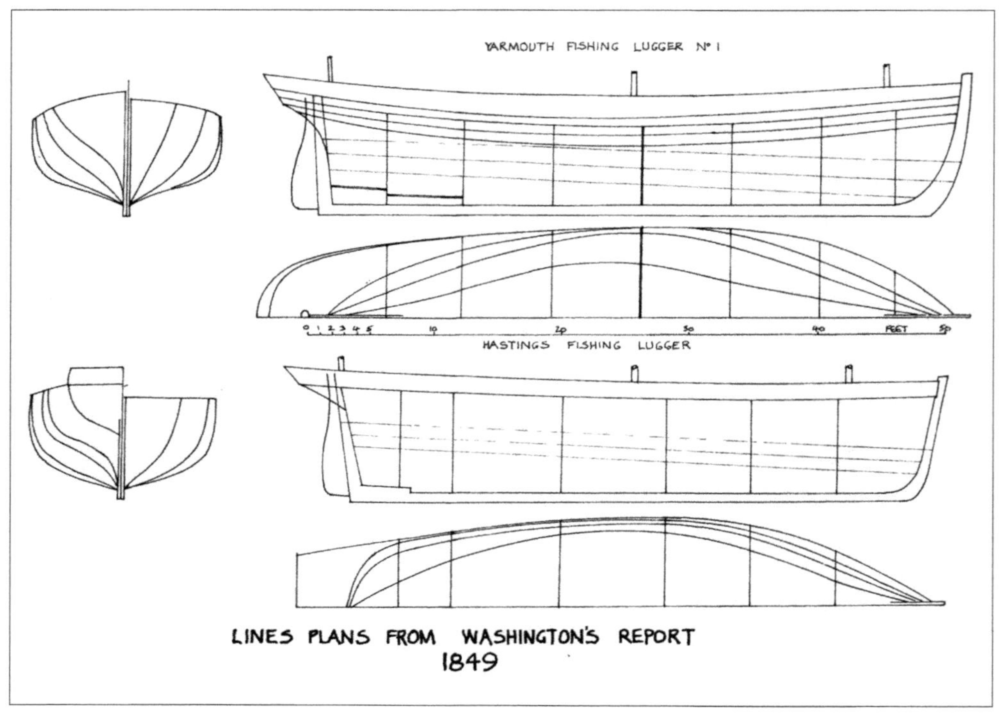
YARMOUTH FISHING LUGGER No 1
0 1 2 3 4 5 10 20 30 40 FEET 50
HASTINGS FISHING LUGGER
LINES PLANS FROM WASHINGTON'S REPORT
1849

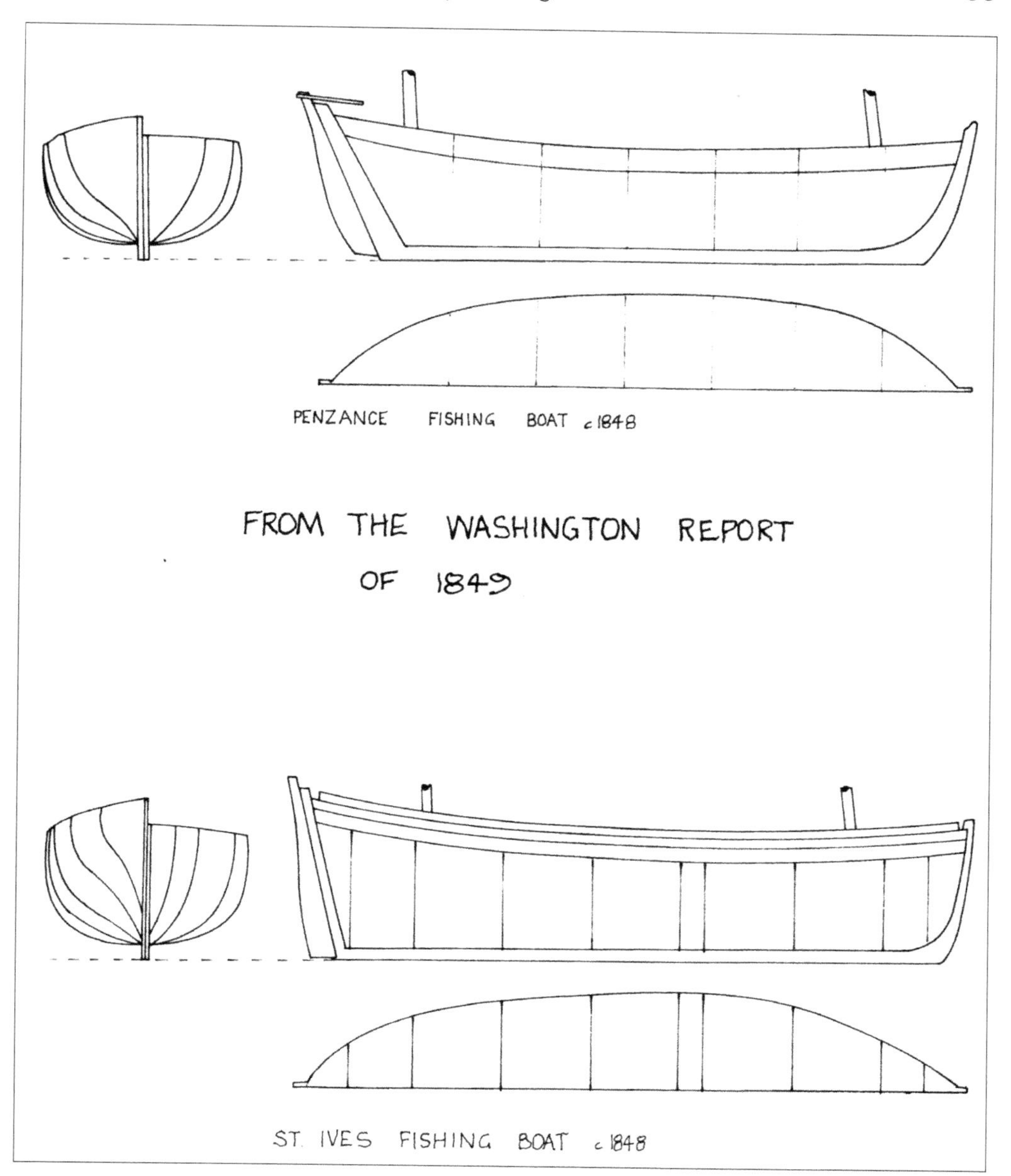
PENZANCE FISHING BOAT c 1848
FROM THE WASHINGTON REPORT OF 1849
ST. IVES FISHING BOAT c 1848

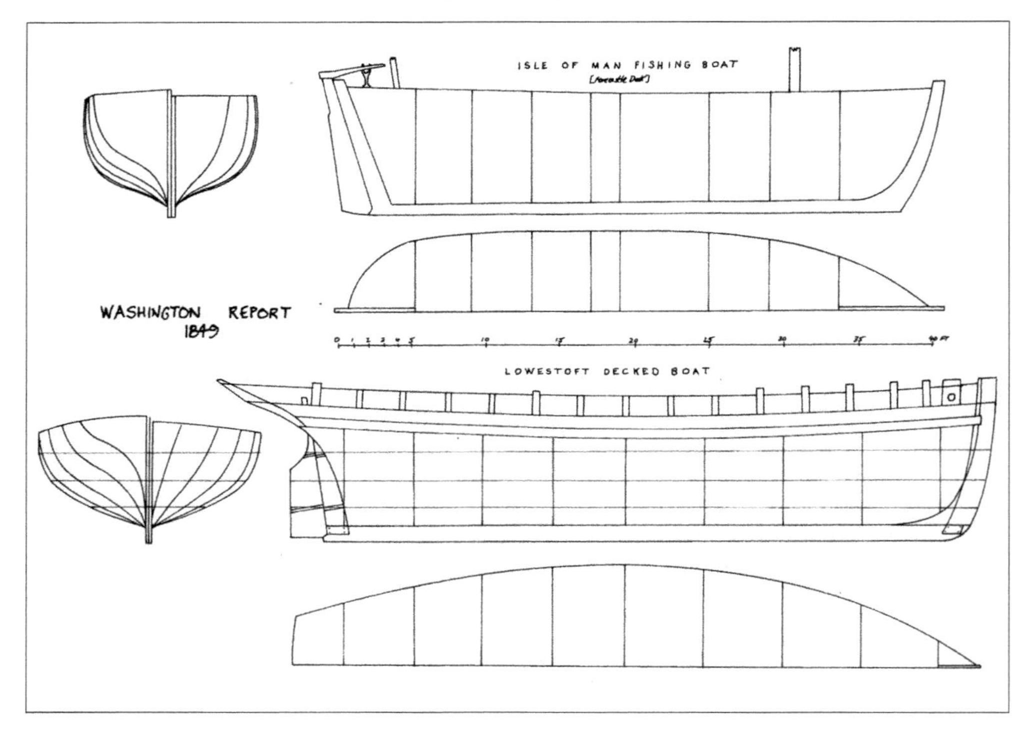
ISLE OF MAN FISHING BOAT
[Forecastle Deck]
WASHINGTON REPORT
1849
0 1 2 3 4 5 10 15 20 25 30 35 40 FT
LOWESTOFT DECKED BOAT

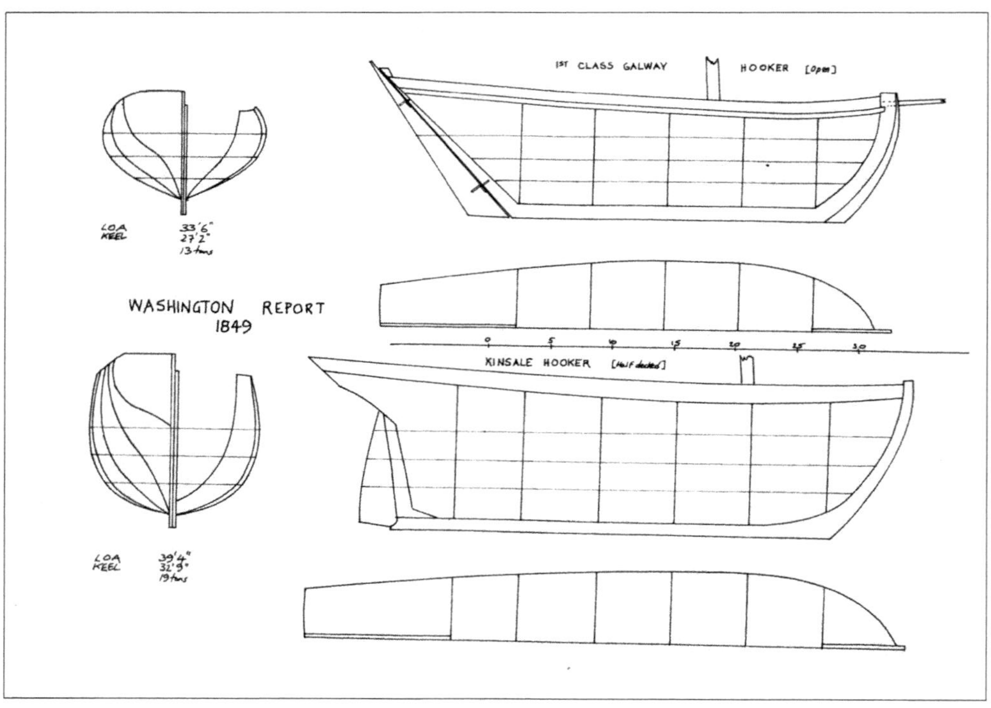
1ST CLASS GALWAY HOOKER [Open]
LOA 33'6"
KEEL 27'2"
13 tons
WASHINGTON REPORT
1849
0 5 10 15 20 25 30
KINSALE HOOKER [Half decked]
LOA 39'4"
KEEL 32'9"
19 tons

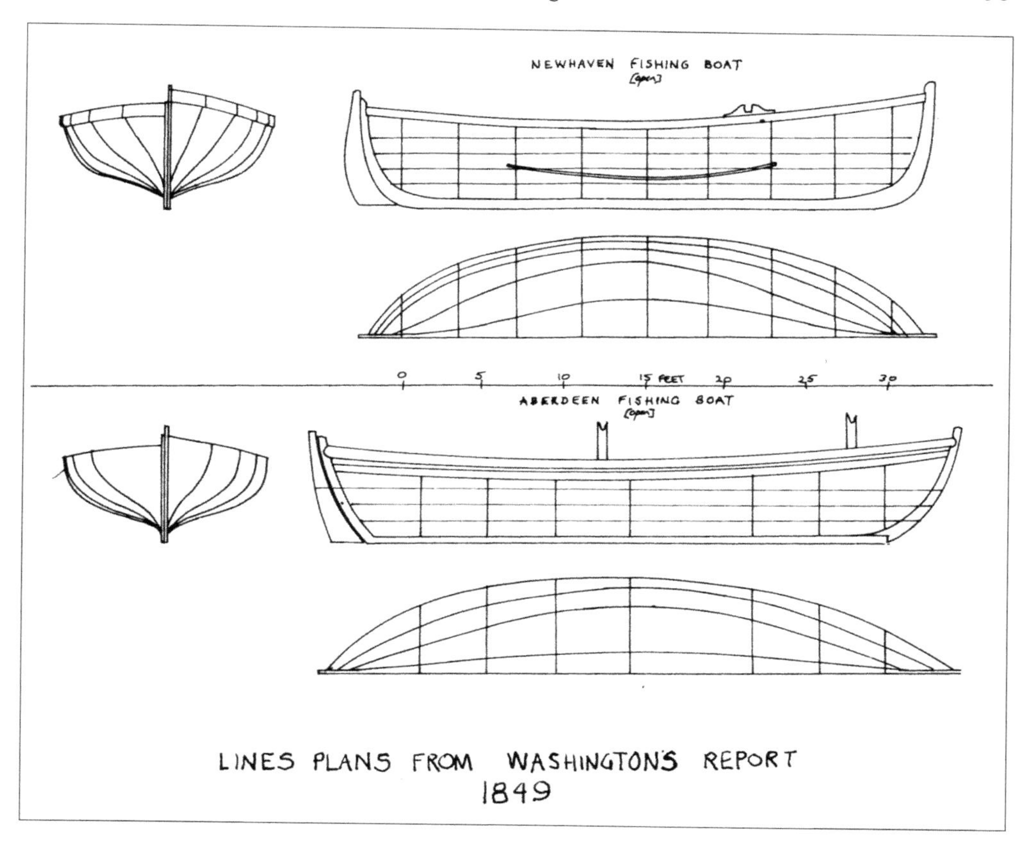
NEWHAVEN FISHING BOAT
[open]
0 5 10 15 FEET 20 25 30
ABERDEEN FISHING BOAT
[open]
LINES PLANS FROM WASHINGTON'S REPORT
1849

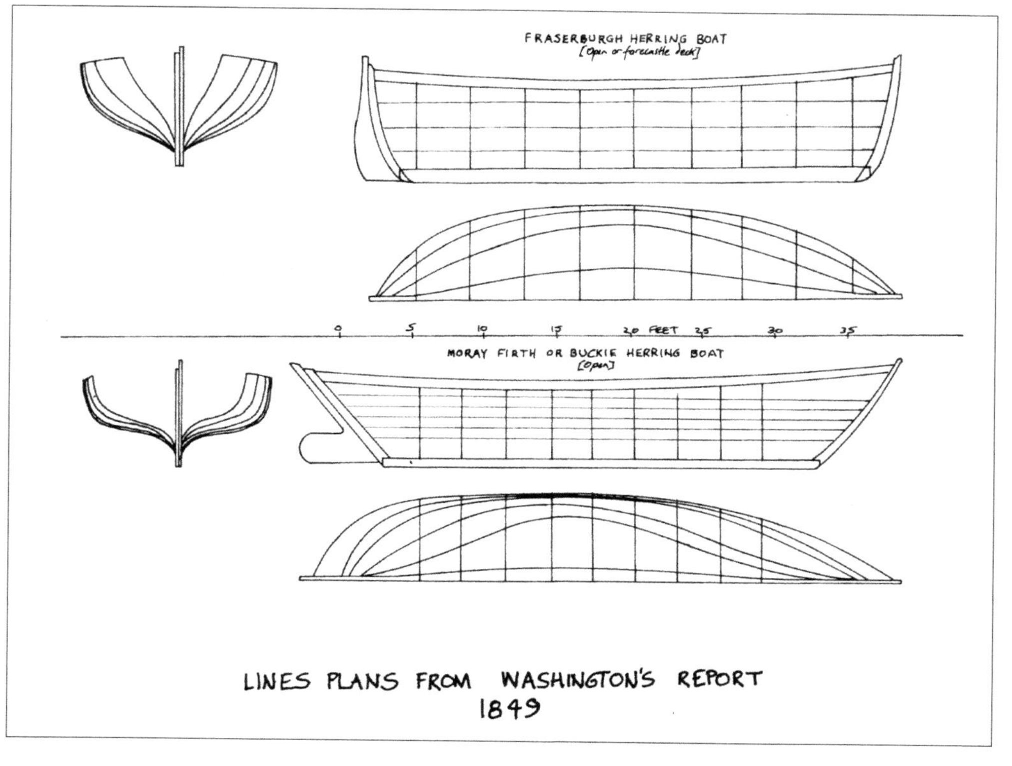
FRASERBURGH HERRING BOAT
[Open or forecastle deck]
0 5 10 15 20 FEET 25 30 35
MORAY FIRTH OR BUCKIE HERRING BOAT
[Open]
LINES PLANS FROM WASHINGTON'S REPORT
1849

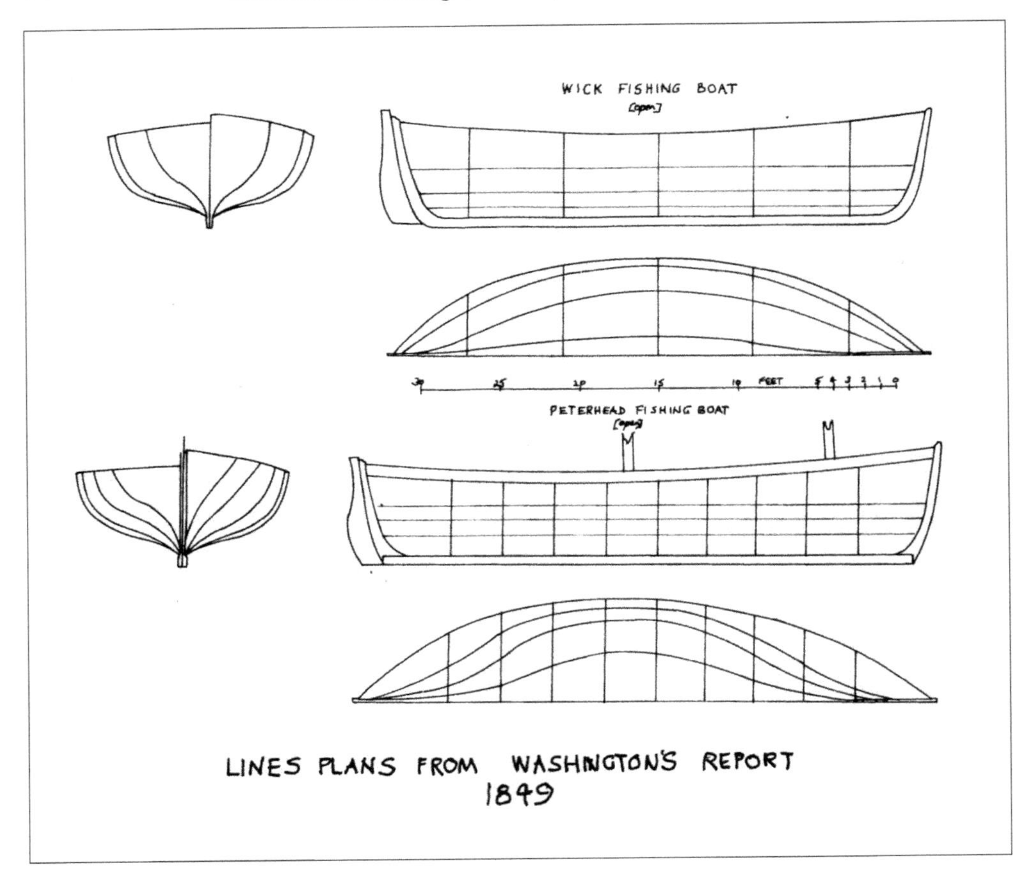

LINES PLANS FROM WASHINGTON'S REPORT 1849

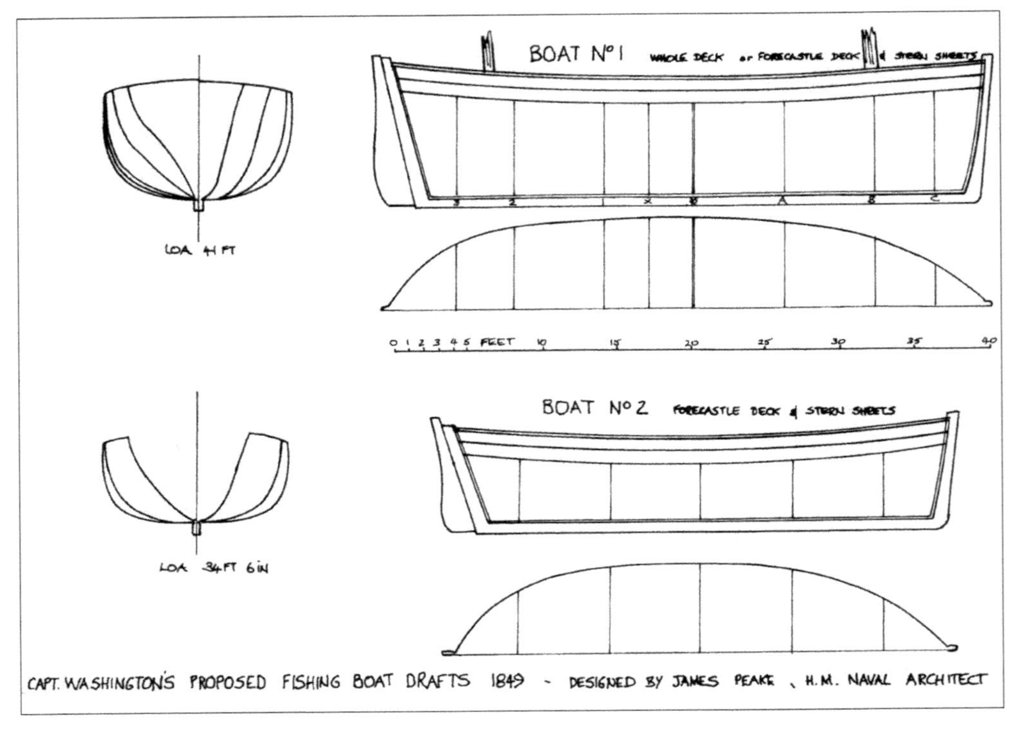

CAPT. WASHINGTON'S PROPOSED FISHING BOAT DRAFTS 1849 - DESIGNED BY JAMES PEAKE, H.M. NAVAL ARCHITECT

CHAPTER 3

The East Coast of Scotland: Berwick-upon-Tweed to Duncansby Head

Five herrings for tuppence.

—The cry of the Fisherrow fishwives

When one thinks of the fishing boats of the east coast of Scotland, one immediately conjures up pictures in one's mind of the large fifies and zulus that, although natives of this coast, eventually ended up in various parts of Britain in different fleets. Moreover, although one thinks of the fifies as the large, upright and powerful fishing boats of the latter half of the nineteenth century that were commonplace along the coast from Berwick-upon-Tweed to Fraserburgh, their origins, in fact, span an earlier two centuries. To the north, in the Moray Firth, another distinct type evolved, completely apart from the fifies, although it can be said that both came from a Viking pedigree. Towards the end of the sailing fishing boat era, these two types were then brought together into one design to produce that most splendid of fishing boats – the zulu. Before we consider these beautiful, strong boats we shall consider the two earlier types in full.

THE SCAFFIES OR SCAFFS (MORAY FIRTH OR BUCKLE HERRING BOATS)

The shores of the Moray Firth abound with tiny coves just big enough for a splattering of fishing boats to gain shelter. Around these havens, small fishing communities grew as the herring fishing developed in the eighteenth and nineteenth centuries. Other places of shelter had substantial piers and workings and could be termed harbours, in contrast to the beaches with only room for a few boats to be pulled away from the grasp of the dangerous North Sea surf. One

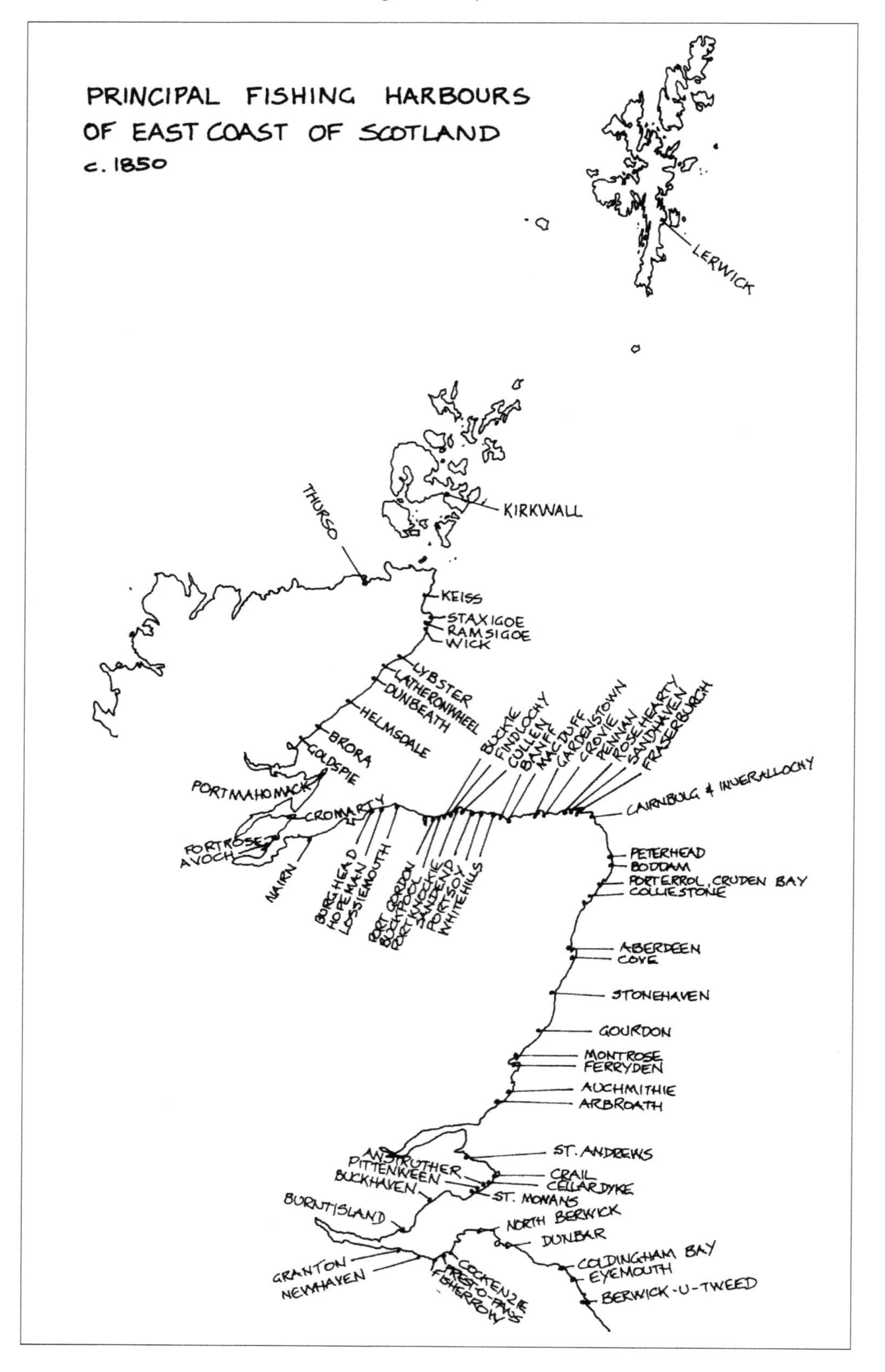
PRINCIPAL FISHING HARBOURS
OF EAST COAST OF SCOTLAND
c. 1850
LERWICK
KIRKWALL
THURSO
KEISS
STAXIGOE
RAMSIGOE
WICK
LYBSTER
LATHERONWHEEL
DUNBEATH
HELMSDALE
BRORA
GOLDSPIE
PORTMAHOMACK
CROMARTY
FORTROSE
AVOCH
NAIRN
BURGHEAD
HOPEMAN
LOSSIEMOUTH
PORT GORDON
BUCKPOOL
PORTKNOCKIE
SANDEND
PORTSOY
WHITEHILLS
BUCKIE
FINDLOCHY
CULLEN
BANFF
MACDUFF
GARDENSTOWN
CROVIE
PENNAN
ROSEHEARTY
SANDHAVEN
FRASERBURGH
CAIRNBULG & INVERALLOCHY
PETERHEAD
BODDAM
PORTERROL, CRUDEN BAY
COLLIESTONE
ABERDEEN
COVE
STONEHAVEN
GOURDON
MONTROSE
FERRYDEN
AUCHMITHIE
ARBROATH
ST. ANDREWS
ANSTRUTHER
PITTENWEEN
BUCKHAVEN
CRAIL
CELLARDYKE
ST. MONANS
BURNTISLAND
NORTH BERWICK
DUNBAR
COLDINGHAM BAY
EYEMOUTH
BERWICK-U-TWEED
GRANTON
NEWHAVEN
COCKENZIE
PREST-O-PANS
FISHERROW

A print by William Daniel showing the harbour at Burghead (Brugh-head?), Morayshire *c.* 1821. the harbour here was enlarged in 1809, and by 1828 there were some ninety-three boats working from it.

Wick harbour in 1865, with hundreds of scaffies lying after returning home from fishing. In the foreground are the farlanes with the herring being gutted. *(Johnstone Collection/Wick Society)*

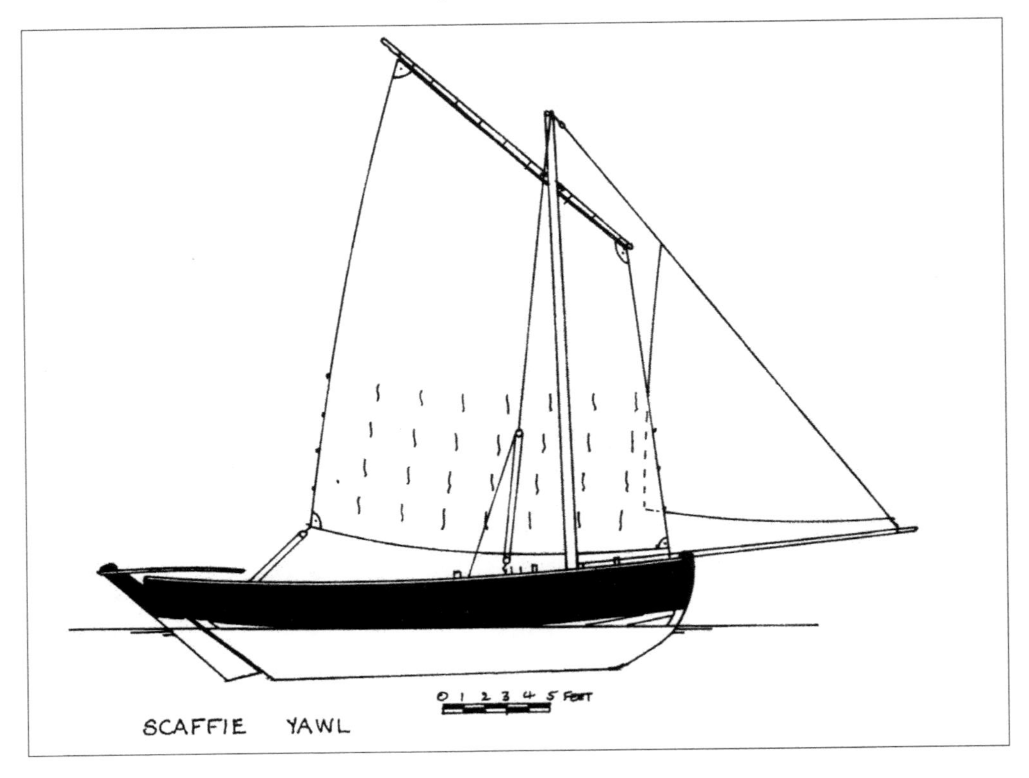

glance at the map shows how many of these settlements there were along the shores of the Moray Firth. Wick, in the north, was the biggest of all harbours, and became the herring capital of Europe in the early nineteenth century after being built by the British Fisheries Society between 1790 and 1803. On the other hand, Ramsigoe, just to the north, had but one boat registered in 1848.

The earliest of the documented inshore boats along this coast seem to be the 'Great boats' of the seventeenth century. At the beginning of that century these were 40 feet overall, two-masted boats with a six- to seven-ton burthen. Influence most certainly must have come from the Viking boats because of their clinker construction, the raked stem and sternposts – the keel length was 30 feet – and because of the two square sails, the latter being about the same size. Other than the sails, the boats were propelled with six oars. Most had a very small decked-over forecastle for stowage. Although generally an Orcadian boat that traded with Norway, they do seem to have travelled south to fish.

The next stage of fishing boat development appears around 1650, again a double-ender with two square sails, and again about the same size. These were said to be unseaworthy and hence did not venture far from land. With the transition from square sail to lugsail in the eighteenth century, the boats themselves became bigger and more seaworthy. During this period, probably more towards the end of the century, the foresail became larger than the mizzen. These yawls, as they came to be called from the Norse *jol* or *yol* meaning small boat, became localised so that each area adopted its own type to suit conditions, although all were open boats.

The small scaffie *Gratitude* at Portknockie in 1936. (*National Maritime Museum*)

By the time of Washington's report in 1849, two types were working in the Moray Firth. The Wick boat was the predecessor of what is now generally referred to as the Stroma yawl or yole as it's more commonly known, and examples can still today be found working along the coast from the south of Wick around to Scrabster and beyond. These are akin to the Orkney yoles and are still today completely open, clinker-built boats.

The Buckie, or Moray Firth boat, was more common along the southern shores of the Firth and was also termed a '*scaith*'. Washington deemed this type of boat to be one of the most seaworthy of all the Scottish boats at that time, and costing only £6 in 1815, they were cheap for the fishermen. They were generally of flimsy build because of the low cost, but consequently were easy to pull up the beaches. They were also said to be simple to operate and could be loaded with a large catch.

The craft evolved after the middle of the century into larger, more seaworthy boats, perhaps combining some of the design features of the Wick boats, so that by the 1870s, the more commonly known scaffie was the only native herring boat. These scaffies were instantly recognisable by their raking sternpost and rounded forefoot. These features became more pronounced in an attempt to improve seaworthiness, and coincided with an increase in length up to about 45 feet, on a 35-foot keel. When the profile of the Buckie boat from Washington's report is compared with the lines plan of the scaffie, one sees that the stem has been made more upright above the water to increase sailing ability, but that it has been rounded sharply to decrease the length of keel, hence reducing the cost, as the general Scottish tendency was to price a boat by

SCAFFIE

its keel length. The sternpost is raked further to increase the overall length and hence the deck space. This was done at the expense of windward sailing ability.

Builders of these boats were to be found in every harbour or settlement. Often as not the fishermen themselves were responsible for building their own boats in the fishing off-season. In the larger harbours, builders sprang up and found enough work building new boats to keep themselves in full-time work. Each builder's boats were different from the next one, and were instantly recognisable by the trained eye. Each new boat was an improvement on the last one, and they were all built by eye. This was, in fact, the case throughout Britain where work boats were shaped more as a result of experience and reality than as objects of beauty. They were built as workhorses, and treated as such, although they were tended with care and attention. Each fisherman and boatbuilder had his own ideas and these augmented traditional methods.

As we travel around the British coasts, we shall see that there were many regattas where the fishermen annually raced their craft, taking the competition extremely seriously, and having a great pride in their craft so that they often spent weeks preparing their boats – as long as there was no fishing to be done.

Even today, we see annual trawler races where spick-and-span fishing boats cruise out and compete on the best of terms. Fishermen always have been a community apart from the rest of society, and even with its upheavals of today, the same remains true.

The scaffies were generally two-masted boats, each with a dipping lugsail, although many boats had a standing lug mizzen. Several of the longer, older scaffies – or scaiths – had three masts with two huge dipping lugs forward. This was as a result of the boats increasing in size, becoming heavier to row, and the desire to sail further out in search of the herring. Another result of Washington's report was that around 1855 the first half-deckers appeared, the scaffies adopting a foredeck,

The scaffie *Gratitude* at the same location, showing various other types of boat behind, and with bows of steam drifters on the extreme right of the photo. (*National Maritime Museum*)

as he suggested, with a forecastle below that created a cuddy with berths and a stove for cooking. But it was a slow process to persuade the fishermen that decking over did not hamper their ability to carry the maximum of fish, as was previously believed. They also believed that decked boats were more expensive to build and that decking the boat over made it more arduous when hauling the nets as they had to stand higher up above the water-level. When one takes into consideration the cold, rough conditions of the North Sea, then they might have had a point, but considering other benefits such as safety, it is hardly surprising that these decked craft were eventually adopted. Buckie, for example, in 1872, had sixty-four decked vessels registered there.

The introduction of cotton nets enabled the fishermen to carry much larger nets – the previous hemp nets were three times heavier than the cotton ones. This, then, gave the fishermen three times more catching power, which, in turn, meant three times more fish, which would not load into a scaffie. There are pictures of scaffies returning home so heavily loaded that the waterline was only inches away from the deck. Sometimes they even had to throw away their catches for lack of room. Coupled with the introduction of the zulu in 1879, and the introduction of steam power for winches, the scaffie's demise was inevitable.

Furthermore, with the coming of the motor in the early twentieth century, it was found that they were totally unsuitable to receive motor units as this necessitated removing too much of the sternpost. They were beached and left to rot, being replaced with larger boats from the south. By the First World War, the majority had disappeared, although the *Gratitude*, BCK252, did survive as an example into the 1930s. She was in fact a scaffie yawl, being only 25 feet 4 inches overall, 16 feet 2 inches keel, 8 feet 6 inches beam. She had been built by George Innes of Portknockie for local fisherman David Mair, and was entirely decked over. More recently, a replica of her has been launched.

One can imagine the magnificent sight of the fleets of these boats. The east coast fishermen tended to paint their boats in vivid colours. A typical scaffie in the latter part of the last century would have been varnished above the waterline up to a brightly painted gunwale in white, green or blue. On deck, they were tarred with blue bulwarks and dark-blue coamings, hatches, mast trunk and thwarts. The upper parts of the masts were painted white, traditionally said to prevent the ingress of rainwater into the wood, and they would have sails tanned from the barking process to protect them. Although the same can be said of many of the different fleets, the sight of these open boats sailing out would, today, have been a feast for anybody's eyes, but it is all the more sad when one considers that none of the original big boats have survived. *Annie*, a small scaffie, can be found among the ISCA Collection and, as mentioned, there is the replica of *Gratitude*, which was built at Portsoy.

THE FIFIES

The fishermen of Fife, that ancient kingdom that lies between the Firth of Forth and the Firth of Tay, were supposedly one of the first people to undertake fishing on a large scale. While the Norwegians and Danes were arriving on the coasts to the north and the south in the eighth century, the Fifers are said to have been busy catching herring.

Fifies leaving Wick towards the end of the nineteenth century.

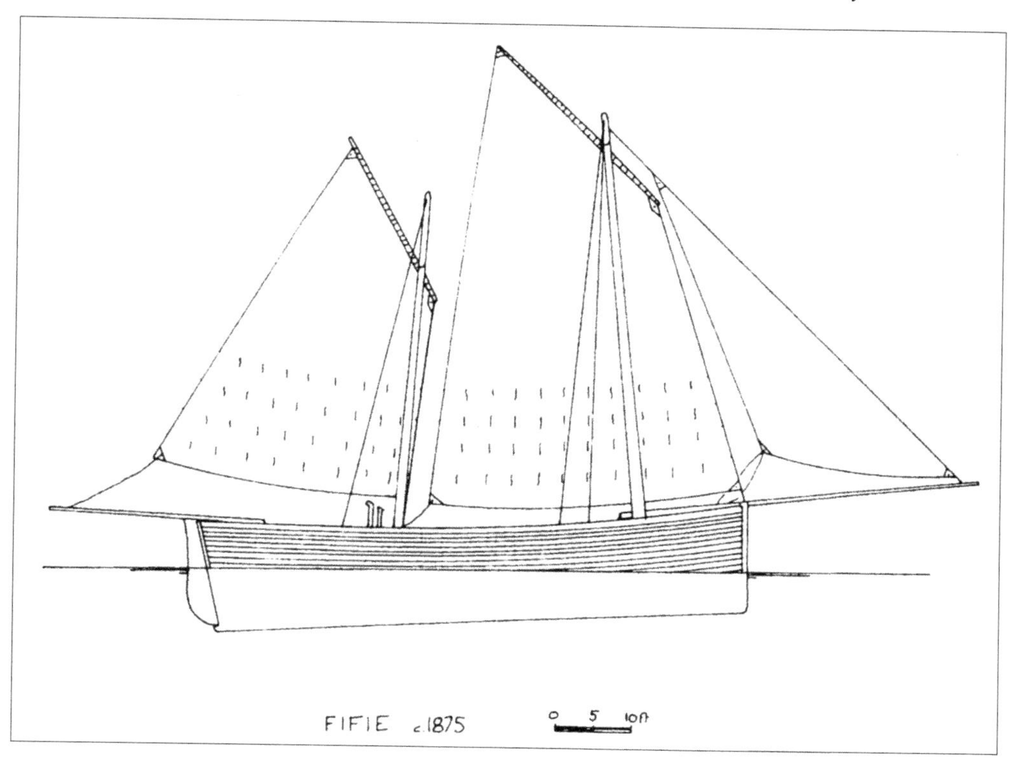

Up to the time of the Dutch dominance of the North Sea herring industry in the early seventeenth century, the Fifers had managed to export herring to Holland. Three decades earlier, when the Scots government woke up to the potential of the herring as a medium for raising tax, the Fifers had deserted and took their families and boats to Holland, where they were welcomed with open arms.

Anstruther is the fishing capital of the present-day Fife, its association with the herring going back to pre-Reformation times. Today it is joined with Cellardyke and Anstruther Wester to form the largest harbour on the Fife coast. Nearby St Monans, Pittenween and Crail all have long associations with the fishing as well.

The uprightness of the Fifers' boats is immediately apparent, and it is believed to have come from Dutch influence, probably after their travels there. Comparisons can also be drawn with some of the English boats from the south, although the English boats did not venture north until towards the end of the eighteenth century. Some say, furthermore, that the long hulls of the Newhaven boats resemble the Viking long ships. As we have seen before, Washington was severely critical of most of the Fife-type boats, although he did recognise the sea-keeping qualities of the Newhaven boat, but attributed it more to the skill of the fishermen than to the design of the boats. He likened them to the Wick boat, which he regarded as being totally unfit for sea!

The two proposed designs in Washington's report had more upright stems and sterns than previously because he reckoned that the raking stems and sterns gave little

The fifie yawl *Alexander*, 600A, owned by George Wood of Portlethen, Aberdeen.

grip on the water. But, as in the north, the fishermen ignored his advice, and continued to fish in their lightly built existing boats.

In 1849, the builder Alexander Hall & Co. of Aberdeen submitted a design of a boat 45 feet overall, with an 18-inch sheer, two watertight bulkheads and buoyancy compartments. The proposed cost was £150, which, as existing undecked boats were costing only £50, was its downfall. A few were built, Fraserburgh getting its first decked boat in 1850. Eyemouth had one in 1856, costing £130. But there was only a trickle of conversions.

Two decades later, however, in 1867, the National Lifeboat Institution sponsored the building of several new boats that closely resembled this design. These decked boats were built and lent to suitable fishermen for one year at several major fishing stations, where, it was hoped, the other fishermen would notice them and adopt similar boats. Furthermore, the Duke of Sutherland ordered a boat of the same design to be built for use at Helmsdale. The fishermen must have been persuaded of the advantages, because within a few years the numbers of decked boats had sharply risen. As in Buckie, Berwick in 1872 had forty boats of about 45 feet overall, some even 56 feet – and these were costing £200. At Leith they cost £250–£300. Newhaven had thirteen such boats built in eighteen months. Others lengthened existing boats and had them decked over. Later in that decade carvel-built boats began to replace the clench-built ones. Although the cost was greater than before, the fishermen soon realised higher earnings to offset this. These boats had higher-peaked sails than before, which increased the speed, had no bulwarks and thereby facilitated the hauling, and the hulls were generally painted in black. The fifie as we know it had arrived.

The fifie yawls remained much the same as they were in Washington's time, although they did adopt the forecastle, and the stems became more upright. These boats have certain characteristics similar to the boats from the island of Stroma in the Pentland Firth, and are said to have been influenced by these Stroma yoles when the Fifers fished and traded with the North Isles. A typical fifie yawl of 1870 was 30 feet overall, clinker-built and double-ended like all the east coast boats. They were used for inshore long-lining – the sma'line as it was known. They rarely fished for herring, and never ventured far out to sea. The bigger boats were for that. They were frail boats because of their lightness of construction, and generally did not survive much beyond six working years.

The large fifies were crewed by five or six hands, and the largest boats were nearly 70 feet in length. Crewing these huge boats was arduous work. The cotton sails were huge, the forward lugsail being nearly 200 square yards in area, while the mizzen was another 150 yards or more. Again the nets were huge, up to 3,500 yards in total length

Wick-registered fifies at Wick *c.* 1890.

Above: Wick harbour – first boats going out. (*National Maritime Museum*)

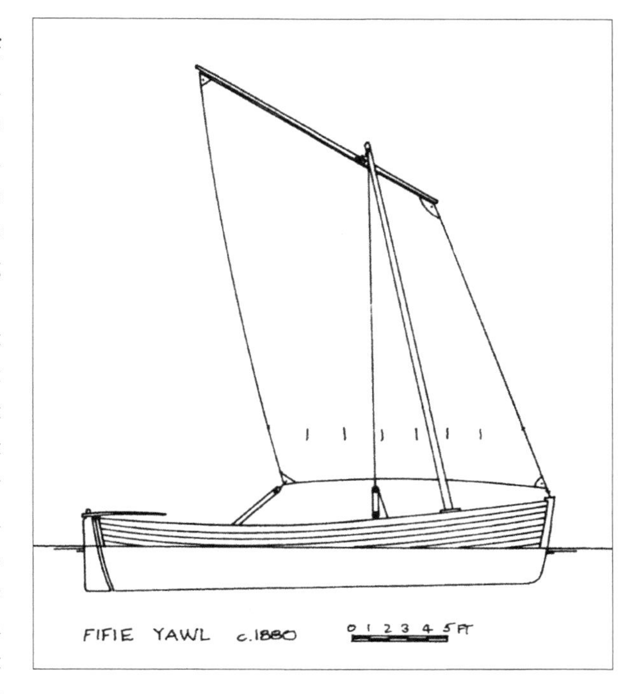

when set, and before the days of the steam winch it was no mean task to get them aboard. But the rig was powerful once up, and the fifies sailed right down to the south of England to participate in the great autumn herring fishery off East Anglia.

These fifies were built at various locations. The most famous of all the yards is that of James Miller & Sons of St Monans, although many other yards grew up from Eyemouth right up into the Moray Firth. That huge numbers of these strong boats were built is shown in the 1883 returns: in that year there were 3,665 first-class fishing boats of 15 tons or over, in Scottish waters. Of these only sixty-two were on the west coast, although it is likely that some of these were fifies. It must be assumed that the majority of these on the east coast must have been fifies, as the smaller scaffies were rarely more than 15 tons, and the new zulu boats had only been around for three years. In Shetland, the fishermen adopted the dandy rig, and then swapped the fore-lug for a gaff rig, and even built a few themselves, calling them the Shetland herring boats. Many were sold south to Yorkshiremen, who preferred these keel boats to their cobles. Some even sailed permanently out of Lowestoft and Great Yarmouth in direct competition with the local yawls and luggers.

The 'Baldies' or 'Bauldies' appeared around 1876 during the Italian civil war, being named after the patriot Garibaldi. These were basically small fifies in that they had similar lines to the bigger boats, but were only around 45 feet overall. They often had a raking sternpost, similar to the scaffies, and although they probably were technically fifie yawls because they were used for the inshore fishing, they were a mule-type breed. The name baldie was only really used along the Fife coast. In Fraserburgh, locally built boats of this type were called Fraserburgh yoles. Along the southern shores of the Moray Firth they were fifie skiffs – or just plain 'skiffs'. I've even heard one man refer to them as Lossiemouth cod boats! Trying to label designs and to identify distinct types tends to be difficult with all fishing boats, as each boat is influenced in different ways. This complicates research when different types sometimes do not follow a logical path to their recognised name. This also reinforces the belief that designs simply evolved through the building process, and not as drawings upon paper. Boats were merely built by craftsmen who knew what was needed from the boat, and different versions emerged for different ways of fishing in different areas, or sometimes even the same area. Concrete designs didn't appear until the twentieth century, where individuality was lost, but that must wait for now.

THE ZULUS

As already mentioned, the first zulu (named after the wars of the same name) appeared in 1879. There is no doubt whatsoever that the design incorporates the best of both the scaffie and the fifie – having the scaffie raking stern which contributed to speed, yet retaining the upright stem and deep forefoot of the fifie with its good sea-gripping qualities. How the design actually came about is a point for contention. Whether it came about as dowry for the lass of a Lossiemouth skipper who was betrothed to a Moray Firth man who worked on a scaffie, or whether it came about when a family from Lossiemouth wanted a new boat and the woman favoured the fifie whereas her husband preferred the northern types, I do not know. Another tale tells that it was simply because the boat's designer, William 'Dad' Campbell, decided he wanted a bit of both. That he designed it seems to be definite, but why and for whom seems unclear. I have read that the boat *Nonesuch* belonged to him, and I've read the opposite. Whatever the reality, it seems she was built at the yard of W. Slater of Lossiemouth, although others say it was at Asher's yard in Burghead or by Alexander Wood, also of Lossiemouth. Some also say that the design was copied from an earlier zulu that had been built by George Innes of Portknockie.

However, it emerged, there is no doubt that it was an instant success among the fleet, especially in Banffshire, Morayshire and Nairnshire, where it outsailed and tacked better than the other boats. There immediately began a surge in building these boats. Every yard, it seemed, was rushing to launch the 'new' zulus, and the fishermen were as keen to buy them, unlike years before when they were slow to deck over their open boats. Around Lossiemouth there were five yards all building zulus. This was the same all along the east coast. Unfortunately, this was at the cost of building new fifies, although a few were built. Other owners sold on their fifies around the country and built zulus. This is one reason why the fifies were quite often seen in many other parts of Britain, as the second-hand price was low.

ZULU

The size of the zulus was impressive. P. J. Oke drew up plans of *Fidelity*, BF1479, in 1936 as a part of his work with the coastal craft sub-committee of the Society for Nautical Research. He lists the keel as being 58 feet long, 8 x 8 inches

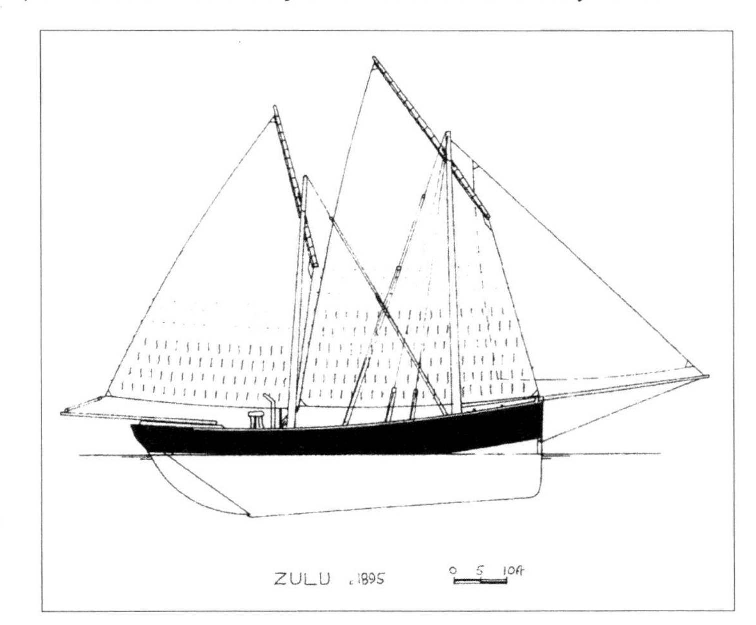

Below: The zulu *Smiling*, BF772, coming into harbour. Note the massiveness of the construction.

Below: Zulus bringing the catch into Peterhead harbour, *c.* 1880.

Below: A salmon coble at Balintore, 1997. These boats are unlike other Scots craft. They were developed specifically for salmon fishing at or around the river estuaries, and resemble craft from Scandinavia rather than Britain. (*Author*)

with a 7-inch keelson. Both the stem and sternpost are 8 inches wide, the former being 14 inches deep, and the latter 15 inches deep at its head. Frames are 4 inches x 7 inches to 9 inches, spaced at 12 inches. Deck and hull planking was 2 inches thick, bulwarks 3 inches thick. She had been built by William McIntosh in 1904 at Portessie, and was 78 feet overall, being one of the biggest ever built. She took about eight weeks to build! The actual largest zulu ever built seems to be the *Laverack*, BF787, in 1902. She was 84 feet overall, except for the bowsprit and bumpkin, and had a keel length of 54 feet.

The rig of these zulus, although similar to that of the fifie, was even more powerful. The hulls grew up to around 80 feet and the masts grew accordingly. The mainmast had a diameter of nearly 2 feet and a typical zulu set 400 square yards of canvas. This needed 30 tons of stone ballast to counteract the rig! The sight of these craft sailing out in large fleets is best summed up by somebody who actually saw it. In his book *Mast and Sail in Europe and Asia*, published in 1906, Herbert Warrington Smyth writes, 'It is truly one of the finest sea sights of modern times to see this great brown pyramid come marching up out of the horizon and go leaning by you at a ten knot speed, the peak stabbing the sky as it lurches past some seventy feet above the water.'

By the turn of the century there were 480 zulus registered at Banff alone. They were being built all along the coast where some yards such as George Innes built fifty in total, while others only built a handful. It was said that they were easier to build than a fifie as the long overhang at the stern required less bend in the planks at the tuck. We have already seen how small fifies were built for the line fishing, and the same was the case for the zulus. Small zulus – zulu skiffs – were built at about 30–35 feet in length, and these continued to be built long after the demise of the zulus and fifies. The last zulu skiff was built around 1920, whereas the last zulu, the *Winsome*, had been built in 1906.

Surprisingly, the zulus have not survived well. There are constantly stories of zulus being burnt on beaches or allowed to fall apart. Today, one of the largest, *Research*, sits in a permanent exhibition inside the Scottish Fisheries Museum at Anstruther. Luckily there have been a few restorations, though of the smaller 50-foot types. *Violet* was built by Nobles of Fraserburgh in 1911, and now is in the USA after a lengthy rebuild. Her sister ship *Vesper* was less lucky and was scrapped a few years ago. Closer to home the small zulu *Nellie* was relaunched in 1996 after a three-year rebuild. More recently (997) the 41-foot *Kate*, LK126, clinker-built by Hay & Co. of Lerwick in 1910, has been salvaged from a beach. Several others remain in commission duly converted as pleasure craft, and a few zulu skiffs have survived the decades. However, many of these were later built and hence don't have the true zulu sternpost that rakes 40 degrees, sometimes referred to as half-zulus.

MOTORISATION

Ten years or so after the introduction of the steam drifter (see Chapter 15), the fishermen had to cope with the advent of the motor. This idea seems to have originated in Denmark,

Above: Fifie *Onward*, PD292, at Peterhead. (*National Maritime Museum*)

Above: The motor fifie ML33 at St Monans in 1936. (*National Maritime Museum*)

Below: A creel boat at Anstruther 1997. (*Author*)

where in 1900, a smack's small boat was fitted with a motor made by Messrs Mollerup of Esbjerg. One year later, the drifter *Pioneer*, LT368, was built at Reynold's Yard at Oulton Broad. Within a few years she returned higher earnings than similar sailing drifters, whereby her critics were silenced. The first Scottish boat to have an engine was a fifie – the purposely built and similarly named (rightly) *Pioneer*. She was built at Anstruther in 1905, measured 72 feet overall, 21 feet beam and had a draught of 8 feet aft, and was fitted with a 25-hp single-cylinder four-stroke Dan engine. Although she experienced what could be termed teething problems in her first year of operation, the motorisation was generally regarded as being successful. One or two sailing fifies had installations made on a trial basis, but these seem to have been surprisingly unsuccessful. The first really successful Scottish conversion was in 1907 with the fitting of a 55-hp Gardner 3KM engine into the 1901-built fifie *Maggie Jane's*, BK146, from Eyemouth. She proved to be a good sea boat, and was financially worthwhile. By the following year, four more local fifies followed her example. As well as the fifies being converted, the zulus began to have engines fitted, although, with their raking sternposts, they were never as successful. In 1909, the 48-ton Arbroath fifie *Ebenezer* had a 60-hp Blackstone heavy-oil engine fitted.

By the time the new decade arrived, there were many boats on the east coast of Britain sporting engines. Even the smaller line boats were having them fitted: for example, the fifie yawl *Vanguard*, 19AH, built for the sma'line fishing out of Arbroath and hailing from the Miller's yard arrived that year with a 7.5-hp Gardner. The Kelvin engine was launched a few years earlier by the Bergius Launch & Engine Co. in Glasgow, a company that was to dominate the motor installations within the Scottish fishing industry, and especially on the west coast.

The only development in design that conversion to motor power had on the fifie was the modification to the stern to allow a propeller aperture to be formed. This proved easy with the upright sternpost, which was not weakened by the cut-out. As with other powered boats, they also had the small wheelhouse that was to become so familiar in years to come.

The number of motorised vessels grew up to the war in 1914. But the increase in power brought about realisation that the larger vessels were no longer desirable. An engined smaller boat was as successful as a large sailing boat. And so the 'new' motor fifie was born, as we shall see in Chapter 18.

Another type of fishing boat worth a mention is the typical, what is now termed, east coast creel boat. Although based on the fifie design, they continued to be built up to the 1950s. Generally open except for a small foredeck, and often narrow side-decks, these boats worked their pots close inshore. Their length hardly exceeds 20 feet, yet perhaps they are the final evolution of those early fifies.

And so, from early origins among the early fishing fleets in the seventeenth century, the fifies (and zulus to a certain extent) became the most prolific of all British fishing boats. Still today many different examples of the fifie can be seen. From the well-known *Reaper*, FR958, at Anstruther or the Shetland herring boat *Swan*, LK243, to the many small craft such as *Olive Leaf*, PD39, these lovely boats are as varied as the coasts they fish off.

CHAPTER 4

The Northern Isles: Shetland & Orkney

Frae rocks an' sands
An' barren lands
An' ill men's hands
Keep's free
Wee loot, weel in
Wi' a gueede shot.

—North-east Scots fishermen's saying upon the launching of a new boat

SHETLAND ISLES

The Shetland Isles remained part of the Norse kingdom until 1469 when they were returned to Scottish control as part of the dowry of a Danish princess to her betrothed Scotsman. This link that began in the early eighth century, and the fact that geographically the Shetland Isles are quite remote from the Scottish mainland, created a bond with Norway that was to survive through generations.

From the early times of their invasions, the Vikings brought ships and maritime influence as well as boatbuilding experience to the islanders. They also brought desperately needed raw materials to the windswept Shetlands where few trees grew. This led to their influence permeating all aspects of sea life on the islands; an influence that was not to be quickly lost or forgotten.

Unfortunately, as in many parts of Scotland, the majority of the island dwellers were at the mercy of a few wealthy landowners. Most of the islanders rented their homes from these lairds, as the landowners were called, and in return were expected to work for them either on the land or at sea. Thus they found themselves increasingly beholden to their landlords to the extent that if any young man left the islands to avoid the work, it was quite usual for his parents to be evicted from their homes. The fishing that they were expected to take part in was completely under the landlords' control, and took place annually. This was the 'haaf' fishery for which Shetland is

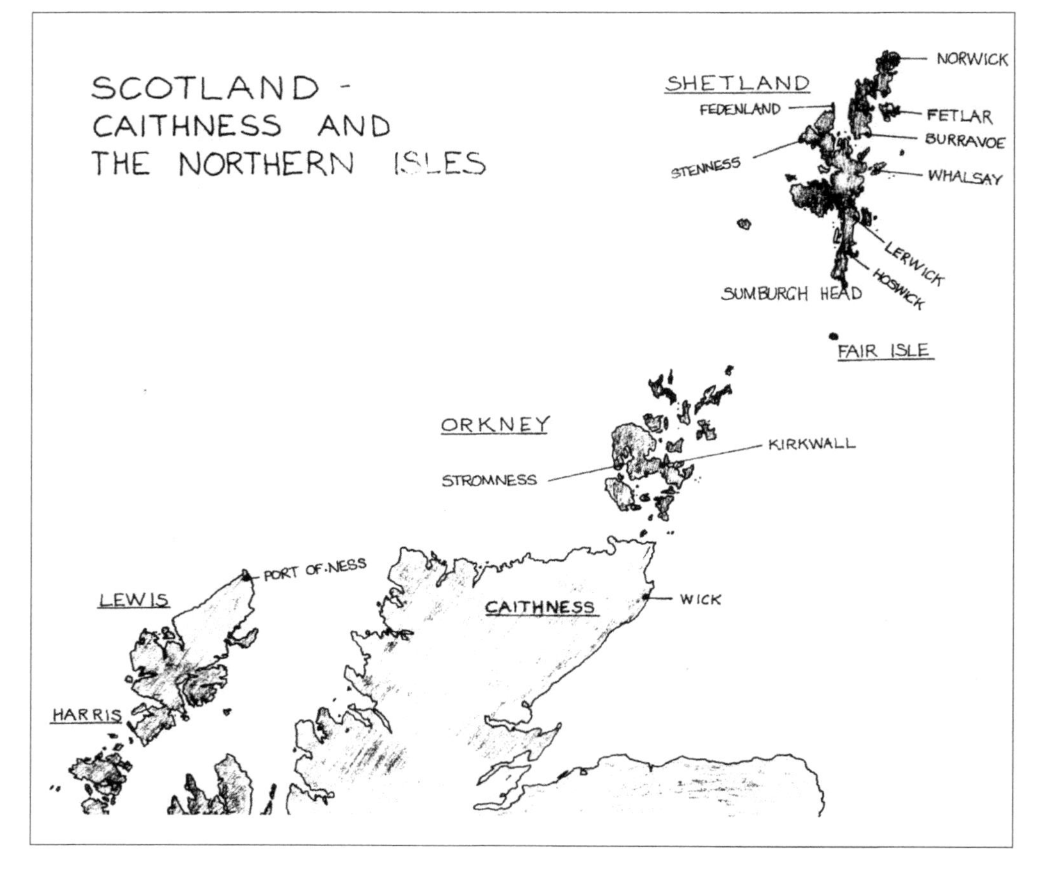

SIXEREENS ANCHORED AT FEDELAND

COPIED FROM A PHOTOGRAPH OF MR. J.D. RATTAR

renowned – 'haaf' meaning ocean in old Norse – in which cod and ling were sought. This great long-lining fishery began on Beltane Day (1 May) and continued up to Lammas Day (1 August). During this period the fishermen were based at one of the many 'haaf stations' or lodgings that were set up around the coasts of the islands by the landowners. These stations consisted of numbers of stone huts around a suitable beach where the boats could be hauled up well out of the water. Up to seventy boats might operate from one station with 400–500 fishermen and curers sleeping ashore at times. The fishermen would aim to make two trips a week out to the fishing grounds that generally lay 30–40 miles out to the west of the islands. Records tell of some boats venturing 60 miles out, and one or two appear to have fished within sight of the Norwegian coast, although these seem to be an exception to the normal practice.

Prior to setting out, the fishermen would use their smaller 'eella' boats to collect bait, usually using young saithe and, later on, herring. They would set their longlines, which were up to 6 miles long and carried 6,000 hooks, as soon as they arrived at the chosen fishing ground. It could take up to three hours to set the lines and another six to recover them, and in a single trip the process of setting and hauling might be done four times. Once they had sufficient fish aboard they would head home to get the fish ashore as soon as possible for curing.

In the 1880s, the first herring curers moved into the islands and at once the landlords' stranglehold over the fishermen was loosened. Similarly, the truck system, the means of payment in the form of vouchers that had to be redeemed in the laird's own local shop, was abandoned, giving the fishermen more freedom and cheaper supplies. With this freedom and the higher earnings that were got from the herring fishery, the fishermen soon turned their backs on their gaolers, and the haaf stations too were abandoned to become idle.

THE HAAF BOATS

The boat that evolved for use in the haaf fishery was the six-oared sixareen (sixern, saxereen, sixtreen or sexæring). In 1774 these open boats were about 24 feet long on an 18-foot keel with a small mast and square sail, although the earliest of sixareens were supposedly sailed over from the Norwegian fjords and were 40 feet overall in length. The boats, were imported direct from Norway in pieces ready for assembly, and cost the princely sum of £6 complete. At that time the Shetlanders imported all of the wood they needed for domestic purposes and other uses as Norway was a great deal more accessible than the Scottish mainland where transport in the north was scarce.

By 1830 the fishermen were constructing their own boats, still using Norwegian timber and having adapted the design to suit their own local conditions. But the boats still retained their Viking appearance – flaring topsides, narrow at the waterline, low free-board amidships and long overhangs at each end. This shape gave them ample room for six men and a full load of 2 tons of fish. Traditionally the length of the keel of a sixareen was two thirds of the overall length of the boat. By the beginning of the nineteenth century they were being built on a 24-foot keel, with an overall length of 35 feet and beam of 8 feet 6 inches. By the middle of the century the cost of these boats had risen by about 40 per cent to £1 per foot of keel and it was about this time they adopted the dipping lugsail.

In a similar way to other Viking boats, the sixareens were clinker-built using wide, green planks of Norwegian fir on sawn oak frames, or bands as they were called in Shetland, fastened with iron. In Viking tradition the bands were not fixed to the keel. These boats were very lightly built and did not last much more than seven or eight years before being replaced. Their remains were salvaged to build the much smaller eella boats. These were 15–17 feet overall on a keel length of 9–11 feet. As well as being used for bait gathering, they were also used as tenders to the bigger boats.

The average scantlings of a late-nineteenth-century sixareen were:

keel	8″ x 3.5″
stem/sternpost	8″ x 3″
bands	6″ x 3″ tapering to 5″ x 3″
planks	8″ x 0.75″
gunwale	3.5″ x 3″
reabands	2.5″ x 0.75″ spaced at 3″ centres
tafts (thwarts)	10″ x 3″
mast	same height as keel was long, max. diam. 6″ at taft

The sail area was also directly related to the length of the keel and the sails were made of untanned cotton. The boats were built at various locations around the coast, often by the fishermen themselves. Being light, they were quick to build, and were what would now be termed a basic boat. Knotted wood was avoided as this was deemed unlucky – these fishermen being just as superstitious as many of their counterparts elsewhere.

The boats were split into eight individual compartments or 'rooms'. With the exception of the division immediately forward of the aftermost compartment, the bulkheads were of open construction, which allowed water to flow along the length of the boat but prevented the fish from doing the same. These open bulkheads were constructed by fixing narrow strips of wood – reabands – from the keel to the underside of the taft or thwart that demarcated each compartment. The aftermost compartment was called the 'shott-hole' or cockpit and was the domain of the helmsman with the sheet deadeyes and cleats close to hand. In later boats a compass was also fixed in this compartment. Immediately forward of this was the 'shott' or 'run', which was the fish hold where some of the stone ballast was stowed until being thrown over the side to compensate for the fish being brought aboard. Here, too, in later boats was the rudimentary bilge pump. Next in line, moving forward, was the 'owse-room' where the bailer worked in the days before the bilge pump. The floor was slightly higher and smooth, so that the bailer could use his 'ouskerri' – water scoop – to bail the water from the boat. One scoop alone could shift two gallons of water

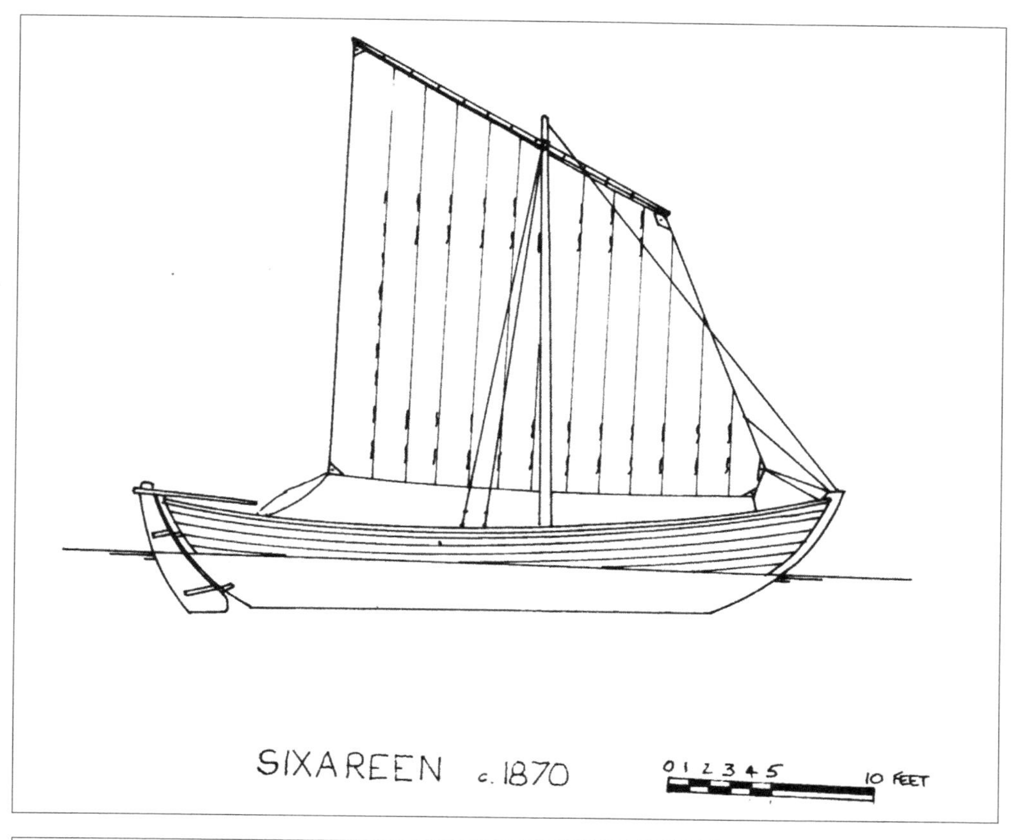
SIXAREEN c. 1870
0 1 2 3 4 5
10 FEET

SIXAREEN

and one can imagine this stalwart working constantly when the weather was bad, standing alone between the might of the seas and a watery grave for all aboard. The 'midroom' was used for the setting and hauling of the lines and here too they would be baited, and the catch removed and tossed into the hold. Immediately forward of this was the 'ballast room' where the majority of the ballast was stowed around the base of the keel-stepped mast. The next compartment, the 'foreroom', served as the galley and was equipped with a peat-burning stove, basic cooking utensils, peat for fuel and a sea-chest in which all the food was kept. One man worked as cook and was responsible for preparing all the rations.

The diet of the men at sea consisted mainly of ling and potatoes, the fish being cooked in various ways. One favourite was 'liver muggies' – the stomach of the ling being stuffed with liver, salt and pepper, and either roasted or boiled. Basics included bread, oatcakes and butter, with tea to drink. The 'headroom' was the store for the water container, usually a big barrel, and also provided space for some of the fishermen to sleep, as did the 'lineroom', which was right forward in the eyes of the boat where the sail was stowed.

Each compartment had its own taft, and those that were used for rowing had 'kabes' or thole pins of wood and a 'humbleband' of cowhide or rope to hold the oar into the kabe when rowing. The main frames of the boat, which coincided with the thwarts, were called bands and were individually named. The 'beating band' was the foremost and was followed by the 'fore band', the 'mast band', the 'mid band', the 'owse band', the 'foreshott band', the 'midshott band', which was an extra half way across the shott room, and the 'futtock band'. There was also an additional frame called the 'tack band' positioned close to the bow to prevent the planking from splitting in heavy seas.

From its early entry into the Scottish fisheries, the sixareen was perhaps one of the most tested of the sailing craft around British shores. Who today would row out into the North Atlantic in a small open boat for a few days at a time, completely at the mercy of the elements? That these men were among the finest and bravest of fishermen had as much to do with their confidence in their craft as it did with their skill. Their seamanship was first class; they were said to have been able to navigate back to land in misty conditions by watching the waves alone, and determining the direction of the 'mother' wave. Combine this with the best of the Viking influence, and it produces a finely tuned team, a partnership between endurance and ability, between man and sea. An influence that was to leave its mark, not only around the Shetlands, but around the rest of Scotland and, indeed, much of the west coast of Britain, as we shall see. While the English fishermen looked to the Dutch and French for their influence, nearly all the later craft of the north and west were modelled on Viking designs, and the sixareen was possibly the ultimate in this regard. The best example of a sixareen is Duncan Sandison's *Far Haaf*, built in 1993, and which today can be seen at the Unst Boat Haven at Haroldswick in Unst.

Another craft that was common to the Shetland fishery was the fourereen or fourern. This boat was a smaller, 26-foot, four-oared version of the sixareen. They

Below: Duncan Sandison's new sixareen *Far Haaf* in 1988. This boat was subsequently damaged and a new boat built. (*Duncan Sandison*)

were mainly used for the winter haddock fishing and tended not to sail more than 20 miles or so offshore and returned home every night. Their construction was similar to that of the bigger boats and, like them also, they were later rigged with a small dipping lugsail on a mast mounted further forward than when a square sail was set. This forward movement of the mast also gave the boat a better performance when running before the wind. The fourereens were crewed by three, and although sturdy little craft, it appears they did not survive at the fishing much beyond the decline in the winter haddock near the end of the nineteenth century. It was found that there were other jobs available that were more congenial than working out in the North Sea in winter in a small open boat.

Around the southern parts of Shetland, in Dunrossness, a different type of craft was to be found and it was said to have been the most elegant of the Shetland boats. The Ness Yoal (jol) was popular for fishing with hand-lines for saithe in the inshore waters, especially between Sumburgh Head and Fitful Head to the north-west. Here the current runs swiftly between the islands and a reliable, well-found craft is still a must to ensure that it is not thrown onto one of the many hazards. At 22 feet overall, they had keel lengths of around 15 feet, a beam of 5.5 feet and a draught of 18 inches. The construction was similar in that the frames or bands were not through-bolted to the keel, fixed only to the garboards in true Viking fashion, and they were built with very wide planks using only five per side in comparison to a sixareen, which normally had nine.

These Ness Yoals were often raced by the fishermen owing to their turn of speed, due in the main to their narrowness. In races with the other Shetland boats it has been said that the Ness boats were often disqualified because of their different shape from the sixareen and fourereen – possibly an excuse as it seems that the Ness boats always won the races. Today, although some boatbuilders build these yoals in the traditional style, others have developed it for racing purposes so that many villages have commissioned their own boat to join in with the rowing regattas that take place most weekends during the summer.

Several storms ravaged the Shetland boats with loss of life. The death knell of the traditional Viking boats came when a north-westerly summer storm hit the fleets in 1881, and in the ensuing dash for home on a lee-shore some fifty-eight men were drowned and ten boats destroyed. But at that time the fishermen were already beginning to adopt the larger fifies and zulus that were already established on the east coast. As the herring fishery grew and hundreds of vessels were packed into Lerwick for the spring herring season, the Shetland fishermen bought in the bigger boats, fifies especially, and the day of the fishing sixareen was over. Those remaining were used to transport people and animals around the islands, and as peat boats.

Some fifie-type boats were built in Shetland. Although built originally as luggers, many of these vessels were converted to dandy rig and then to smack rig as this was found to be not only more powerful but easier to handle. One such vessel was the *Swan*, built by Hay & Company at their Freefield yard at Lerwick. She was launched in 1900, but by that time the steam drifters had begun to make a serious impression

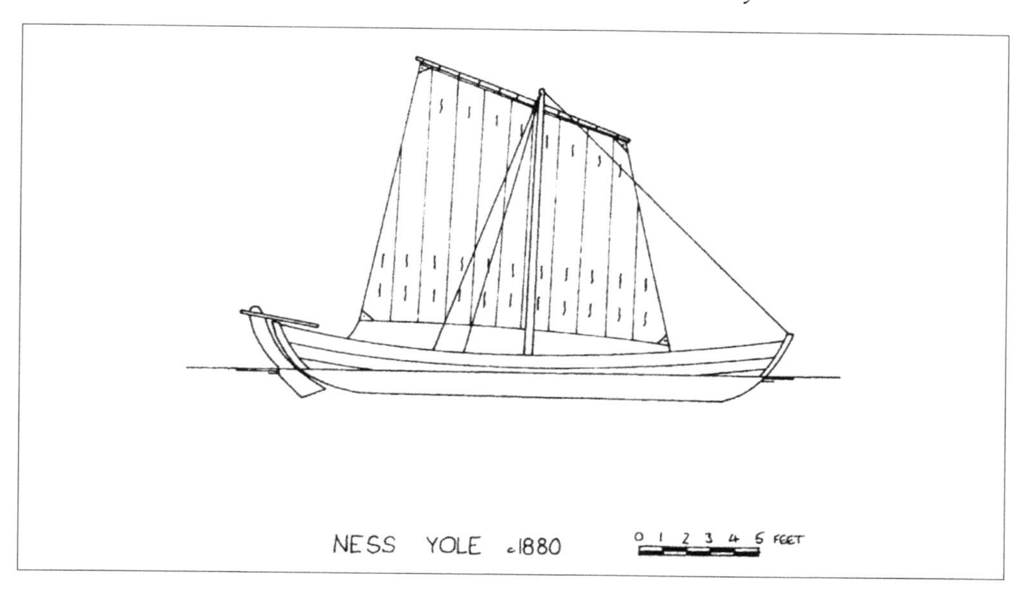
NESS YOLE c1880
0 1 2 3 4 5 FEET

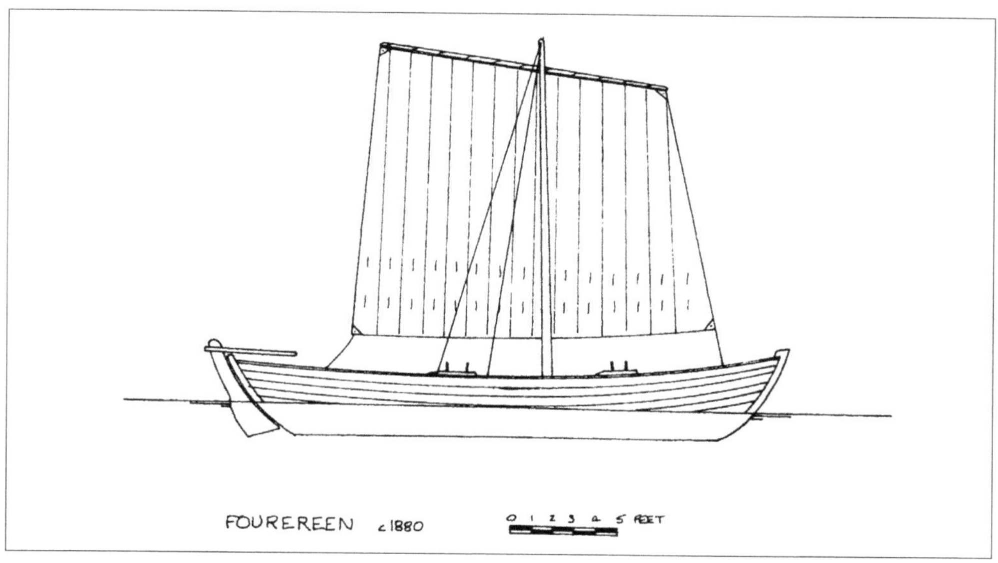
FOUREREEN c1880
0 1 2 3 4 5 FEET

NESS YOLE

Above: A Ness Yole in Dunrossness shored up by carefully placed stones. (*Jack Peterson*)

Lerwick's *Swan*, LK243, which was completely rebuilt with help from various grants, including the lottery fund. (*Author*)

upon the fishermen. The *Swan* fished as LK243, and she remained fishing under sail until having a motor fitted in 1935, after which she seine-netted until 1950. Today she has been rebuilt by the trust of her name and is based in Lerwick. Hay & Co. also built various zulus – *Kate,* LK126, previously mentioned, being one of the last in 1910.

THE ORKNEY ISLANDS

In the Orkney Islands, less emphasis was attached to the fishery; crofting was a better alternative due to the greater fertility of the land. Nevertheless, Orkney did produce its own type of fishing vessels that showed some Viking influence, although this was probably derived from the mainland – only a few miles away.

The largest of the Orkney boats was the Orkney yole, sometimes called the South Isles yole. These boats had a much fuller stern and greater beam than the Shetland boats, and were similar to the very early scaithes or scaffies, and to the mainland yoles. They were clinker-built in Norwegian fir on oak frames, iron-fastened and were built locally or on Stroma. They were generally 16–20 feet long – typically 18 feet 3 inches x 7 feet 3 inches. Unlike the Shetland boats that were often rowed, these yoles were primarily sailing boats, and had two masts with spritsails carried on opposing sides of the masts. Motors were introduced in the early part of the twentieth century, but it was found that the propeller slipstream tended to suck down the stern too low in the water. This was compensated by adding another strake. Later boats were built with an even fuller stern to increase the buoyancy aft. These yoles remained working until the 1960s, both fishing and generally transporting goods around the islands, after which time they were rigged with a gunter sail and used for pleasure purposes. Many remain for that purpose today.

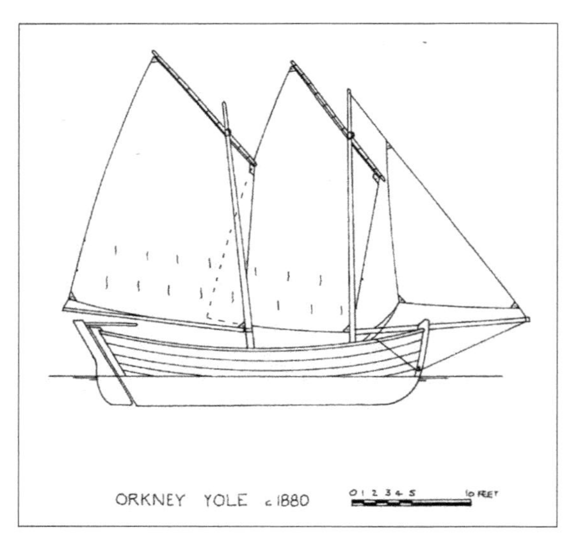

ORKNEY YOLE c.1880

ORKNEY YOLE

On the north shores of Orkney, similar yoles were to be found, but these were rigged with two standing lugsails. These North Isles yoles, or Westray yoles, had very subtle differences both in their shape and in their method of construction.

Two small Orkney yoles in the foreground and the Kirkwall-registered firthie (fifie) *Rose of Holm*. (*Orkney Library Archive*)

The Custom house whill or quill in Stromness harbour, *c.* 1930. (*Len Willson Collection*)

Orkney yoles landing at Skippigeo, Birsay. These are two-masted boats with main masts lowered for the run onto the beach. (*Orkney Library Archive*)

Another type of craft, in reality a small yole it seems, was called a 'quill' or 'whill', and these had a similar shape at both ends.

The Orcadians also fished with much smaller boats close inshore. Their traditional dinghy had a wineglass transom with a slightly flared bow and pronounced sheer-line. They were rigged either with one lugsail or a spritsail. Another small local type is the 'flattie' which seems to have been introduced to the islands from Canada around the turn of the century. As their name implies, they are flat-bottomed, anywhere between 6 and 12 feet long, and were essentially used as rowing boats. Len Wilson of Kirkwall, to whom I am indebted for much of the information on these smaller Orkney craft, built a flattie as recently as 1964, and several examples still remain.

The herring fishery had existed for centuries in the Orkneys, often executed by outsiders – the Dutch being the first to base a fleet there in the twelfth century. Stromness, Stronsay and South Ronaldsay were the three bases for the herring, and up to 750 small boats of 12–17 feet were said to fish for the early summer herring in the first years of the nineteenth century, half being local boats. These yoles drift-netted some 10 miles offshore. By the middle of the century these craft were up to 30 feet overall. During the 1870s, the majority of these yoles were superseded by the larger imported herring luggers as in Shetland, although many yoles survived in individual hands. Some luggers were even built on Orkney, such as the fifie *Pioneer*, K219, built at Stanger's Yard in Stromness. By the beginning of the twentieth century several zulus had been brought into Orkney in addition to the hundreds that fished annually from the mainland. These local zulus became known as the Burray Boats, of which they

The South Isles yole *Family Pride*, which belonged to the coxswain of the Longhope lifeboat that was tragically lost with all hands in 1969. Dan is at the helm in this picture. (*Dennis Davidson*)

Spritsail yoles from the Island of Graemsay, lying at Stromness pier, 1920s. (*Orkney Library Archive*)

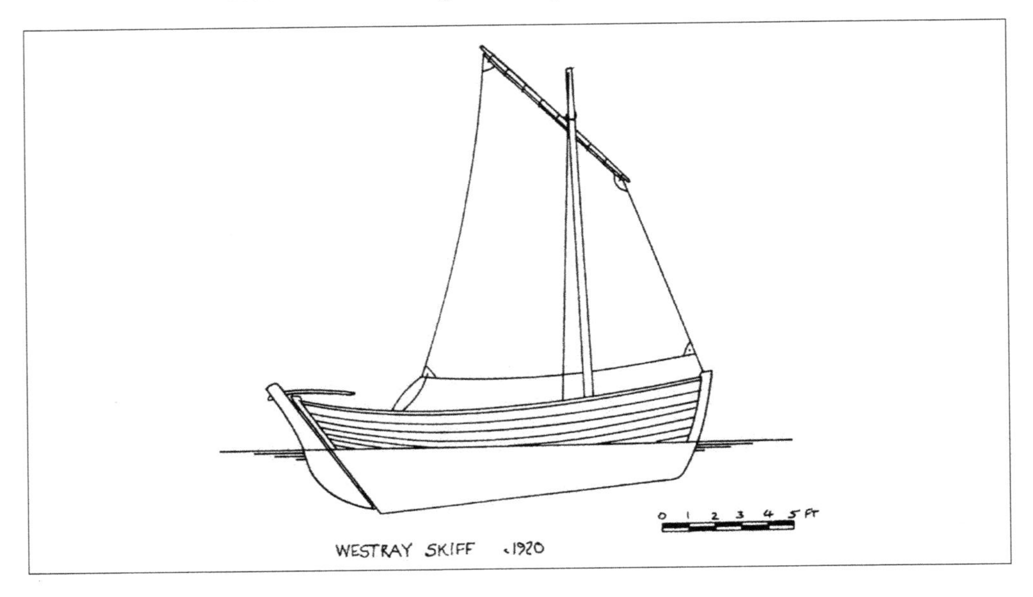

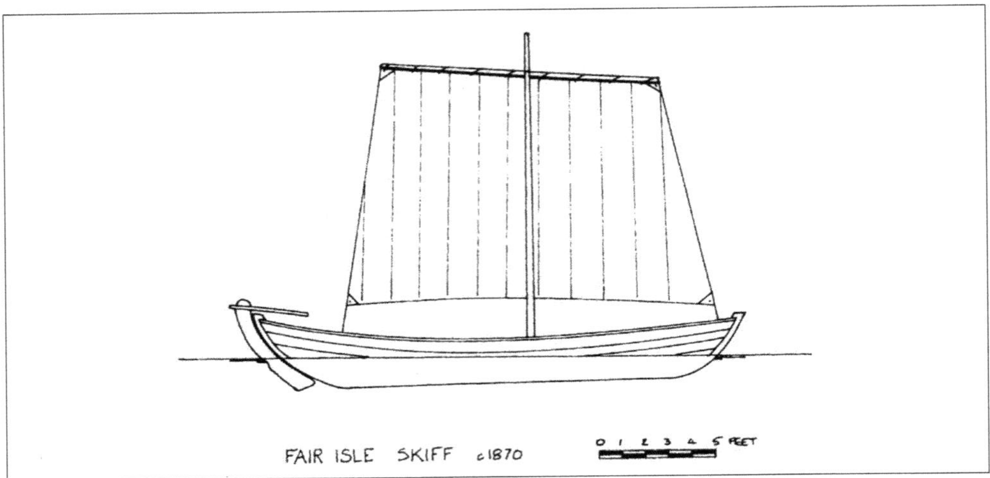

were twenty or so. Although boats had been built on Burray, these zulus were all built south of the Pentland Firth.

As we've seen, the yoles continued to be used by the local inhabitants, although primarily as transport between the islands and only for a limited amount of fishing. Several new boats have been built over the last few years and are raced in the summer. Smaller versions of these yoles were the Westray skiffs, from the most northerly island. These small boats were said to have been used for fishing since the eighteenth century and probably not least because they were easily hauled up the beaches. Their use was centred particularly on that island, and they were used for creel fishing for lobsters and ward fishing for coal fish. *The Marys*, built by James Rendall, Queensbrechan, Westray in the 1920s, is 15 feet 8 inches, and is rigged with a standing lug and oars.

Fair Isle skiffs lying at South Harbour in Fair Isle. (*Jack Peterson*)

Firthies were boats that were generally built on the mainland side of the Pentland Firth or on Stroma, and resembled the Stroma pilot boats. Many of these vessels can be seen today in Keiss, Scrabster or Thurso. They will be discussed later. Furthermore, it seems that the fifies were also referred to as 'firthies'.

THE FAIR ISLE SKIFFS

Living between Shetland and Orkney, the Fair Isle fishermen developed their own craft for fishing. Their skiffs closely resembled early Viking craft, more so than did the Shetland boats. They were 22 feet overall, on a 15-foot keel and with 5.5 feet on the beam. Unlike other types they had a keel that sloped to give a few more inches of draft at the stern. In the 1880s these boats cost £7 each. They were crewed by three oarsmen, renowned for their distinctive short chopping rowing strokes of up to 45 per minute. The boats also had a small square sail which was generally used when fishing. The fishermen were great seamen in the Shetland tradition and were some of the only ones who held an annual regatta to race their fine boats. This took place from South Harbour – the only harbour on the island – every summer. The boats themselves were built locally, there being four boatbuilders in the South Harbour around the turn of the century. How many boats were built is unknown, but they seem to have survived longer than many similar craft, being used for line fishing for haddock until after 1900, prior to which they were catching saithe in the summer. Today, Ian Best works on the island, and he is keeping the tradition of building these boats alive.

CHAPTER 5

West Coast of Scotland: John O'Groats to the Solway Firth, including the Western Isles

The Perseverance just outside,
the fastest sleeve-valve in the Clyde.
—Clyde fishermen's ditty, *c.* 1930

The same Viking influence is immediately apparent in the boats of the western coast of Scotland, and along the northern coast. This north coast, having some of the most inaccessible parts of the British Isles, has fewer fishing stations, and these are tiny in comparison to other parts of the country. The area suffered serious social upheaval in the early part of the nineteenth century during the Clearances when local population movement put great pressure on the coastal communities. Prior to 1786, fishing was only for home consumption, but after that date fishermen were allowed to sell their catch and commercial herring fishing therefore flourished. After about 1820, with the increase in local population, most of the commercial fishing had moved to the east coast communities. Over the next eighty years or so the fishermen from Farr, Tongue and Durness worked the herring fishery based on the east coast, away from home. When the herring failed at the end of the nineteenth century, it caused great hardship among the northern settlements. Lobster fishing remained the only reliable fishing, with huge amounts reportedly being sent to London in 1842. Herring was used to bait the pots.

The harbour at Sandside Bay was built in 1830 for the local fishing, and is typical of the small harbours to be found along the coast. Durness, Loch Eriboll, Tongue, Farr, Portskerra, Scrabster, Thurso and John O'Groats all had similar quays, such as those we've already seen on the east coast. There were two fishing stations on Loch Eriboll, the remains of which can still be seen. At Rispond, on its western entrance, there is a rubble pier and two-storey fish house, sheltered from all directions. This was built in 1831 by a Mr Anderson for the sole purpose of catching the early 'rich' herring at the beginning of June. These fish he cured in the Dutch way and readily exported them to London aboard one of the many smacks that visited this area. He employed two sloops in 1834, and another twenty boats crewed by four men. At Portnancon, a few miles along on the same side, there is still another old pier and the remains of a smokehouse can be seen.

The boats that were based here were typical of those all along the coast. Looking at prints by William Daniell in 1821, we see many views of the small early scaffies or Pentland yawls, as the particular model from the north was called, that were common along the east coast. These Pentland yawls – or Stroma yoles – were chiefly built on Stroma by a number of builders. George Simpson began building in 1865, and continued right up to 1913. These small boats were either 12 feet for lobstering, 14 feet for the cod or 18 feet for transporting cattle. They were built in a Viking traditional style, although every other floor was fixed to the keel. The boats under 15 feet were sprit-rigged, whereas those over were lug-rigged.

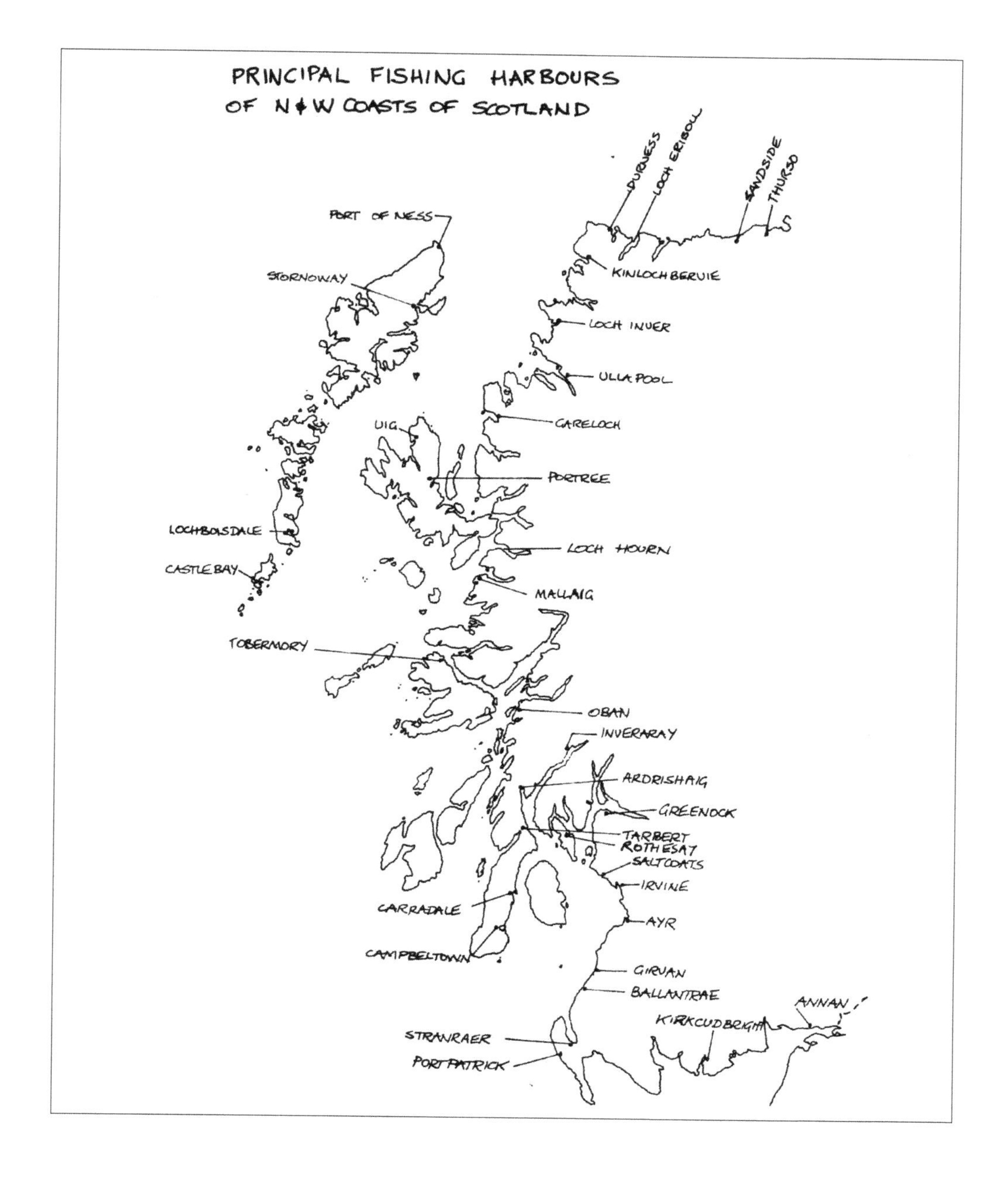

Along this coast, however, many of the fishing boats did not have any sail at all, although we must surmise that some did; indeed a view of Rispond shows a small two-masted vessel that could easily be a scaffie yawl. Of Loch Inver, on the west coast, Daniel tells us, 'Some small boats of a beautiful shape were seen in Loch Inver. They are built in Norway, and are occasionally to be purchased here at the moderate price of thirty shillings.' The practice of importing craft from Norway, as we shall see, was not only confined to the Shetland Isles.

Norwegian-based boats are also to be found among the Outer Hebrides, the most notable being the Ness Sgoth (skiff) from the Port of Ness on the northern tip of Lewis. Ness was the second biggest fishing station on the island, and the boats evolved from early Norwegian boats. They were built in four locations around the port, and were constructed in true Viking fashion in the same way as the Shetland sixareens. These boats were among the largest of the British open boats at 32 feet overall on a 22-foot keel. Also like the sixareens, they were often rowed great distances to fish for cod and ling. They had the obvious Viking look with long overhangs and low freeboard until the end of the nineteenth century when they adopted the zulu-type stern that raked further than before. This type of stern was common on the western skiffs, and most probably was an influence not from the east coast but from the boats from further south along the Scottish coast, or perhaps both. Many north-west boatbuilders trained on the Clyde.

The last of the fishing Ness Sgoths was built in 1935, and as recently as 1995, another boat was built by John Murdo MacLeod, grandson of the builder of the

The pier at Tanera Mor, Loch Broom – a print by William Daniell, *c.* 1821.

The harbour at the Port of Ness, Lewis, *c.* 1890, with Ness Sgoths lying around on the quay. Note their similarity to the Stroma yoles. Two fifies lie at the pier.

1935 boat, as a replica. *An Sulaire* was modelled on the same lines, and is a beautiful example of these fine boats.

Similar boats were to be found at Stornoway, the biggest fishing harbour on Lewis. The east coast fishermen contemptuously referred to these craft as 'pikers', although the reasoning behind this is unclear. By the end of the nineteenth century most of these craft had been superseded by the fifies and zulus, so great was the impact of these bigger boats all over the coasts.

The scaffie yawls were also to be found further south: Daniell shows a print of the pier at Tanera Mor in the Summer Isles which has one of these vessels lying alongside.

The British Fisheries Society, set up in 1786 to promote the fishing on the Scottish coast, built three harbours on the west coast as well as Pultneytown in Wick. Ullapool, Lochbay and Tobermory were all harbours that were established by the society before the end of the eighteenth century. Although Wick prospered, the western settlements never developed into what the society had hoped. This was for several reasons: the outbreak of war with France delayed the onset of the fishing due to French men o'war being seen in the Minch; emigration to America after the Wars of Independence led to a loss of population; and the threat of eviction after the Clearances when at the beginning of the nineteenth century sheep were all the rage for landowners. At Ullapool, it was said that the herring deserted the coasts immediately because of the noise the new settlement created!

Nevertheless, the fishery did develop in the nineteenth century, with Ullapool and Tobermory having fleets of small craft, and harbours such as Oban, Castlebay on Barra, Mallaig, Broadford on Skye and other similar settlements being home to considerable fleets. Other harbours prospered under 'foreign' boats – usually from the east coast or the Clyde – Castlebay on Barra was inundated with up to 500 boats during the herring season between May and June after about 1870. There were between fifteen and eighteen steamers carrying the herring to Liverpool a few years earlier, the fish reaching the Liverpool market within twenty-four hours. The same steamers would tow the boats out of the harbour in groups, sometimes even as far as the fishing grounds. But by this time, the majority of these boats were fifies, zulus and Clyde skiffs. The bigger boats were able to carry larger catches and became more profitable.

Line boats at Barra worked the rich fishing grounds mainly to the east of the island. These small boats were up to about 22 feet long and had a single dipping lug. It has been suggested that they had been imported originally from the east coast, when it was said the east coasters sold them to the Barra men instead of sailing them home after the herring season. Many were later built here by the islanders, and further north through the Uists and into Harris. Many lie scattered around the islands like silent memorials, even today.

Very similar to the Barra line boats are the Grimsay boats, which were used mostly for fishing from the Monach Islands. The small island of Grimsay is sandwiched between North Uist and Benbecula and these boats developed purely as a result of the need of the fishermen to sail to and return with lobsters from the Monach Islands, which were rich in lobsters. These boats were built largely by the Stewart family, who originally came to the island in the 1840s from Argyllshire, bringing their design of boat with them. Their design incorporated the need to ride out over surf into deeper water, sail over to the Monachs with enough gear for a week's fishing, haul creels while there and return with fishermen and the catch. Rigged with two gaff sails, the

The pier at Uig, Skye, with a collection of fifies, zulus and Lochfyne skiffs andf a couple of other trading smacks.

The Lochfyne skiff BRD115 lying in Portree Bay, Skye, *c.* 1910. Although many of these boats were imported from Loch Fyne, some were built by local boatbuilders.

fishermen sometimes raced each other home on the Saturday, but while working they left the mizzen mast ashore. The craft were renowned for their seaworthiness, their lightness and their fineness, especially at their entry, and they were double-enders (*eathar*) until the transom-sterned boats (*geola*) were built.

As usual, motorisation had its impact upon the type, bringing about a fuller hull shape and a transom, although some boats retained the double-ended influence that is longstanding hereabout. Five generations of the Stewart family worked until recently, and it has been suggested that they built in excess of a thousand boats. Some of the Stewart boats still survive at the fishing.

'*Bata*' is the Gaelic term for 'boat', and around parts of the mainland coast these same small 16- to 22-foot double-enders were so called. These were used for fishing and general transport, and were undecked with a single lugsail. Several examples remain working around Port Henderson and Gairloch.

WESTERN SKIFFS

Although these sgoths are indeed skiffs, further south another type of boat evolved over two centuries or so. In the eighteenth century small line boats were working close to the shore, catching mostly for home consumption. Each small community or sheltered cove had its own fleet of these boats, mainly double-enders normally rowed, but occasionally having a small lugsail, or even a spritsail. Lobsters, cod and ling were the common catches until the herring fishery gained in popularity around the middle of the century. These small skiffs were open boats, clinker-built, and were a development of the Norway yawls, as all the imported boats are generally known, adapted for local use. As we shall see with the Irish fishing boats, some of these came aboard the trading smacks that sailed through the Caledonian Canal, after its opening in 1822; smacks also brought supplies of Scandinavian pine both for domestic use and for boatbuilding.

THE EARLY FISHERMEN OF LOCH FYNE

The fishermen of Loch Fyne have been catching herring since the Middle Ages. In 1527 Hector Boece reported that in Loch Fyne 'is mair plente of herring than is in any seas of Albion'. In 1603 Sir Walter Raleigh spoke of the Dutch selling herrings to other nations 'of £1½ million' (presumably being the value of the catch) – employing 20,000 men, all Scots, and with all herring being caught on the Scottish west coast, most notably Loch Fyne. Likewise, the herring they caught has been renowned for its particularly tasty quality. Campbeltown, said to have one of the best harbours in the world, was at the centre of the 'buss' fishery in the early part of the eighteenth century. The fishing gave a seasonal alternative to working the land and an addition to an otherwise relatively monotonous diet. As transport improved, so did the fishing, enabling larger boats to be built.

The earliest of the documented boats on the Clyde appear to have been schooner-rigged wherry-type vessels. Influence for these could possibly have come from the east coast of Ireland where many schooner-rigged boats fished. Then early in the eighteenth century, according to the *New Statistical Account of Scotland* in 1845 for Glassary

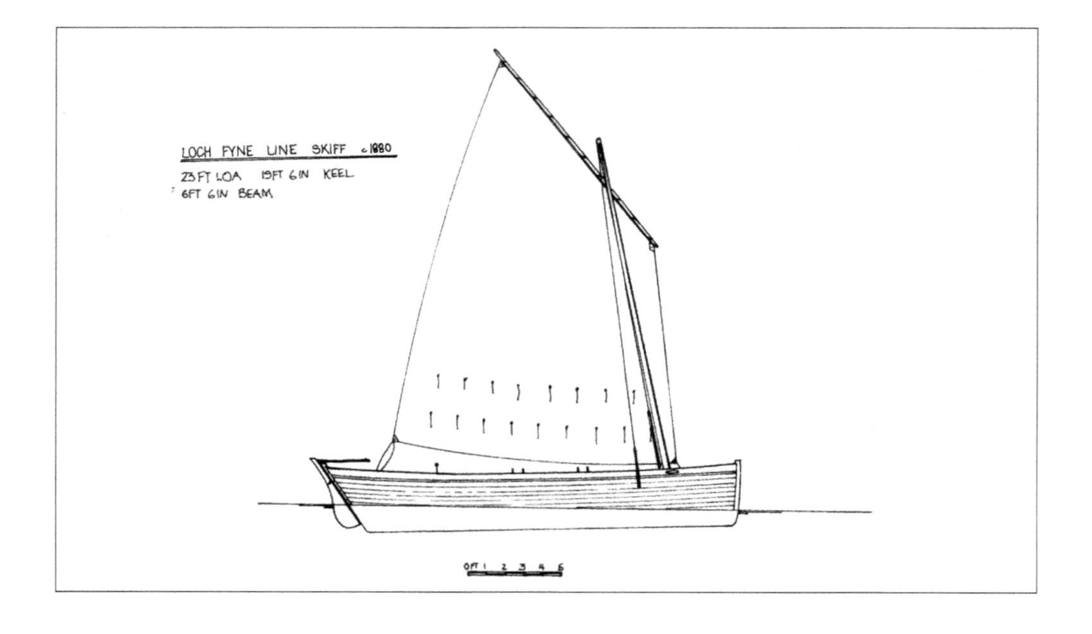

Parish (Lochgilphead to Inveraray), 'their boats are becoming larger and better, and the Ayrshire fishermen have brought in a good style of skiff, with a single lug sail'. When the herring weren't in, they fished for cod and ling with lines. Hence these boats became known as the Loch Fyne line skiffs. They were 14–16 feet long in the keel and all had rounded sterns from the Norse influence of centuries before. There were four men to a boat, which was individually owned, with one share going to the owner, and one to each of the crew. Some 200–300 boats were seen of an evening in 1790 around the lower part of the loch. It was said that the lack of decent harbours and the scarcity of good, cheap timber prevented the building of bigger boats, which in turn 'prohibited a more successful prosecution' of the herring fishery.

THE NABBIES

The Ayrshire fishermen who introduced the lugsail into Loch Fyne had been fishing with their line boats since the seventeenth century at least. By the eighteenth century they were in search of the herring, although it was said at the time that the boats were so small that it was 'not possible to go any distance to look for herrings in boats of this description'. These boats were only 14 feet in keel length, with a round stern and had a raking mast with lugsail and jib.

As these line skiffs worked further out to sea, they grew in size up to 24–28 feet of keel. These larger great-line boats became known as the 'nabbies' – possibly a Scots variation of 'nobby', meaning smart or elegant, from the trim, stylish appearance. They soon became the favourite boat on the east side of the Clyde and were described as 'one of the prettiest, smartest and handiest forms of sea boat to be found'. It must have been the early nabbies that so impressed the fishermen of Loch Fyne that they copied the rig.

These nabbies had fuller sterns and finer lines forward than the earlier skiffs. In addition to the mast being raked further aft than any other boat, they were one of the first to set a jib. Furthermore, some nabbies had small mizzen sails but this addition

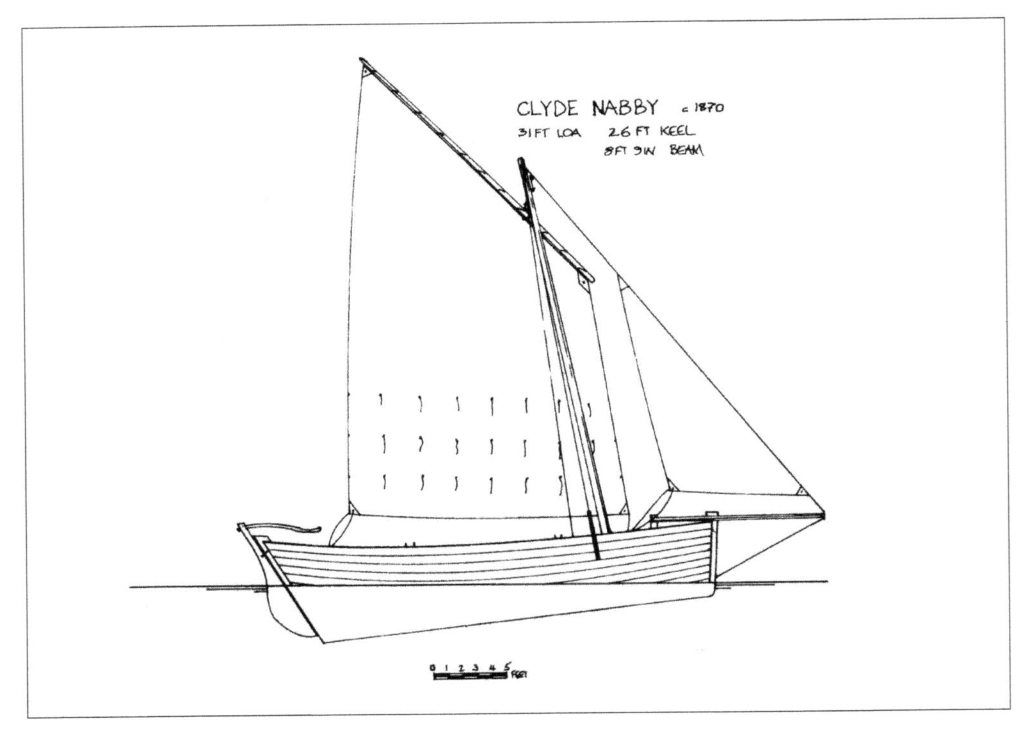

did not seem to survive long. They were generally open boats, the later ones having a small forecastle for storage and they became popular as they were deemed very sea-kindly. This was particularly necessary in the often stormy approaches to the Clyde where shelter was sparse and the wild Atlantic seas swung in around the Mull of Kintyre to create a violent swell.

The nabbies were built in the Ayrshire boatyards of men like William Fife of Fairlie, who built fishing boats when the demand for yachts was low, and the Marquis of Ailsa at the Maidens. Both men could make five times as much money from building a racing yacht, for which they were both renowned, as from a nabby or smack, yet both were renowned also for building good fishing boats. Other builders on this coast were John Thomson of Ardrossan, James Adam of Gourock and Hugh Boag, one of William Fife's best shipwrights, who set up on his own account at Fairlie. John Thomson continued building these nabbies after the First War, one typical vessel, albeit among the largest ever built, being the *Annie*, which was launched in 1916. She was 44 feet 4 inches overall with a keel length of 37 feet and a beam of 11 feet 6 inches. She was rigged with a lugsail and jib, but also had a Gleniffer 18–22 hp engine as was common at that time. According to the register of fishing boats, she was used for 'nets and lines'.

A word of warning: the term 'nabby' must not be confused. Some of the older fishermen continued to use the term after they had adopted the motorised ring-net boat in the 1920s and 1930s: it was only a colloquialism, and in no way reflected any similarity except that they were all used for ring-netting.

Above: The harbour at Irvine with clinker-built nabbies, the majority of which were registered at Ayr (AR).

THE LINE SKIFFS

As the nabbies developed from the early line skiffs, so did the line skiffs themselves evolve into perfected open boats. The Largs line boats had a dipping lugsail only at first, this being later replaced by a standing lugsail like the Clyde and Loch Fyne boats. Further south, the Ballantrae and Girvan skiffs were bigger and, in all reality, it is extremely difficult to differentiate between the line boats and the bigger nabbies.

The smaller boats were made by builders such as Davidson of the Maidens, William Harbison and Hugh Edgar of Dunure and Jas Kirkwood of Girvan. Others came from Dunoon, Garleston, Ayr and from Dan Fyfe of Stranraer. As well as the line boats of Loch Fyne, similar craft were to be found north of the Crinan Canal. In Oban, one typical skiff was the *Gylen* built by MacDonald of Portbeg, Oban, for local fisherman D. Currie and registered as OB5. She was 18 feet 4 inches overall, 15 feet 8 inches on the keel and had a lugsail and a jib on short bowsprit. She had fuller lines than the Clyde boats and a much fuller stern, the maximum beam being well aft of amidships, more similar to the Loch Fyne boats than any of the others. These vessels were to be found out to Tobermory and among the inner isles.

Around the early part of the nineteenth century Loch Hourne was a centre of the herring fishery, it being one of the busiest stations in Scotland. When, however, the railway arrived in Mallaig in 1901, all the business was transferred there, and today nothing remains at Piper's Island by Arnisdale that can be called a port. Hundreds of people worked from here more than a century ago, and now not a building or ruin can be seen. This really does reflect the nature of the herring fishing, especially on the west coast of Scotland where it was procured from the most basic of sites ashore in the most inaccessible of places, yet which might be moved to another site without inconvenience or delay or, indeed, remains. All along the coast, only vestiges of rubble piers and fish curing houses act as telltale signs of a history of the herring; apart from the fishing harbours that still remain, nothing else is tangible evidence of that great era.

THE INTRODUCTION OF RING-NETTING

At the beginning of the nineteenth century, drift-netting was the common method by which herring were caught off the west coast of Scotland, indeed Britain. Some 600 boats were said to have been engaged in fishing around Loch Fyne. During the 1830s, irregular use of the drift-net as a 'trawl' began seriously to increase the size of herring catches. The first documented haul of fish caught in a trawl-net is thought to have been in a bay near Tarbert, Loch Fyne, in either 1833 or 1835. It seems that the net was stretched across the bay and then dragged inshore so that the ends met, enclosing the fish. Soon the fishermen were using the technique, still from the shore, but with the help of a boat rowing one end of the net out around the bay. The method developed into what is now termed ring-netting: two boats drag the two ends of the net around the shoal of fish and back to each other. At first, the length of the net was 150 yards,

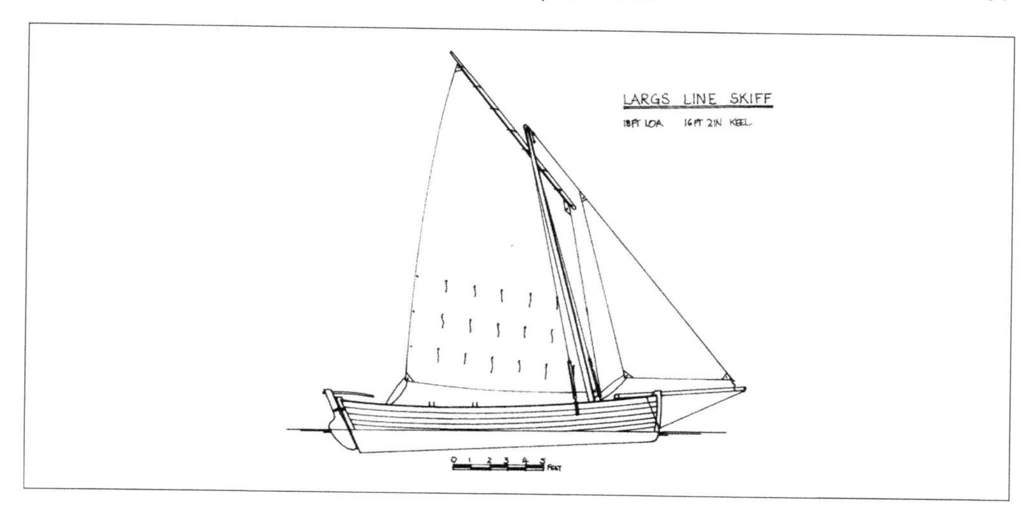
LARGS LINE SKIFF
18FT LOA 16FT 2IN KEEL
0 1 2 3 4 5 FEET

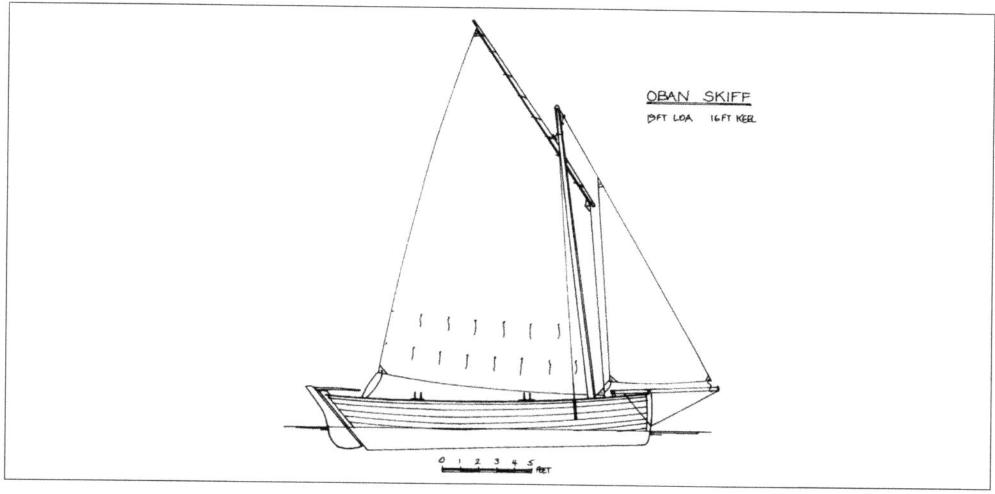
OBAN SKIFF
19FT LOA 16FT KEEL
0 1 2 3 4 5 FEET

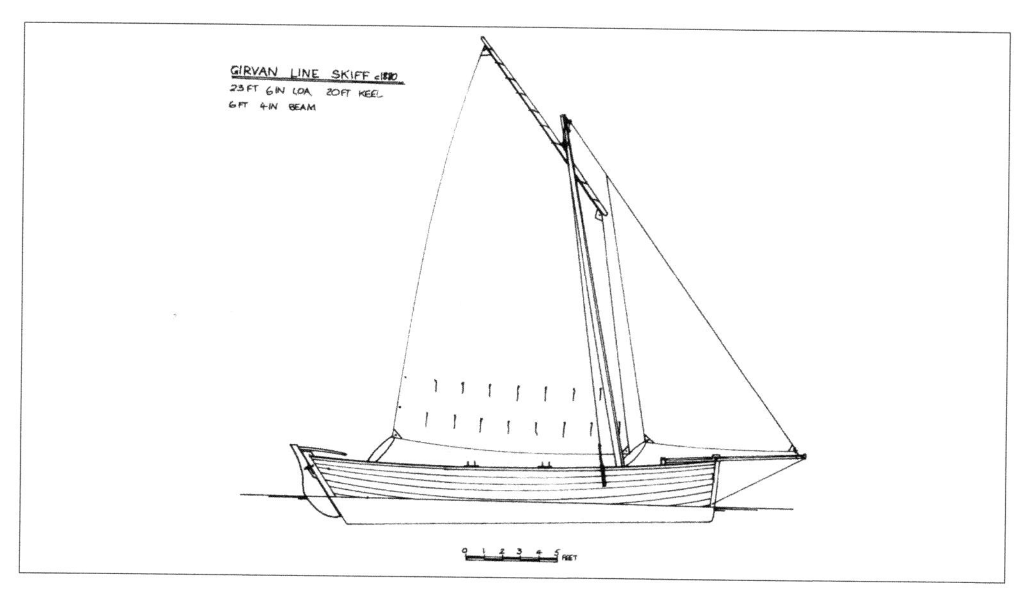
GIRVAN LINE SKIFF c1880
23FT 6IN LOA 20FT KEEL
6FT 4IN BEAM
0 1 2 3 4 5 FEET

but as the size of hauls increased and the cost of net decreased after the introduction of cotton, the nets became bigger so that by 1910 they were more than 300 yards long.

The introduction of ring-netting at Tarbert gave rise to serious arguments among the fishermen, with those from the upper reaches of Loch Fyne particularly in favour of retaining the traditional drift-netting and believing that the huge hauls of the ring-netters would all but empty the loch of herring within a few years.

Under pressure from local drift-netters, the Fishery Officer in Inveraray, at the north end of the loch, soon began to crack down on the ring-netters, believing the method to be illegal; he first seized a trawl-net in 1842. It set the scene and petitions were soon being drawn up against the new method, not only by the local drift-netters, but also by local lairds, and later by the Glasgow fish curers.

In 1851 an Act of Parliament was passed, making the use of any net other than a drift-net illegal. In 1852, sensing objection from the Tarbert fishermen, the Fishery Board sent two ships to patrol Loch Fyne: its own *Princess Royal* and the Admiralty vessel HMS *Porcupine*. The move had the desired effect of stifling the ring-netting, and those fishermen who did persist and were caught had their nets seized. In 1853, a Tarbert fisherman, Colin McKeich, was shot and wounded by the crew of HMS *Porcupine* while they were searching his skiff. The men responsible were barely reprimanded and thus, to the ring-netters, it appeared that such wounding was acceptable to the authorities.

The next year HMS *Porcupine* was recalled to join the war with Russia, and at the same time a cholera epidemic broke out around Loch Fyne. The disease had been brought to the area by French traders, but the villagers of Tarbert believed it to be a judgement from 'on high' against ring-netting and the fishermen were prevailed upon to give up the practice.

The method was revived a few years later, however, due mainly to a report being published which advocated the repeal of the Act of 1851. By 1858, there were 119

The motorised skiff *Nellies* underway. (*National Maritime Museum*)

boats ring-netting in Loch Fyne. Later that same year the tempers of the drift-net fishermen finally snapped, and there were many confrontations between the two factions – angry words were mixed with threats but, surprisingly, there was barely any violence.

SKIFF

In 1860, two amendments were made to the 1851 Act which further repressed the action of the trawlermen. These were brought about following considerable pressure from the Glasgow fish curers, who felt threatened by the increased supplies of fresh herring.

In 1861, a further amendment was passed, and by then HMS *Jackal* had been stationed in the loch. Then, in June, young Peter McDougall from Ardrishaig was shot dead while fishing off Otter Spit. The *Jackal*'s two crewmen, who were responsible for the shooting, were found not guilty of murder, and the ring-netters' resolve plunged to new depths in the face of a well-entrenched opposition. Later that year a fourth amendment was passed. Feeling heavily repressed by the strengthened laws against them, some of the Tarbert men reluctantly gave up trawling and bought drift-nets. Others who expressed a will to continue were confounded by the fact that the selling of illegally caught herring was now illegal itself, and thus any market to which they had previously sold their fish was closed to them.

By 1863, families were starving, some fishermen were imprisoned, and a Royal Commission report was published that condemned the repressive legislation. The

SETTING OFF TO THE FIELDS OF HERRING

next year, after a further commission again condemned the Acts, the beginning of a rebellion was felt on Lochfyneside.

Four years later, trawling had recommenced in earnest and some of the original objectors had joined in, as had one fish curer who had realised the advantages of the method. In that same year, 1867, under great pressure from all sides, the government repealed all the Acts and amendments, making the use of a trawl-net legal once more: within a year all trace of an era of law enforcement had disappeared.

THE TRAWL SKIFFS

As the practice of ring-netting took a hold on the loch, the fishermen began to adapt their skiffs to suit the new method. With a drift-net, the net is carried to the chosen spot and set in the early twilight, and the skiff lies to the net all night; at first light the net is hauled and the fish taken ashore. The difference with the trawling method is that the net is set and hauled perhaps many times during the night, and that the fishermen need room on the deck to work. The lightness of the boats did not alter, they remained extremely manoeuvrable so that they could easily turn while encircling the shoal. As the boats developed, two characteristics became apparent: the mast was moved further forward so that more deck space was created, and the mast was raked aft more to compensate for moving the centre of effort forward.

At the time of the repressive legislation, the skiffs cost between £12 and £20 to build. This would give the fisherman a boat of about 18 feet of keel (Scottish boats

Right: A painting of Tarbert Harbour, *c.* 1880, Loch Fyne, by David Muray ARA.

Below: Various double-ended skiffs on the beach at Iona.

Clinker-built Rothesay-registered trawl skiffs at Loch Ranza, Arran. (*National Maritime Museum*)

were generally priced according to length of keel) and a small lugsail and a net. These boats were rarely ballasted, so that they could maintain a good speed while evading detection by the fishery protection vessels. Once the fishing was legalised, they tended to ballast using stones from the beach, white ones being deemed unlucky by the superstitious fishermen and being thrown overboard. The size of boat grew thereafter up to 30 feet overall, 24 feet on the keel, and these vessels had a tiny forepeak that was used for storage under the short foredeck.

Most of them were built locally and James McLean of Ardrishaig was their most prolific builder; the fishery records show that out of an Ardrishaig-registered fleet of some 418 skiffs between 1887 and 1902, he built 140 of them. Other builders who were responsible for more than just a few were McTavish of Tarbert, Munro of Ardrishaig, James Fyfe of Rothesay, and, in the upper part of the loch, Donald Munro of Inveraray, who later moved to Blairmore to build a new generation of skiffs. The McLea yard in Rothesay also built several before being demolished in 1880 to make way for the new promenade. Other small skiff builders were Donald Brown in Torrisdale, Angus McLachlan at Crinan Ferry on the Crinan Canal, and Archie Smith at Tighnabruaich on the Kyles of Bute. Others who merely built a few skiffs included John Wardrope of Campbeltown and John McEwan of Ardrishaig. Lachie Lang of Campbeltown started building skiffs as they were developing into fast seagoing boats. Today, the skiff *June Rose*, at 30 feet, is perhaps the only example of a smaller trawl skiff, although she was built around 1900 by James Fyfe, and could be termed one of the new generation of skiffs.

Carvel-built Lochfyne skiffs, again at Loch Ranza, these being bigger than previous trawl skiffs. (*National Maritime Museum*)

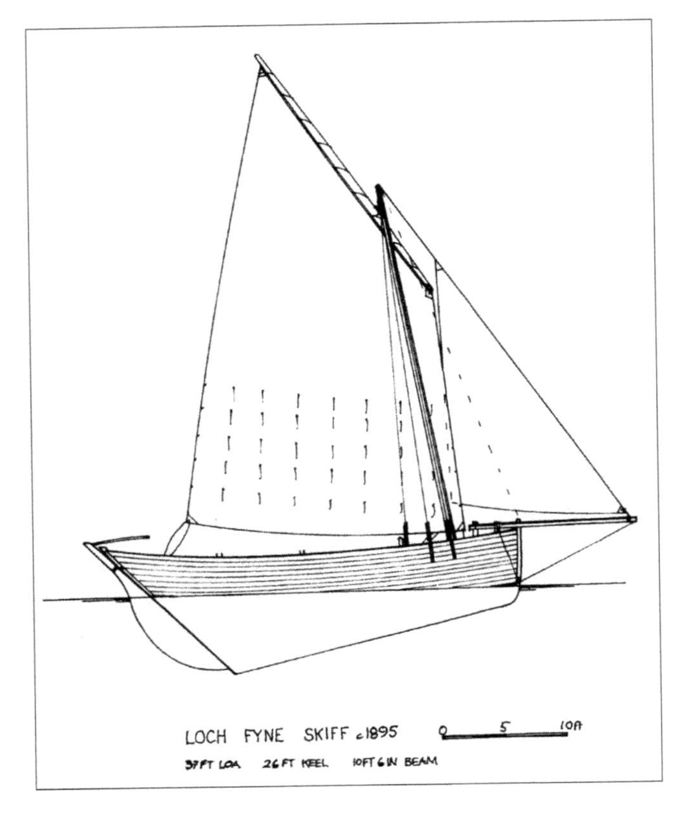

THE LOCHFYNE SKIFFS

In 1882, two new skiffs that were instantly recognised as being different from the rest of the fleet arrived in Campbeltown. These two boats, *Alpha* CN185 and *Beta* CN184 were launched in Girvan for Edward McGeachy of Dalintober, and had 25 feet of keel. The main feature that gave them superiority was their forecastle in which it was, for the first time, possible to sleep and cook aboard. Previously, the fishermen had either camped ashore while fishing away from home, or had taken with them a smack for the whole group of men to sleep aboard. Although the price of these boats had risen to £72, this was deemed worthwhile because of the greater ability to fish away from home. Like the trawl skiffs, these Lochfyne skiffs, as they became known, set a standing lug and a foresail on bowsprit.

The hull was split into three compartments, the stern sheets, the hold and the forecastle. The forecastle, or den as it was referred to, was entered through a small sliding door on the starboard side of the hold. It had room for four men to sleep – two on the lockers and two on hinged shelves – and the boy generally slept on the floor.

Over the next forty years some 540 skiffs were built around the Clyde. As well as those builders mentioned earlier, Robert Wylie of Campbeltown was renowned for his superior skiffs. Tarbert and Ardrishaig had five yards at one time. Several boats were built by the fishermen themselves – Matthew MacDougall of Carradale building some in the field alongside his house. Murdo MacDonald of Alligin, Loch Torridon, built the *Queen Mary*, UL138, in 1910 on similar lines, and she can still be seen at the Gairloch Museum. She had been built for fishing out of Badachro, where herring were cured in 1886. There were two other sister boats built for local owners: *Lady Marjorie* and *Isabella*. Large amounts of herring were reportedly caught in the Gairloch and John Watson was a curer there.

The Lochfyne skiffs have been described as one of the prettiest of British fishing boats, yet their lifespan was short due to the introduction of engine power. It was in 1907 that Campbeltown fisherman Robert Robertson became the first fisherman on

The skiff *Fairy Brae*, 55CN, under full sail on a still morning off Davaar Island, Campbeltown, *c.* 1913. (*National Maritime Museum*)

Similar skiffs alongside at Campbeltown, *c.* 1913. (*National Maritime Museum*)

the west coast to install a motor into his boat, the *Brothers* CN97. He chose a Kelvin 7.9-hp single-cylinder engine as did many fishermen in years to come.

The Kelvin engine came from the Bergius Launch & Engine Co. of Glasgow. Walter Bergius had set up the company in 1904 to manufacture motorcars, but within a few years launched his marine engines.

They were an instant success on the west coast, and the name Kelvin was to be synonymous with the fishing industry over the next fifty years or so. The 1908 engines were of the poppet valve type and cost £50 with another £20 to have them installed. By the next year there were a dozen Campbeltown skiffs with engines, and by 1911, no pairs of sailing boats remaining. New boats were having engines fitted before launch – the author's old boat *Perseverance*, CN152, being built by Robert Wylie in 1912 and having a 15.20-hp Kelvin. At 40 feet overall (29 feet keel) she was the largest boat in the fleet at her launch. Unfortunately, she was lost off the Portuguese coast in 1995, three months after being sold to new owners. By 1913 there were seven different models of engine available to the fleets.

Although the Lochfyne skiff was capable of having a propeller shaft put through the hull, offset to starboard because of the practice of setting and hauling nets over the port side and thereby avoiding any entanglement between net and prop, conversions were never as successful as on other craft such as the fifies. Many more boats were built during the second decade of this century, although not in such numbers as in previous decades – the fishing was just not as viable with the herring in decline, and with the war in Europe taking its toll on the lives of the fishermen. After the short period of only forty years, the skiffs were superseded by a new breed of boat around the early 1920s. Today only a very few skiffs have survived, the majority emanating from yards on the east coast and built during the 1920s. The *Sireadh* was built in 1922 by James Miller of St Monans, Fife, for the Monroes of Minard, and fished until 1936. Although no longer rigged with a lugsail, she has survived the passage of time by sailing all over Europe until returning to Britain and is now undergoing a full rebuild.

Another boat is the *Fairy Queen*, CN196, built in 1926 by Nobles of Fraserburgh for James Robertson to replace the earlier Lochfyne skiff of the same name that had been built in 1885. After being sold to Lybster in 1947, she was subsequently taken to Ireland's west coast, where she remains today. The Ardrishaig 1908-built *Clan Gordon* sails from Ullapool and at present is the only one of these vessels to have a lug rig.

The design of the Lochfyne skiff was deemed so successful that various other pleasure versions have been built over the years, and the type remains one of the best small fishing boats that Britain has ever seen. These yachts include *Miranda, Kirsty, Nel, Rowan* 4, *Nighean Donn* and *Craignair.*

THE CLYDE BOGGLES, FIRTHIES & SMACKS

Finally, mention must be made of some other boats that fished the waters of the Clyde. As we've seen in Orkney, 'Firthies' is a general term applied to all vessels typical to

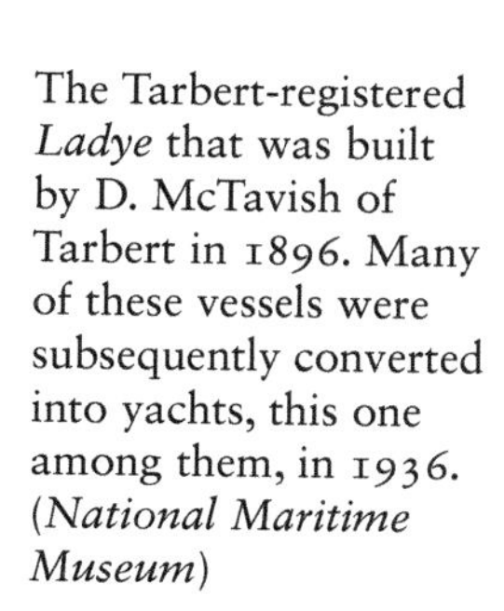

The Tarbert-registered *Ladye* that was built by D. McTavish of Tarbert in 1896. Many of these vessels were subsequently converted into yachts, this one among them, in 1936. (*National Maritime Museum*)

Motorised skiffs rafted up on the beach at Waterfoot, Carradale, in 1913. Note the propshaft on the starboard side. (*National Maritime Museum*)

The Campbeltown-registered skiff *Good Hope* in regatta rig. (*Angus Martin*)

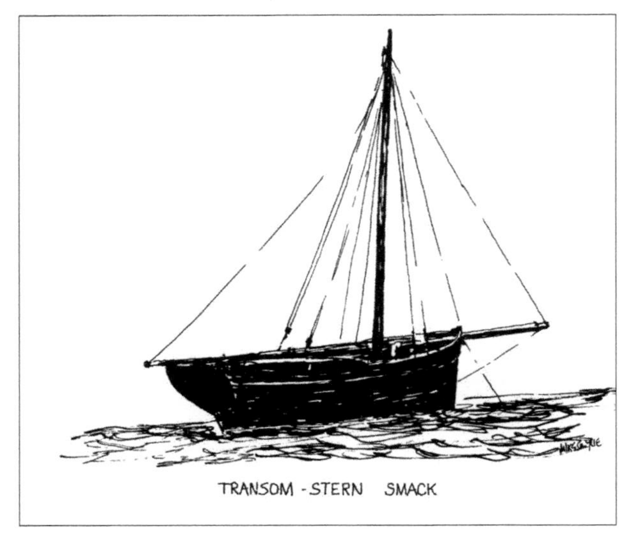

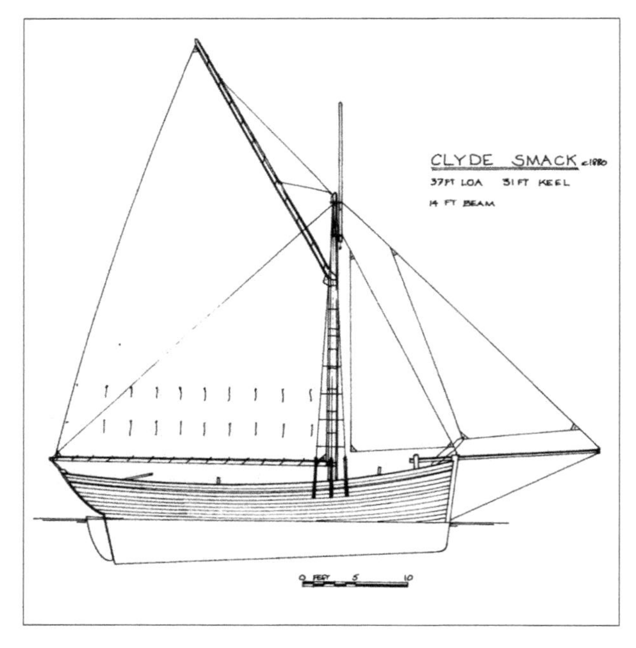

a firth. With Glasgow rapidly becoming an important port for the import and export of goods to and from Scotland, the Clyde was full of large sailing vessels. Shipbuilding was growing in the nineteenth century on the upper reaches of the river to supply the ever-increasing need for boats. Among the islands and lochs of the Clyde there were even more vessels supplying goods to the more outlying areas. The gabberts were a type of trading vessel designed to pass through the Forth and Clyde Canal to trade with the east coast. All three Scottish canals – the Forth & Clyde (1709), the Crinan (1801), and the Caledonian (1822) were built to encourage and promote sea trade, and all three were used by the local fishing fleets in their quest to find supplies of good fish.

The boggies were luggers of the area, more often than not brought into the region from away. The word 'boggie' is reputedly said to come from the word 'bogue' – to fall off the wind – as these luggers tended to do. A similar distinction can be found among the Hastings luggers called bogs.

In Campbeltown there were twenty of these luggers in 1880. More than half of them had been brought in from the Cornish fleets that we shall learn about over the next few chapters. Two were from the local fisherman Matthew MacDougall, aforementioned. After the Kinsale mackerel fishery declined in 1895, most of the luggers were sold off and converted for trading, a few even being set up on the foreshore at Campbeltown to house destitute families.

The smacks were single-masted gaff-rigged boats that were common all around the British Isles so that no particular type was typical of any one area. There were transom-

Maggie Campbell, BA472, at Portpatrick, *c.* 1935. This small 22-foot nabby had been built in 1890. She was one of the last to survive, their lightness tending to mean they had short working lives. (*National Maritime Museum*)

sterned smacks, counter-sterned smacks and, to a lesser degree, round-sterned smacks. They seem to have evolved from Norse influence, similar to the gabberts. When the first smacks arrived in Campbeltown in 1840, they cost £120 for a 22-foot overall boat. They were full-bodied with a 9-foot beam. Although expensive when compared to the skiffs, they became popular among some of the fishermen, who also used them for trading when the fishing was poor. They became larger, and local builders soon began to build them for the local market.

The smack's introduction marked the beginning of the use of the fore and aft rig, which was gaining favour around the British coasts. The smacks survived through periods of great change, and were still in use up to the First World War, although their numbers were few. With the decline of the herring fishery and the introduction of the motor, they suffered the same fate as many other types of sailing fishing boat and were either left to rot or sold on. The rig, however, never did really catch on among the Clyde fishermen, most skippers preferring the standing lug for which the fishermen of the west coast were famous.

THE PORTPATRICK LINE BOATS

A different type of boat evolved out of Portpatrick for the near-shore cod fishing. Here, alongside the fast flowing waters of the North Channel, the fishermen had to cope with very different waters from those of the shallower Clyde. The harbour itself grew up in the seventeenth century as the crossing point to Ireland, as its name

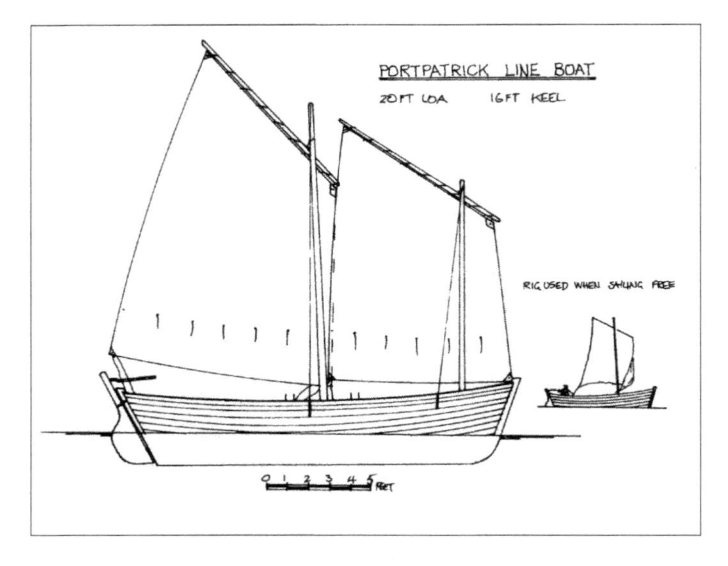

suggests (Patrick's port), and was built in 1820. By 1862 the railway had arrived from Stranraer, but by then the latter had already taken the business of the Mail boats. In 1875 the harbour was abandoned altogether, leaving the few fishermen alone. It enjoyed a brief revival in the 1930s when the Clyde ring-netters landed herring there.

The Portpatrick line boats show an amazing likeness to the Orkney and Stroma yoles, possibly because of the nature of the waters they all work in. The waters of the North Channel are perhaps similar to those of the Pentland Firth, which just goes to show that boats evolved through their natural working surroundings more so than because of any other influence. This can also be seen in other areas – the fifies developed on similar lines to the Cornish drivers because of a similarity of working habitat. The line boats are beamier and rounder than the Clyde skiffs, and generally set two lugsails because of the need for more power in these stronger tides. Thomas Muir's *Brothers*, BA318, was built by J. MacDowell of Portpatrick and she fished for cod between November and May, to begin with 1 mile offshore, but then gradually further out as the season progressed. They worked the lines with 800 hooks per line. In the summer, they trawled for plaice and turbot. Further south, a few boats worked out of Port Logan, where it was once hoped to base a cross-Channel harbour. Being extremely open, it was never really an option.

THE SOLWAY FIRTH

Between the Mull of Galloway and Annan, there are several good harbours that sprang up in the seventeenth and eighteenth centuries. Mostly they were built for the coastal trade, as numerous smacks plied these waters and afar. Coal was imported and agricultural produce taken away. Many of these harbours grew up on the rivers leading to towns, and it was not a rare sight a hundred years ago to see a trading smack or schooner sailing upstream to towns such as Gatehouse of Fleet, Dumfries or Dalbeattie, navigating in waters with only inches between riverbed and keel, and only yards between riverbank and gunwales. This could probably be said of many coastal towns all around Britain. Some of these harbours sprang out of fishing communities that have existed for centuries. Port William in Luce Bay is one example, being

On the beach at Rothesay, showing clearly the propeller of the 40-foot carvel-built *Psyche*, built in 1915 by John Fyfe of Port Ballantyne. (*National Maritime Museum*)

founded in 1776 and described as 'a small, neat village consisting of low houses, well built facing the sea'. Although now devoid of many fishing boats, it is believed that the port grew up out of the ancient fishing settlement of Killantrae.

Annan has had a harbour since the eighteenth century, and in the early nineteenth century the quay was 250 feet long, increasing to 400 feet during the next century. The first fishermen, however, only arrived from Morecambe in 1854 after they were driven there during a gale. They immediately discovered rich fishing grounds so they sailed home and returned soon afterwards with their boats and families. Initially there were four families – Holmes, Baxter, Woodman and Woodhouse – settling there, but others followed. They took with them their 30- to 35-foot clinker-built, half-decked, square-sterned, cutter-rigged trawl boats and their smaller 13- to 15-foot sprit-rigged drift-net or whammel-net boats. Before long the trawl boats became counter-sterned, narrower, carvel-built craft, while the smaller boats adopted the lugsail, and became double-ended.

Although fast tea clippers were built in the harbour, the four local builders, Messrs Shaw, Neilson, McGubbin and Wilson relied upon the building of fishing boats for their basic work in 1870, mainly local whammel-net boats. Whammel-netting involved the staking out of one end of a net inshore and rowing around in a circle paying out the other end, and then hauling it in. Salmon were caught in this way on the Scottish side of the Solway. The whammel boats were originally 15-foot clinker craft, but they soon grew in the latter part of the nineteenth century into 20-foot boats, and were carvel-built. The *Dora* was built by James Wilson in 1900, and was 19 feet 3 inches overall. They appeared similar to the Clyde skiffs, with well-raked sternposts and upright stems. Their planking was ¾ inch on 1¾ x 2¼ oak frames, and they had a deep keel. To protect this against chafing on the coarse local sands, they had a cast-iron keel bolted on, with a

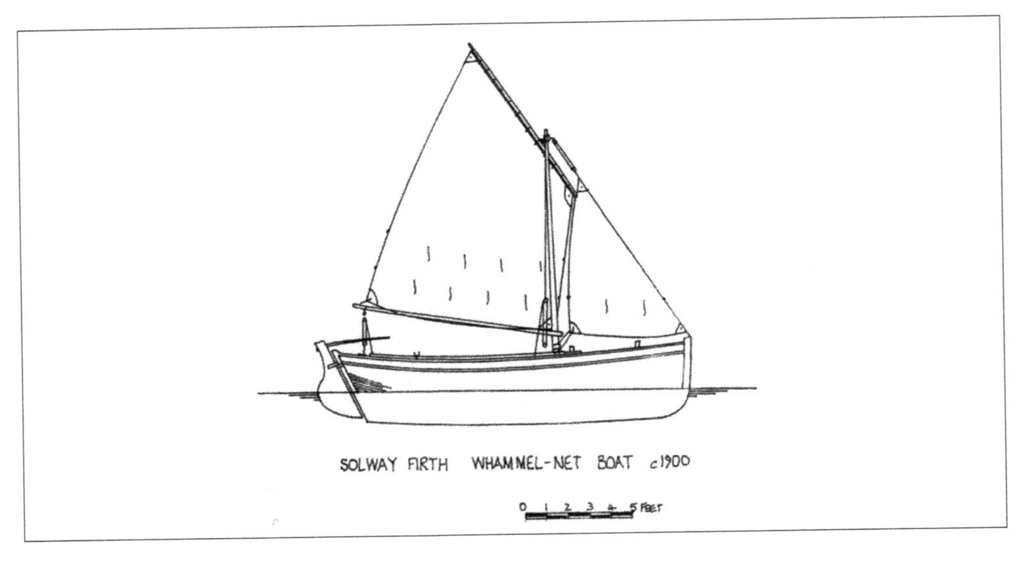

rockered bottom edge. They were decked except for a narrow cockpit, and had a small cuddy forward with seats, lockers and a stove, which two crew could squash into while awaiting the tide. The mast was carried well aft, unlike other Scots boats, said to be a leftover from the rowing days. Engines were fitted in the 1930s, and today unfortunately no examples remain. Similar boats worked out of the other ports already mentioned along this part of the coast.

One example of a boat built in Waterfoot, Annan, is *Noreen*, BA63, a salmon coble from 1926. She originally had a standing lugsail, and in 1928 she had a Kelvin 7.9-hp engine installed. Today she still fishes for salmon out of Ballantrae, where she is kept, and is only one of a few surviving boats of this age that were built there.

The design of fishing boat altered drastically in 1922 with the coming of the canoe-sterned boats *Falcon* and *Frigate Bird* into the Campbeltown fleet. These were introduced by that most innovative of fishermen, Robert Robertson, who years earlier had introduced the first engined skiff. Within a few years, after initial scepticism, the fishermen 'realised the advantages of the new types – easier conditions and higher earnings – and adopted them. Immediately, generations of sailing fishing boats had come to an end, and the birth of the Scottish fishing boat, which nearly all recognise these days, had arrived.

CHAPTER 6

Ireland: The Emerald Coast

Oro, mo bhaidin, ag snamh ar an gcuan...
('Yoho, my little boat, floating on the sea...')

Given the proximity of Ireland to Britain, and the history of British control over the island up to the early part of this century, it is inevitable that British influence played an important part in the shaping of the Irish fleets. However, the British fleets were not alone in their influencing, much as they would like to have been. In the same way that Scottish and Welsh fleets (and indeed those from many parts of England) evolved from the Norse control over large tracts of Britain in the ninth and tenth centuries, so did those of the Irish. Furthermore, especially on the south and west coasts, a completely different influence produced a design unlike any other British craft.

The Irish markets were peculiarly inaccessible to the home fleet. What caused this is uncertain: possibly the British dominance, the lack of a coherent infrastructure, or the nineteenth-century emigration to the USA. It nevertheless resulted in a large amount of cured herring being imported up to the 1850s. In 1844 Ireland imported 127,770 barrels of herring and 17,683 cwt of cod, ling and hake. Nearly all this came from Scotland, only 12 miles away at the nearest point. This compares with a total annual Scottish production of about 400,000 barrels of herring in 1830. On the contrary, however, 180,000 barrels of herring were exported alone from Downings in Donegal in 1775; this just shows some of the injustices of the Irish food chain.

The huge shoals of pelagic fish off the southern coasts, which were the subject of various reports in the eighteenth and nineteenth centuries, were never capitalised upon on a scale that could possibly have compared favourably with the North Sea. Consequently, the design of boat never matured enough past the basic crude forms until well into the nineteenth century. The markets that did thrive were mostly under foreign control with most of the fish disappearing into export with hardly any benefit to the local population.

Contrary to popular belief, Ireland does have a diverse collection of traditional fishing boat types even if its fishery suffered from neglect, under-investment and British

A Drontheim at Whitecastle, Lough Foyne, County Donegal, with an unusual pleasure or racing rig, *c.* 1952. (*Donal MacPholin*)

control over centuries of its sometimes tragic history. That the terrible famine and the causes affected fishing in the mid-nineteenth century goes without saying. While some of the regional fishing craft have developed from similar roots to some English, and particularly Scottish, vessels, many are peculiar to Ireland and others have their roots from different Continental Europe directions. In two cases, established boat types have been introduced into the fleets direct from Britain. The boats themselves can loosely be split into two groups – those coastal and deep-sea craft and those confined exclusively to river fishing.

THE NORTHERN YAWLS

The north coast of Ireland has much more in common with the south-west coast of Scotland than it has with the rest of the UK and many other parts of the Irish mainland. The trading links between Galloway, Kintyre and Islay have existed for centuries and have seen goods transferred from coast to coast. The same has been true of the population, with many in Kintyre being of Ulster extraction, and these links can be traced right back to the tenth century when the Vikings were in firm control. The Drontheims, or 'Norway yawls' as they were often called, were imported into Ireland in kit form direct from Norway in much the same way as the Shetland sixareens were brought into the Northern Isles. In this case, the boats came from Trondheim, and hence their name as a variation of that place. When the Caledonian Canal opened in

1822, trade increased, and, instead of carrying kit boats, the trading boats began to import more timber for boatbuilding and domestic use. The fishermen then began to build their own yawls, adapting the Norwegian design to take account of their own unique conditions. The main difference between the Norway and Shetland boats and these home-built Irish boats was the rig. The Irish adopted the sprit rig and mostly with two masts. This arrangement was common in and around the Irish Sea in the eighteenth century where, from right down in Cornwall and along the whole of the Welsh coast, sprit-rigged boats worked. Indeed, the sprit was the most popular rig for small boats at that time on all the North European waters.

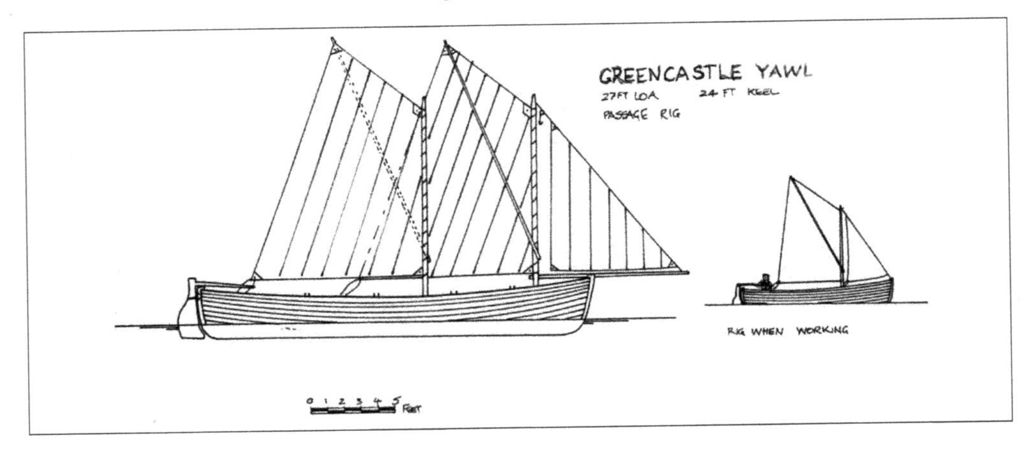

The Drontheims were to be found all along the north coast, and locally they took their name from the fishing communities they served, hence 'Greencastle yawls' and 'Skerries yawls' – the word yawl, as we have seen, coming from the Norse 'yol'.

The boats were about 25 feet overall, clinker-built with a sandstroke – a tapered board fixed to the keel as the first plank. This enabled the keel to be replaced at frequent intervals with ease as the keels were prone to rapid wear from being dragged up and down the stony beaches. The cost of a typical boat in 1890 was £11.

Further to the west, the Drontheims – or skiffs – were found along the Donegal coast and out to Inishbofin, and at Achill, where they were called Achill yawls. These were also referred to as the 'big salmon boats'. They are one of Ireland's indigenous working craft that almost disappeared in the twentieth century, although their importance in the lives of the people of Ireland's biggest island was unimaginable in terms of today's standard. They were the workhorses of Achill. From the mid-eighteenth century until the mid-nineteenth century these boats were the means of transportation and trade. They fished, collected seaweed, carried limestone for fertiliser, turf for fuel, stones for building and timber for new boats. Their fishing was seasonal again, as elsewhere, with herring fishing, line fishing and oyster dredging: in December 1867 there were 100 boats dredging oysters in Achill Sound, the shellfish which were later carried to Galway.

Original Achill yawls were about 18 feet in length, carvel-planked, rigged with one dipping lugsail, which was sewn by the women of the island and double-ended, the design said to have come from the west coast of Scotland. Nowadays they are often longer – up to 28 feet – and some have square sterns to take outboard motors. Their build was said to be relatively basic, due to the shortness of finance to pay for them, and the hull tarred. Some had, indeed, been built with money from the Congested Districts Board, and in 1892 the CDB reported that there were 226 yawls and rowing boats on the island. Though this doesn't specify the actual numbers of yawls, it is presumed to be substantial. However, their numbers dropped to three surviving vessels in about 1980 when Dr Jerry Cowley bought an old yawl and subsequently learned how the families raced their yawls every summer. Because of this a resurgence in the building of new craft has come about so that some thirteen yawls often race during the annual Cruinníu Bádóirí Acla or Achill Sailing

A 26-foot Drontheim at Inishowen, County Donegal. Built in the 1950s, it was owned by Michael Doherty of Portronan, who fished it until the 1970s, after which it passed to the Ulster Folk & Transport Museum. (*Donal MacPholin*)

Festival held between July and August. Some of these new yawls have been built on the island by builders such as Poraigh Owen Pattern of Saulia and the O'Malleys of Corraun, though not as cheaply as they used to be when money was in very short supply. The oldest afloat is the *Cutty Sark*, said to be coming up to 100 years old. Today's rigs, though, are a far cry from those of a century ago: modern sail cloths with radial cuts on aluminium spars. Some boats have kept their traditional black colours, though many others are brightly painted, and some come from Clare Island to the south.

Today, the old Drontheims and yawls rarely fish; more often than not they are seen racing at local regattas. But Drontheims are still being built at the yard of James MacDonald & Sons of Greencastle, previously of Moville, where they had been building them since the 1750s, albeit mostly recently in a larger half-decked form of about 40 feet overall. Each trawler owner tends to have one of these vessels to supplement his normal fishing with salmon between about May and July.

EAST COAST YAWLS

The Greencastle yawls were also taken over to the Scottish islands of Islay and Colonsay, as well as parts of Kintyre. Here they were called the 'Greenies' or 'sgoth Eireannch' in Gaelic – literally 'Irish skiffs'. They were generally used for long-lining for cod and saith, and they usually took the catch back to Ballycastle to sell it there. A number of these craft were delivered aboard the steamer to Campbeltown, many being dropped off near Sanda Island after the fishermen had rowed out to collect them.

The Groomsport yawls, described as whale boats, fished with longlines, and sometimes acted as pilot boats. They were similar to the Drontheims, retaining a strong Viking resemblance in being double-ended and clinker-built. They were also said to have been imported from Norway in kit form, but the similarity ends there. They ranged in length between 20 and 30 feet, the larger ones setting two dipping lugs, the forward being the smaller. These boats were fitted with Kelvin 7/9-hp engines after about 1908. Similar boats were found at Donaghadee, and many of these were built on the Copeland Islands by James Emerson. The small boats, called variously Killough yawls and Newcastle skiffs, set only one sail, and the rig was considered to be extremely crude, sailing to windward being virtually impossible. They fished with longlines, catching haddock and codling in winter, plaice, turbot and sole in spring and mackerel and herring in the late summer. The Newcastle boats were referred to as 'skiffs' as they tended to be smaller, because Newcastle only had a small drying harbour, and the rig was also different in that the lugs had booms that were lashed to the mast. Nevertheless, they were renowned for their speed and ability to cope with the short, sharp sea of the North Channel. At the Skerries, similar craft were called Skerries yawls, and the same for Kilkeel yawls. In fact, similar craft all the way down the coast to Dublin and beyond were referred to as yawls though there were localised differences due to the builders' preferences.

Not a lot seems to be known about the Arklow yawls that were transom-sterned, fully decked dandy-rigged craft built specifically for herring and mackerel fishing with a length of anywhere between 25 and 35 feet. It seems they were a development of the boats that dredged the oysters offshore from the river mouth. The yawls appear to have been first built in the 1860s by John Tyrrell and became popular in usage. It is possible that Arklow hookers, a term occasionally come across, are the very same craft, as the mackerel were taken on longlines as well as in drift-nets.

There are still a few Ballyhack yawls in this small village on the east side of Waterford Harbour, though today's are motorised whereas the original yawls were gaff rigged. These transom-sterned half-deckers were fished in the main for herring and mackerel, though some set lobster pots according to John Carroll, whose family had been building them at their yard alongside the harbour there. In the back of the yard is one of the oldest yawls, the *Star of the Sea*, which John hopes one day to restore. Either that or build an original yawl.

Groomsport yawl *Lissey*, *c.* 1901.

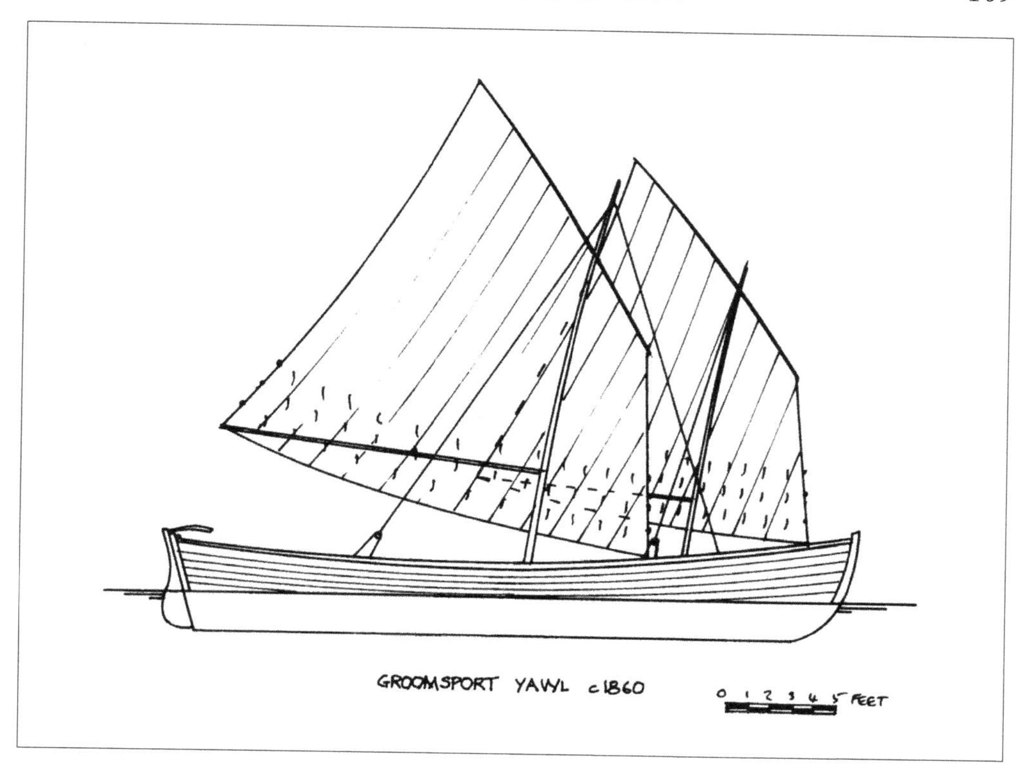

A Newry-registered nickey *Uncle Tom.*

NICKEYS, NOBBIES & ZULUS

When the well-known Cornish builder William Paynter left St Ives in 1875, he chose Kilkeel to set up a boatyard. By this time, his Cornish luggers were to be found all over the west coast. The Manxmen had adopted the design to replace their earlier dandies in the 1860s, and had called them 'nickeys' (see Chapter 7). Many of these nickeys sailed over to harbours such as Kilkeel, Ardglass and Annalong where they found favour among the local fishermen.

Sensing a market for Irish-built boats, William Paynter moved his entire operation over from St Ives, and built a series of the luggers. Costing about £700, they were 50 feet overall, 45 feet of keel and 15 feet beam approximately, and had two lugs – a dipping lug main and standing mizzen – as well as driving a kicker eighty, staysail, topsail and balloon jib for use in differing weather conditions. The nickeys were renowned throughout the Irish Sea for their great speed – up to ten knots – and were considered a good, well-proven design.

Paynter's venture, however, was not a success. He was undercut by the then-established Manx builders, and a disastrous fire gutted his premises a year after arriving. Because of a decreasing market, he found that he could not sustain the business he needed to, although he did manage to recover sufficiently after the fire to build several more vessels. Realising that his intended market would not materialise, he returned to his native Cornwall in 1883; his apprentice John MacKintosh continued in his place after his departure.

Ardglass harbour with nobbies and a couple of Scots boats. Note how cluttered the decks are. (*William A. Green*)

These Irish nickeys fished mainly for herring all along the coast, and many sailed south each year to join the Kinsale mackerel fishery. When motor power came along, the nickeys were ideal for conversion, most receiving Kelvin units such as the 13/15 model and later the 26-hp engine. Some nickeys survived the First World War to be converted as motor fishing boats. One such vessel, the 51-foot *Mary Joseph*, N55, was built by Paynter in Kilkeel in 1877 and survived fishing for nearly 100 years, until she was retired in the 1970s and now lingers in the ownership of the Ulster Folk & Transport Museum.

In much the same way as they adopted the nickeys, the Irish also took to the Manx nobbies (see Chapter 7). Smaller than the nickeys, they set two standing lugs and no topsails. They were chiefly built in and around Portavogie, becoming known as Portavogie nobbies, although some continued to be brought in from the Isle of Man.

On the west coast it was the Connemara nobby that was introduced by the Congested Districts Board from the Isle of Man in the late 1890s to encourage full-time fishing. Between 1898 and 1905, twenty-one were built in Kilronan, Mweenish, Galway and Lettermullan.

The CDB also introduced fishing boats from the east coast of Scotland into Donegal and Mayo in the very late nineteenth century and into the twentieth. These were zulus, though smaller at about 50 feet in length, whereas a full first-class zulu was over 80 feet long. They've often been termed half-zulus. One of these was the *Leenan Head*, built in Banff, which came to the small island of Inishbofin, off the north coast of Donegal, in September 1909 to fish, after a spell in Loch Swilly. This

Portavogie nobbies at Ardglass, *c.* 1901. These half-deckers had accommodation in the forepeak in the same way as the Clyde boats. (*Kenneth Smith*)

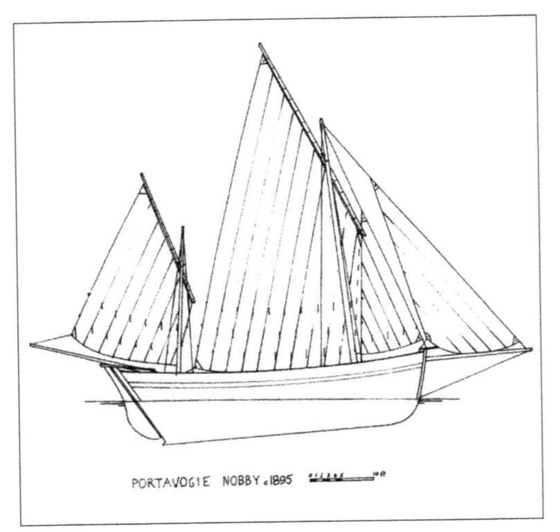

The Newcastle skiff *Jane* at Dundrum around 1930. (*Author*)

boat, like many others, had been built with finance from the CDB and leased to those fishermen willing to fish at a time when the authorities were trying to pull the fisheries of the north-west coast out of the Middle Ages. Whereas the southern mackerel fishery had produced a boom to local economies, this coast received little investment and remained primitive throughout most of the nineteenth century. With a vibrant Donegal and Mayo herring fishery fished mostly by Scottish boats towards the end of that century, the CDB decided to introduce these boats to enable the local inhabitants to compete. Their older boats and currachs were unable to do this. To begin with, they were built in Scotland and brought over but before long Irish boatbuilders and small yards were building their own versions. The *Leenan Head* continued to fish for a number of years until she ended up as the ferry boat out to the other island of Inishbofin off the Galway coast. Today she is based in Paimpol, Brittany, and is probably the only one of these Irish zulus still sailing, although the Scottish, 1921-built *Bracken Lass* still sails. She was built as the *Thomson's*, BCK397, and fished out of Buckie for three years until being sold to Balbriggan. She remains in Irish ownership.

Nine smaller versions of the Scottish zulu were built in Lettermullan and Aran. Some were gaff rigged on the main, retaining the standing lug mizzen sheeted to a bumpkin, because the mizzen mast had to be stepped further aft. The first Irish-built zulu was the 46-foot *Bencar*, built in Mweenish in 1898. In Donegal, where the zulus found more popularity, they were built at Killibegs or imported into the region from Tyrrells of Arklow, Skinners or Mahoods of Portavogie. The CDB had 187 nobbies and zulus built, ninety of which were built in Ireland. A nobby cost £205, a zulu £210, compared with a Manx nobby at £186 and a Scottish zulu at £157.

TRAWLERS

Various types of trawler operated along this coast. Brixham trawlers were common, (as they were off the Welsh coast) and especially around Dublin. Some of them were owned locally, although most travelled in convoy from ground to ground. Belfast smacks worked out of Groomsport and other harbours of Belfast Lough. Smacks evolved all over Britain in the late nineteenth century, and the Belfast smacks were similar to those found in the Clyde. The similarity of the smack all around the British Isles was almost certainly because the gaff rig was powerful enough to pull heavy trawls, yet was easy to handle. It probably originated, from a fishing point of view, in the North Sea, as we shall see in later chapters.

Also found in small numbers working off the north coast in the early part of this century was the clinker-built, yacht-like Antrim cutter, a trawler with clipper bows and a counter or round stern.

THE IRISH WHERRIES

IRISH WHERRY c.1840

One common group of vessels in the early nineteenth century was that of the Irish wherries. These were peculiarly the only schooner-rigged fishing vessels to operate in the waters of the British Isles. Generally they ranged in size between 20 and 50 tons, and were crewed by seven or eight men.

Some were decked over completely, but most originated as half-deckers, perhaps being later fully decked over. They fished mainly for cod, ling and haddock with longlines, and travelled the length and breadth of the Irish coast in search of fish.

Numbers of boats fishing in 1802 were as follows: Ringsted 7, Howth 7, Malahide 3, Rush 16, Skerries 36, Balbriggin 9, and Baldoyle 9. Sometimes, because of their concentration there, they were termed Skerries wherries. They seem to have ceased fishing by the middle of the century. Several suggestions have been put forward as to the derivation of the word 'wherry': some say from 'ferry', others from 'wherret' (to hurry), while others say it has Norse connotations. Another plausible origin is to be found in the Welsh 'chwerw' (pronounced in a similar way to 'wherry' with the throaty 'ch' sound), which means bitter, as in beer, which could have arisen as a result of a trading link between the two countries.

THE SOUTH AND WEST COASTS

Cork Harbour had various fishing craft, among these the Rathcoursey hooker and the Loughbeg shrimper, the latter being a small gaff-rigged open boat that trawled for shrimps within the confines of the bay.

The Kinsale hooker is another type of vessel that only appears in the occasional report such as that by Captain John Washington when looking into the 1848 gale that caused a large loss of life, and the loss of many fishing boats, on the east coast of Scotland (see Chapter 2). Washington shows a 40-foot Kinsale hooker with counter-stern in his report. It seems that these hookers were developed for the longline fishery and are said to be a development of Dutch craft that fished these waters in the eighteenth century, though there could have been other influences at work here. Another interpretation of the term comes from the small fore-and-aft-rigged craft of medieval times called the 'urea' – later corrupted to 'howker'. These were easily handled coastal craft used by the Spanish, who frequented much of this coast in the sixteenth century.

They were generally regarded as unseaworthy from Kinsale, which was at the time the only place on the south coast engaged in the herring fishery to any extent, although huge shoals were reported elsewhere. Dungarvan, further to the east, had 163 similar hookers employing 1,000 men in 1825 long-lining for cod, ling and hake. Dungarvan cod and ling were considered superior in quality to Newfoundland fish. The trawl-net had in fact been condemned locally when it was introduced in 1730. One hundred years later there were sixty-nine decked vessels averaging 13 tons and employing 320 men, forty open boats with 200 men, and 270 rowboats with 1,080 men.

In his *The Scenery and Antiquities of Ireland*, Joseph Stirling Coyne found Youghal in about 1842 as 'a place of little maritime trade; as a harbour it is chiefly resorted to by fishing-boats'. He continues by giving an early description of the typical fishing boat of the time:

> These craft, called *hookers*, are generally 10–20 tons burthen, and are open, with the exception of a few feet of deck in the forepart of the boat, beneath which is a small cuddy or cabin, in which the crew repose during the brief intermission of their toil while engaged in fishing. The boats are rigged cutter-fashion, with a large fore-and-aft mainsail, a foresail, and a jib; and though without the defence of a deck,

Towelsail yawl *Saoisse Muireann*. (*Author*)

it is almost incredible in what a heavy sea they will live. The Youghal hookers are esteemed the best and most seaworthy boats built on this coast.

Coyne also mentions the Nymph Bank and its inexhaustible stock of fish – 'both round and flat' – and notes that the fishermen of the coast take little advantage of this prolific fishing ground but instead prefer to go fishing across the Atlantic at Newfoundland.

The south-west mackerel yawls are better documented thanks to the replica boat that was built in Hegarty's Yard at Oldcourt. This came about after, in the early 1990s, the lines were lifted from an old mackerel yawl, *Shamrock*, which had been built by Skinners of Baltimore about 1910 and was among the most famous of these craft. At the time she lay abandoned on the beach near Schull, just down the coast from Ballydehob, where they took off the sections and measurements to allow a replica to be built. Then Nigel Towse, along with Liam Hegarty, decided to build a replica, this being *An Rún*, built mostly by eye and launched in October 1995. After that three more yawls were built at Hegarty's boatyard and recently one was in the South of France and another in Hegarty's though the whereabouts of the other is doubtful.

The other type of boat from this coast is known as the towelsail yawl, often called the Roaringwater Bay lobster boat. Nigel Towse has the only surviving original boat, *Hanorah*, which was originally built on Heir Island, as many of them were, in 1893. The name 'towelsail' comes from their habit of stretching out the sail over a boathook at the bow for the three crew to sleep under, for they ventured for miles along the coast, being away fishing anywhere between Cape Clear and Dursey Head, up north, for seven weeks at a time. Until relatively recently these boats were unknown outside the immediate area, until, that is, Ballydehob-based teacher Cormac Levis wrote and had his book *Towelsail Yawls – The Lobster boats of Heir Island and Roaringwater Bay* published in 2002. Since then several more new boats have been built by Hegarty's boatyard including *Saoirse Muireann* for Cormac himself.

Hanorah had been built for Con Harte, who passed her down to his son Dan. After Dan ceased fishing her, she was sold on and subsequently had a foredeck fitted and

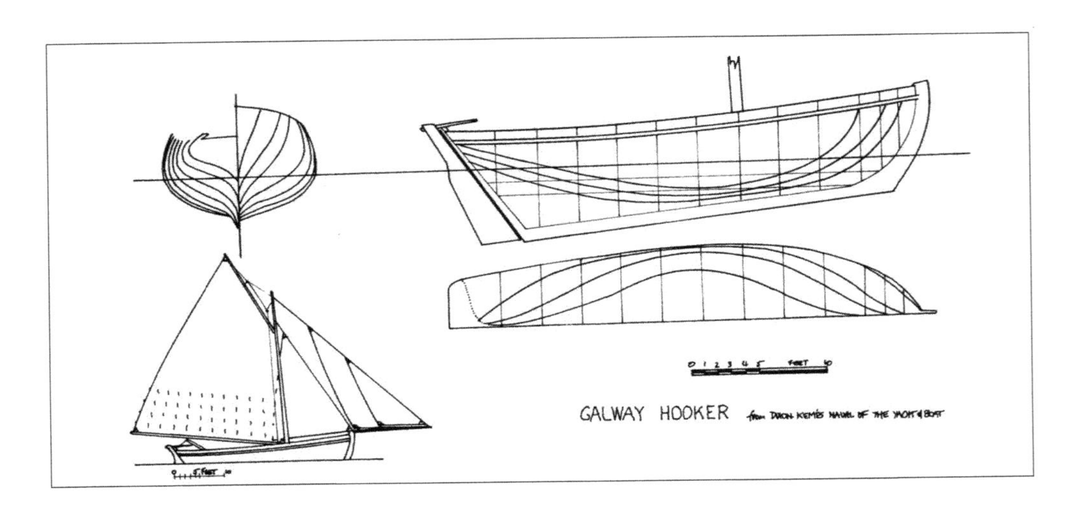

a motor and pot hauler installed, one of the first to have an engine and hauler which enabled one man to control the boat and haul the pots, thus reducing the crew from three to two members. The other steered the boat. By 1984 she'd a few more owners until she was abandoned in Mill Cove, Schull, after having her engine removed. And there she lay gradually rotting away until 1999 when Nigel and Liam, coaxed on by Cormac, pulled her out of her mud berth, after which she was taken to Sherkin Island to be restored. Today she's a pretty boat in her original green and grey colours and a pleasure to sail, as I found out in 2007.

Seine-net boats worked the West Cork and Kerry coasts for pilchards and mackerel, and latterly for salmon. I recently came across a couple in Valentia though they were long past their working days. Nowadays restored craft are raced during the summer months.

GALWAY HOOKERS

Galway hookers have been well documented since their revival back in the 1970s after Dubliner John Healion bought the 1890-built hooker *Morning Star* and restored her to her former glory, not as a converted yacht but carefully to the original layout. Other hookers had been converted but these became yachts with decks and the like but when John Healion arrived in Connemara as the first hooker to return there in June 1976, he had, unwittingly, begun a revival that has resulted in plenty more of the large and half boats being restored as well as various new boats having been built. As to their roots, there's still an ongoing debate, though the obvious choice is to say they evolved from Dutch craft because of their name and the bold tumblehome they display. However there could be some Viking and even French influences brought upon their design over many generations.

The largest of the hookers are referred to as *Bád Mór*, vessels of a length somewhere, on average, between 37 and 39 feet, although the largest built is 44 feet and the smallest 35 feet. In general these are transom-sterned boats with a very full, apple-cheeked bow, a high degree of tumblehome amidships and a sloping sternpost. They are half-decked with a cuddy forward of the single mast, which supports a gaff mainsail and foresails set on a bowsprit. Hookers never set topsails as far as I can make out. They were massively built: 1.25-inch planks on 6 x 3-inch sawn oak frames spaced about 12 inches apart amidships, even less forward. The keel was approximately two thirds of hull length and of oak or beech. The heavy beamshelf, 10 x 3-inch oak, provided strength for the deck. The transom sterns are said to have originated from France.

The half-boats or *Leath Bháid* are so called because they aren't half the size but carry half the load of turf compared to the bigger boats. Otherwise they are almost identical in shape and rig. Sizes of length range from 32 to 34 feet. The third type, although again of a similar shape and rig, was called the *gléoiteog*, pronounced 'glowchug', from the Gaelic *gleoite* meaning pretty, and ranged from about 24 to 28 feet. Being smaller, these were more affordable to the majority of the coastal dwellers and thus were the real workhorses, being used for all manner of work. Richard Scott, in his 1983 book *The Galway Hooker*, which today remains the seminal book on the boats, calls them 'the maids of all work',

a fitting description for a vessel that fished, ferried people and animals, carried turf, gathered seaweed and did any other work that was considered necessary.

The last type – called a *Púcán* and pronounced 'pookaun' – was a bit different in that it set a dipping lugsail. The hull shape was basically similar though smaller at 22–24 feet. They were entirely open yet they were still used in sheltered waters for fishing and carrying. Obviously they were even cheaper to build and were favoured by those not wishing to venture too far out into the bay and they have been described as the finest of all the Connemara sailing boats.

In 1830, Claddagh (Galway's main port) had 105 sailing boats and eighty rowing craft, employing 820 men. The largest group, of up to 13 tons, cost £70 to build, with an extra £10 for a foredeck. The 10-ton vessel cost £55, and the 6-tonners £35. They were built in various locations in the area, boatwrights earning 3*s* a day plus three glasses of whiskey. The fishermen often made the sails themselves, and rigged the boats in the local church!

The hookers were all-round workhorses and consequently many boats fished for only one month a year. They were a necessary part of a family's possessions. Some called the hooker a poor man's boat but this seems unreasonable. These days the hooker revival is apparent all over the coast. Visit any west coast gathering and six or seven will turn up, some only recently built. But 100 years ago the coast from Mizen Head to Slyne Head would positively have teemed with such craft, for roads were virtually nonexistent.

Similar craft are found in mainland North America. The Boston hookers are a direct descendant of these Irish craft, which in one or two instances may well have been sailed over. The majority would have been built by immigrants from Ireland, copying the boats from back home.

Left: The recently restored Galway hooker *American Mor* in 1984. (*G. Hembrough*)

SKIN BOATS

The final and oldest of the indigenous Irish fishing boats are the currachs (curachs in Gaelic). These, like the Welsh coracles, consist of canvas, or skin in the case of the early ones, stretched tight over a simple lightweight framework. Unlike the coracles, Irish currachs were up to 25 feet long and not much over 3 feet on the beam; long and slender craft. The Donegal boats, however, were around 14 feet long.

The currachs are as varied as the harbourless coast that they belong to, their lightness enabling them to be carried into the water and launched into a heavy swell. A currach can equally be swiftly snatched from the frothing sea. The high bow enables them to move with ease over large waves, but the amount of lift varies to suit local conditions.

The *naomhóg* can also be referred to as a Kerry canoe but not a currach. Currachs are inshore vessels used in protected waters while these 'canoes' are often sailed or rowed in exposed parts of the coast such as out to the Blasket Islands off the Dingle Peninsula. Without getting too involved in the ins and outs of these wooden-framed, hide- or canvas-covered craft (there's a specialised subject in itself), suffice it to say that these boats fished as well as being used to transport people and animals, hunt seabirds and their eggs and collect seaweed. They come in various sizes and shapes, and their oars are renowned for having no blades.

Connemara currachs, as mentioned earlier, worked in the largely protected waters of this part of the coast and out to the islands. Similar but larger currachs worked from the Aran Islands where, even today, some are still used for laying lobster pots around the back of the island. Wooden currachs, almost identical to punts, also work this indented Connemara coast.

A typical currach from the Aran Islands.

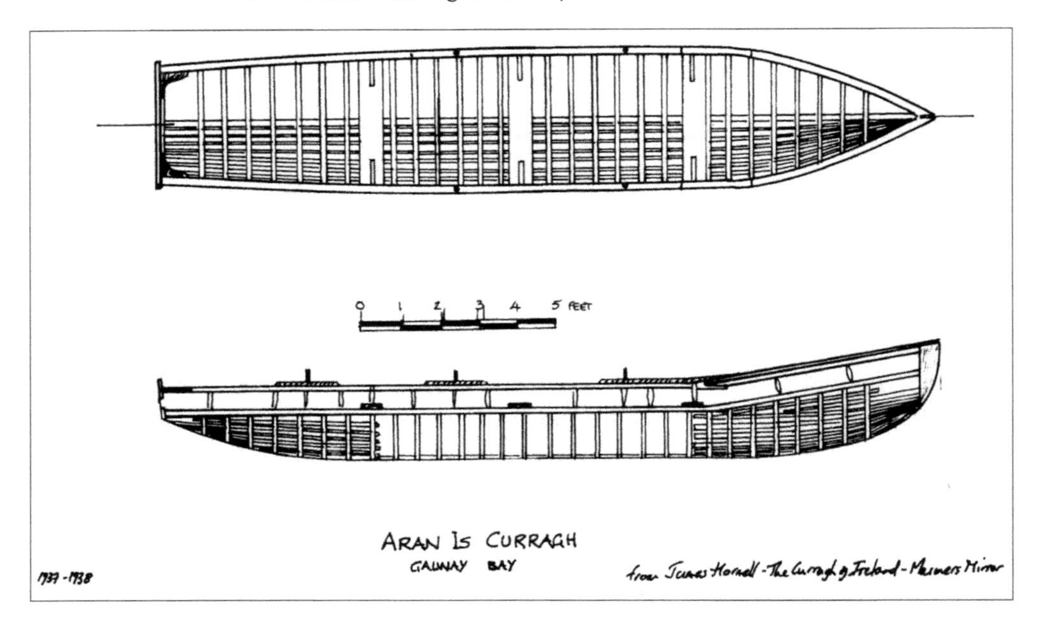

There are five separate and quite distinct versions of the Donegal currach in use over time. The first of these is the Donegal 'paddling' currach – literally *curach céasla* in Irish – which is unique in that it is Ireland's smallest. It's some eight feet in length and is characterised by the fact that it is often paddled by one man crouching in the front of the boat because it lacks thwarts (seats). Furthermore it was said to be crudely built. The second is the Tory Island currach, which was used primarily for fishing from the offshore island, though similar examples were common on the mainland opposite and the nearby Inishbofin Island. These were extraordinary boats in that they were rowed great distances, for Tory Island itself lies some 8 miles offshore. They were characterised by the addition of a shallow keel and rubbing strakes on the outside of the hull. Third is the Bunbeg currach, a small currach that evolved through the need to cross the waterway or inlet that Bunbeg lies upon. These currachs are small enough that they can easily be carried on the back of one man, much as the coracle fishermen of Wales carry their coracles as already mentioned. Fourth in the list is the Dunfanaghy or Sheephaven currach from the waters east of Horn Head. Being one of the largest of the five types, it was renowned for its seaworthiness, though weren't they all? Finally there's the Fanad or Ros Guill currach, Ros Guill being the headland to the east of Sheephaven while Fanad Head lies a bit further east. This currach is basically a different version of the previous type, its differences lying in its construction.

There's also a Boyne currach, which was used for draft-netting for salmon in the lower reaches of the river and was one of the last currachs to be covered in hide. These vessels are almost circular in shape and show some resemblance to the coracles of Wales, though two men would be aboard, one rowing from a kneeling position and the other paying out the net, whereas the Welsh coracle only ever carried one man.

RIVER BOATS

PRAMS

Although the Boyne currach and some of the other currachs can be described as river boats, they've been included above for continuity. However, Ireland has a rich and diverse range of other small open river craft, used mostly for fishing salmon, although some have been adapted for other uses.

For instance, prams were the salmon and mussel boats of the River Boyne. Though looking the same, some bigger than others, subtle differences differentiated those that fished the salmon and those that raked up the mussels. Built mostly by the local fishermen themselves, and most would have one of each, both have the same basic shape with a fairly flat but rockered bottom, are keel-less and clinker-built for local deal in the main. The transom is strangely circular in shape while the bow is almost truncated in that it looks as if it's been cut off in the same way as a pram dinghy does, which is where the name comes from. On the bottom of the hull there are runners along the entire length to ease their pulling in the mud. The salmon boats are easy to recognise as they have a stern platform for holding the net and are between about 14 and 16 feet in length, though none are used for fishing. The mussel boats on the other hand are generally bigger at 19–21 feet to carry the extra load. They only have one thwart, though some have a large platform near the bow upon which the fisherman stands while dredging. Salmon boats have two thwarts. Thus these have more beam and two extra planks either side. With mussel harvesting halted in 1998 due to the river being dredged, and the beds being re-seeded afterwards, it was hoped to recommence collecting in 2003, yet almost eight years later it hasn't due to the lack of reproduction. Such stories of poor harvests occur all over mussel-producing areas, so it is possible that they will come back again.

COTS

Further south, in the expanse of Wexford Harbour and the River Slaney, cots worked both up the river and larger versions out to sea. The word 'cot' comes from the Irish *coite*, which means a log boat. Today's cots, though, are not log boats yet the suggestion is that they developed from such craft. Whether these log boats then became 'extended' by the addition of planking – craft like this are in use in many parts of the world still today – and then became fully planked craft is not known, never will be I suppose, but it is improbable. More likely the term became accepted in the vernacular and continued. Cots are in fact flat-bottomed boats and display many similarities to dozens of other river and lagoon craft in use throughout Europe. What is also quite likely is that they are a combination of two influences upon boatbuilding techniques that have come about over a couple of millennia, these being a southern influence from the Roman times in Britain and a more recent – by comparison – influence from the northern Vikings. This is because they have a flat bottom consisting of planks of

Larger Wexford cot. (*Author*)

timber butted together – in a way similar to the craft of the Veneti tribe of Northern France, as described by Julius Caesar, yet their upper planks are clench or clinker-built in the same way as the Vikings built their craft. In other words, a hybrid of building techniques resulting in a strong yet easily built vessel.

On the outside of the spit, the land that forms the southern boundary of Wexford Harbour and along the coast to Rosslare, bigger cots once developed into seagoing craft, being up to 40 feet in length. Holdsworth writes that these cots were used for the herring fishery and had three masts sporting three spritsails and a jib. He states their normal length was 30 feet and that they were flat-bottomed with a small bit of keel at each end and a bilge keel on either side, though in his illustration these weren't very deep. Forty-six of these large cots were regularly fishing in the 1920s and they travelled a fair distance in this quest, as far as the north of the Irish Sea. As well as fishing for herring, they trawled, potted and longlined. It is also thought they sailed over to the Welsh coast to fish. Their demise seems to have come about after the advent of motorisation when their hulls were entirely inappropriate to receive motors. The smaller cots survived purely because they could be adapted to take outboard motors.

Today, cots can still be seen on the river around Wexford and upstream where Jim Devine still builds them in his shed. The 18-foot boats were used to fish for salmon until licences were put on hold after a ban was imposed on salmon fishing in 2005. Larger 23-foot boats are still raced. They are all flat-bottomed with a differing degree of rocker and many today have small transoms to take an outboard whereas traditionally they were double-ended.

Cots can also be found on the River Blackwater, which emerges at Youghal, working upstream for salmon. A peculiar type of salmon yawl is also found at Youghal: wooden punts that worked drift-nets. A few can still be seen tied up alongside the quay there.

PRONGS

It is believed that the prongs developed from the currachs of the west coast and have been for generations the workhorse of the River Suir at Cheekpoint and downstream, some being seen at Ballyhack. Most families would have owned at least one. At somewhere in the region of 17 feet in length (this varies according to builder), the prong is built in larch planking on oak frames, sometimes even of elm. Wood was sourced locally. The shape of a prong, with its pram bow and rockered bottom, is supposed to have closely evolved over many years of usage on the river so that it could slide easily onto the mud without the bow lifting up and allow the occupant to step out with ease. This is a characteristic seen on many a flat-bottomed river vessel.

Many fishermen built their own prongs though boatbuilders in various locations close to the fishing centres also did. In Cheekpoint there were three builders at Glass House Mill, by Belleview Point just along the river. One of these, John Lonergan, was well known because he had two club feet. He moved about by folding his deformed feet up under his buttocks and used his hands to propel himself over the ground and was as nimble and skilled as any other builder. Two brothers named Aitkins also built there and all three built punts as well as prongs. Another builder was Chris Sullivan, who produced punts, cots and prongs. I was shown a 1950s notebook in which Chris Sullivan kept his records of boats built – small pencil diagrams with a few simple measurements. There's one built for his brother Charlie and another for a local hotel.

Several of the River Suir cots can be found moored in tiny recesses or creeks in the riverbank. These flat-bottomed boats were long and narrow with a very shallow draft and were quite different to the prongs. They are almost primitive in appearance and are said to be an improvement on a log boat, or dug-out, that the fishermen used until a couple of centuries ago. The argument was that the availability of whole trees to use as logs decreased and at the same time planked-up timber became available. These cots have been worked on the rivers Barrow, Suir and Nore, the latter being a tributary of the Barrow, which meet just north of New Ross. They are often referred to as 'the Three Sisters' and are regarded as three of the finest rivers in Ireland. Mooncoin was about as far downstream as the cots were based on the Suir, although a few might venture down as far as Waterford city when the river was low. The fishermen generally built their own cots and because of this it is almost impossible to specify an exact design, each man building to his, and the river's, requirements. Modern materials and techniques have been introduced in some cases. For example, although most use ribs of oak to strengthen the planking, a couple of cots have had ribs made of angle iron incorporated into their building instead of oak.

A snap-net was the net they used to catch salmon until the ban was imposed and this had taken its effect upon the fishermen here as well. The net is used much in the same way as a coracle-net is used in Wales. Fished using two cots, the net is stretched out between the two cotsmen, who drift with the tide – both on the flood and ebb – in the hope of netting a salmon. Their use is documented as far back as 1200.

GANDELOWS

Gandelows were once the workhorses of the River Shannon, being used to transport people and their goods, fishing, wildfowling and seaweed collecting, even to take pilots out to incoming vessels. Built with a rocker in the planked bottom in both directions – i.e. fore and aft and athwartships – the gandelows have raked sides and a tiny transom and are about 21 feet in length, though this does vary. They were thus designed to glide into the mud banks of the river and again be easily extracted from the mud like, I guess, the prongs. Some had a 'leg of mutton' sail while later versions had an outboard well cut into the after end of the boat.

Those built further up the river in Limerick didn't have any rocker across the boat. There seemed to have been three families associated with the building of these craft there. Furthermore, the Limerick Port Authority kept a fleet to be used as tenders to the various harbour workboats as well as the pilots. As the lighthouse keepers and pilots were often local men, I suppose they were well used to these gandelows. At some times in the year they were used for reed collecting among the extensive reed beds of the Shannon just below Limerick, the reeds being used to thatch the houses.

Similar craft, called 'gangloes', had been used since before living memory at the extreme west end of the estuary, around Kilmore strand and the estuary of the river Feale. Although slightly different in shape to the Limerick and surrounding area boats in that they had more sheer for work in the open sea, they are assumed all to have come from the same influence which, it is suggested, was the Newfoundland dories used in the cod fishery there, a possible though tenuous claim.

It is supposed that gandelows were popular around Limerick because of the timber being imported from Canada and Norway. Any boatbuilding of this type needs long, clean lengths of spruce or pine, and northern timber was regarded as the best due, mostly, to the fact that it is not quick grown as much timber is these days, thus giving a knot-free and straight piece of wood to work with. With most folk living by the river owning such a boat, it was often a member of that family that gained the required skills to build their boats. A typical gandelow, if well maintained, would be expected to last twenty years while those dragged over stony beaches would not last half that time. The important thing was that they were cheap to build and I'm sure that on many occasion, the vital pieces of timber were whisked away without anyone seeing.

Brocauns fished the river at Limerick until they disappeared in totality. Shaped like many of the other Irish river boats – especially the cots – many believe they originated from a simple log, largely because of the shape – long, narrow and low in the water. They were used for pair fishing in the river much in the way that the snap-netters worked in the rivers Suir, Nore, Barrow and Slaney.

MOTORISATION

Motorisation had its effect on the fishing fleets of Ireland as it did elsewhere. However, the first purpose-built motor fishing boat built in the British Isles came from the Irish yard of Tyrrells in Arklow. This was the *Ovoca* and she was designed by John and Michael Tyrrell in 1905, with plans drawn and a model made (which still exists). Plans and model were taken to the Chief Inspector of Fisheries at the Fisheries Branch of the Department of Agriculture, who eventually agreed to advance the money for the building of the boat as long as they recommended fishermen to work the boat. Thus construction began in 1907 and the 48-foot *Ovoca* was launched in the January of the following year complete with her 20-hp Dan hot bulb engine. This was innovative stuff in 1908, for the fishing fleets of Britain were just beginning to convert their old sailing boats to motorisation. To build one from scratch was almost unheard of. *Ovoca* had one other distinction – she was the first canoe-sterned fishing boat in the British Isles. She had two masts, ketch-rigged, engine room forward and cabin aft. During her first year fishing in the Irish Sea and off Donegal her gross earnings almost equalled her cost, less the fishing gear. The large sail plan was soon deemed unnecessary and was shortened and bowsprit removed. During the First World War she went from Arklow to Balbriggan and worked until 1966, when her registry was closed and she became derelict.

One other type of craft worthy of a mention here is what has become known as the BIM 50-footer. These boats were designed by James Stafford, a naval architect employed by the 'Comhlachas Iascaigh Mhara na hÉireann' (the Irish Sea Fisheries Association) in 1946. It wasn't until 1952 that this became the less-of-a-mouthful corporate body 'An Bord Iascaigh Mhara', now widely known as the BIM. Stafford had served time with Tyrrells of Arklow from 1931. Subsequently, the BIM had interests in yards at Baltimore, Dingle and Killibegs as we've seen, though the first boat of this type was launched in Killibegs in 1949 where Stafford was based. These boats, designed to be 49 feet 8 inches in overall length – give or take the odd foot – cost about £4,000 at the time and the fishermen were expected to put a down payment of 10 per cent, the rest being paid to the BIM over a period with minimal interest. This had been usual in the fisheries since the days of the CDB, a sort of hire purchase, common today. Far too common, some say. However, it's worth remembering that it was possible to make a packet of money from fishing in that period and some skippers managed to pay off their debt in a couple of years. Eighty-eight of the boats were built in total, twenty-one at each of at the BIM yards, eight at Meevagh, nine by Tyrrells of Arklow, six by W. G. Stephens of Banff, east coast of Scotland, one by Henry Skinner of Baltimore and the final one at Sisk's boatyard in Dun Laoghaire. Furthermore, Tyrrells also built the *Vega*, which was wholly owned privately. By 1952 they were costing £5,500 and over £7,000 two years later. By the time the last one was running down the ways in 1970, its full cost was £22,000. It is estimated that seven boats are still working at the fishing and seventeen have been converted. The BIM were also responsible for the design of, among other craft, the inshore lobster boat fished with widely in the north and west today, a motorised variant of the northern yawl.

CHAPTER 7

The Isle of Man: Manxland

If mist be on the mountains on old Christmas day there be plenty of herrings going.
—Manx fishermen's saying

Because the Isle of Man is situated in the middle of the Irish Sea, it has always commanded a position of great strategic importance, ever since the Vikings settled these parts, and possibly before. When the Vikings came during their conquests between AD 760 and 1100, they brought a maritime expertise that was long to remain, and signs of it are apparent in the fishing boats that once worked these waters.

Around the Manx shores in the past there was an abundance of fish and the people of the island have a history of maritime excellence. So it is hardly surprising to find that it was said that 'in 1883, one in four Manxmen were directly or indirectly dependent of fishing for a livelihood'.

The island's early chroniclers mention the church tithes on all fish caught. But details of the vessels used for catching these fish do not emerge until nearly 1600. However the development of one of island's most popular and still-surviving boats, the Manx nobby, seems to follow a definite track through five distinct stages or designs. As has been the case in other parts, the deepwater herring boats and the inshore line boats developed separately, albeit similarly, so it seems appropriate to consider each in turn.

THE HERRING BOATS

THE SCOWTES OR SCOUTES

Early records refer to the Icelandic *skuta* as a fast vessel using oars and sail. The development of these vessels led to the herring scowte. These were open boats of a positive Viking appearance, with sharp bow and stern, curved stem and sternpost with considerable sheer giving low freeboard amidships and of 5–6 tons burthen. Keels ranged from 20 to 24 feet probably associated with 24- to 28-foot lengths overall.

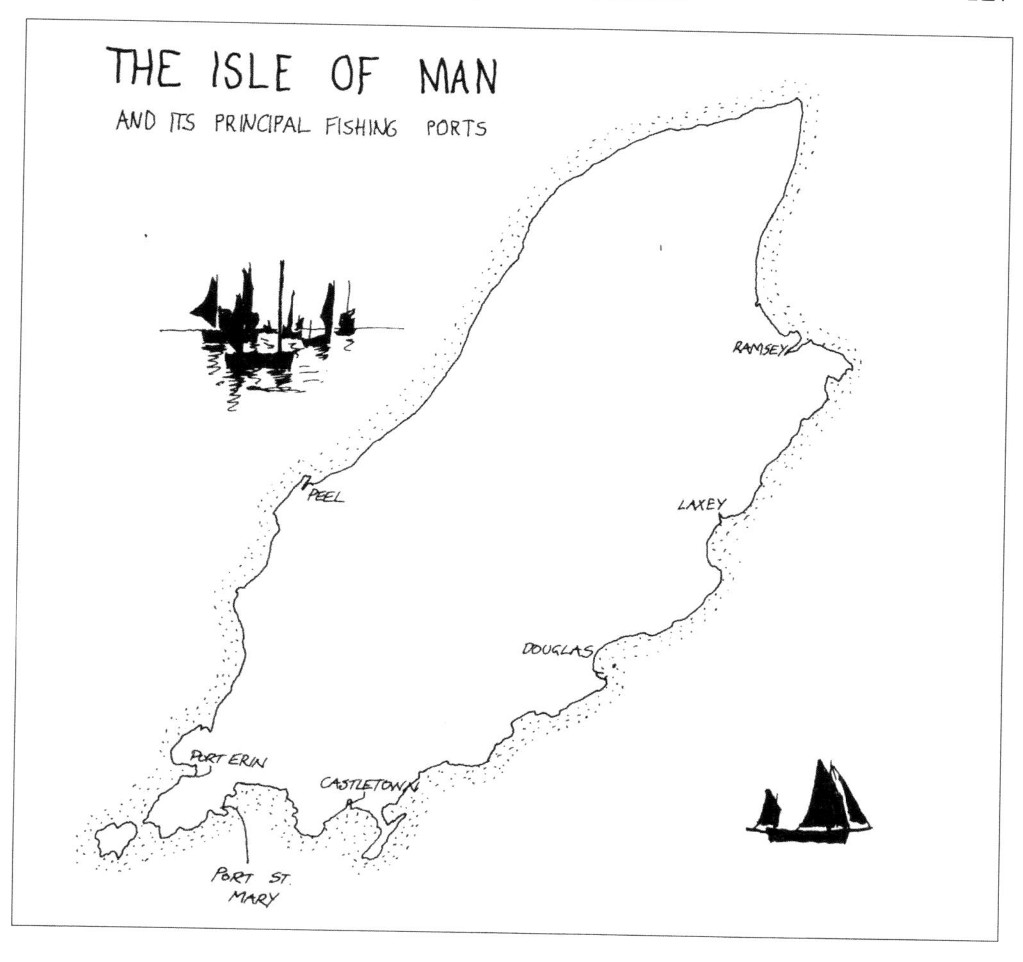

They had a single mast with a large square sail and, usually, four sweeps (long oars) for rowing by their crew of six. They were of clinker construction, and have obviously been compared with the sixareens. It appears that the scowtes were common from the fifteenth century to the late 1700s, although during the eighteenth century they became known as the 'square sails' to contrast the newly emerging fore-and-aft-rigged vessels.

During this period legislation peculiar to the island was introduced to encourage fishing – especially herring fishing. The majority of vessels were engaged in drift-netting. A law was passed in 1610 enforcing a closed season from 1 January to 5 July, within 9 miles of the shore, and prohibiting the shooting of nets before sunset. This was observed until 1823 when visiting fishermen broke it, ensuring that the local fishermen had to follow suit.

By 1775, ten years after the English takeover of the island, duties on fish imported into Great Britain from the Isle of Man were abolished. Simultaneously, a tonnage bounty of 1*s* was paid to deep-sea herring boats, which was doubled in 1801 but abolished altogether in 1820. These bounties encouraged the part-time fishermen to

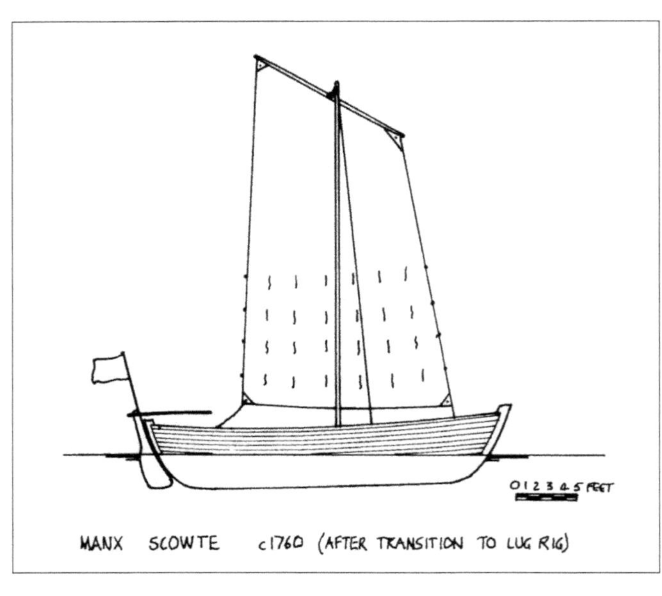

Above: The Manx scowte, 1760. Detail from J. Goldar's engraving of a painting by Richard Wright.

abandon crofting and turn to fishing full time. This in turn encouraged the building of the fishing fleets, and hence the increase of boats on the Isle of Man.

Wherries were also to be found among the fleets at this time, more generally owned by the fish buyers rather than the fishermen, although a few fished. These two-masted craft were rigged with fore and aft sails, and were similar to other vessels found around the Irish Sea and the Clyde. Moreover the largest smack built in Peel was launched around this time: the *Dagan* was 60 tons, and very fast. Supposedly every large boat built for some time was given the same name.

THE HERRING DISASTER OF 1787

In a gale that struck the Isle of Man in November 1786 both the end of Douglas pier and the lighthouse were swept away. Damage was substantial to the pier, with some 84 yards of the structure being destroyed. The loss of the lighthouse was more significant: its seven or eight half-pound candles, with the 8-foot circular tin reflector behind, had given a light that could be seen at a distance of up to five leagues at sea. The beacon to guide the men home was lost.

On 20 September 1787, the pier remained in a similar state, and instead of the lighthouse there was only a lantern which had been erected on a pole near the end of the old pier. That morning an unusually large catch of fish was landed and towards dusk, because it was a fine day, some 400 boats, mainly scowtes, headed out for the fishing grounds off Douglas Head.

At midnight a brisk easterly wind sprang up, and immediately the fishermen turned back and headed for the safety of the harbour. But one of the first boats to return smashed the lantern on the pole, and as the wind rose to gale force, other boats were unable to make the harbour. They were smashed against the rocks, and eyewitnesses spoke of the cries of the drowning fishermen and the screams of their wives standing helpless on the shore in the dark.

At first light the enormity of the disaster became apparent: the beach and rocks were covered in wrecks, and bodies were floating around the harbour. Between fifty and sixty boats were estimated to have been wrecked, and various estimates of death ranged from six to twenty-one.

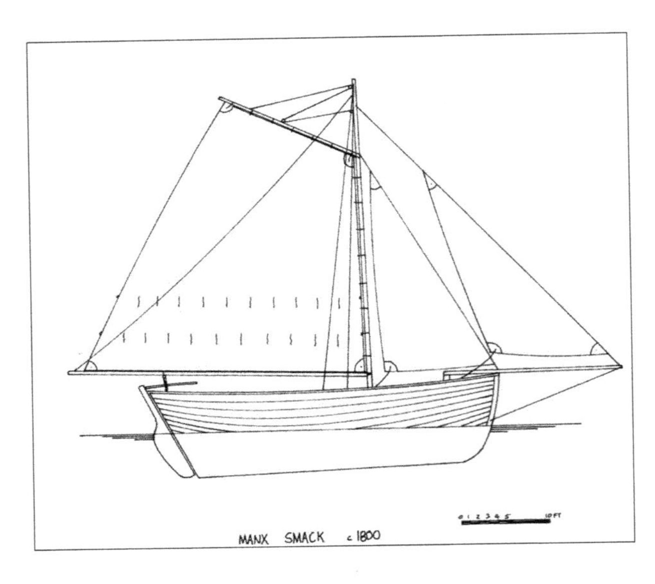

THE SMACKS

After this disaster, there was intense debate on the design of the Manx boats. As with the later Washington report in Scotland, the general opinion was that the main fault with the current design was the lack of a deck on the vessels, although it is interesting to note that this was some sixty years before the Washington report. As a result, decked boats were built, similar to the *Dagan*, yet smaller, and almost immediately they appeared with cutter rigs (fore and aft sails) similar to vessels along the west coast of England and the Welsh coast. These smacks, as they were soon called, gave some degree of shelter in extreme conditions but were generally without living and cooking facilities. The vessels were larger than their predecessors, being 23–33 feet in keel length with a 13-foot beam and a 6-foot-deep hold on the largest of them. They were mainly built on the Island, although some were imported from the mainland.

It was during this period that the fleet increased to its highest numbers, and by 1800 there were more than 600 boats that were locally owned. By 1823, this had reduced to 400 boats, although these still employed 2,000 men.

THE HERRING LUGGERS OR DANDY SMACKS

In the same year the first Scottish and Cornish boats appeared. Their catches were so successful that they returned year after year. The Cornishmen came with their powerful 'drivers', lug-rigged boats with two masts. They would tend to ride their nets with the foremast lowered and only a mizzen set.

Around 1835, when there was a decline in the fishing, the Manxmen adapted their gaff rigs by shortening the boom and retaining a shortened gaff main, but they stepped a jigger mizzen mast to convert to dandy rig with a lugsail on this mizzen. They also stepped the mainmast in a tabernacle on the deck so that they, too, could lower the

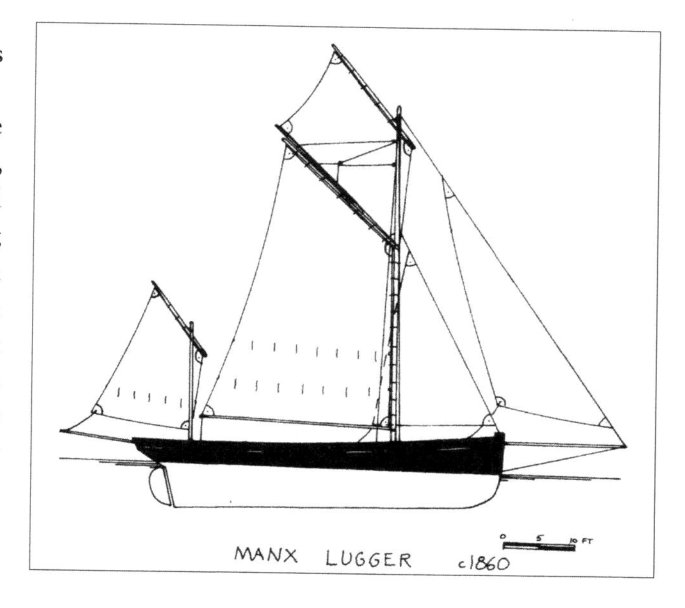

mainmast to ride the nets more quietly than before.

These vessels became known as the Manx luggers, and between 1840 and 1860 almost all the herring fleet was made up of such working boats, ranging in size from 37 feet to 41 feet overall, with keel lengths of 32 feet to 34 feet. The forecastle was enlarged for accommodation, and in 1848, the basic price of a boat was £155. Now being able to sail further and gain larger catches, the vessels were built bigger all the time. Within ten years their size had increased to 38–47 feet on the keel, and they were half-decked. The largest of all the boats were fully decked. Generally they were crewed by seven or eight men.

THE NICKEYS

During the 1860s the Manxmen began in general to adopt the boats of the Cornishmen. These were drivers from St Ives mainly, the same luggers that were common in Campbeltown, and were delivered to the Island. They became known as the 'nickeys' because, it was claimed, of the high number of Cornishmen with the name Nicholas, although an alternative reason could be that the first boat to arrive in Manx waters was called the *Nicholas*.

Within a few years, the Manxmen began to build their own nickeys. The first was the *Alpha*, from the yard of William Qualtrough in Port St Mary in 1869. Another boat from the same yard was the *Expert*, launched in 1881. She was 52 feet 2 inches

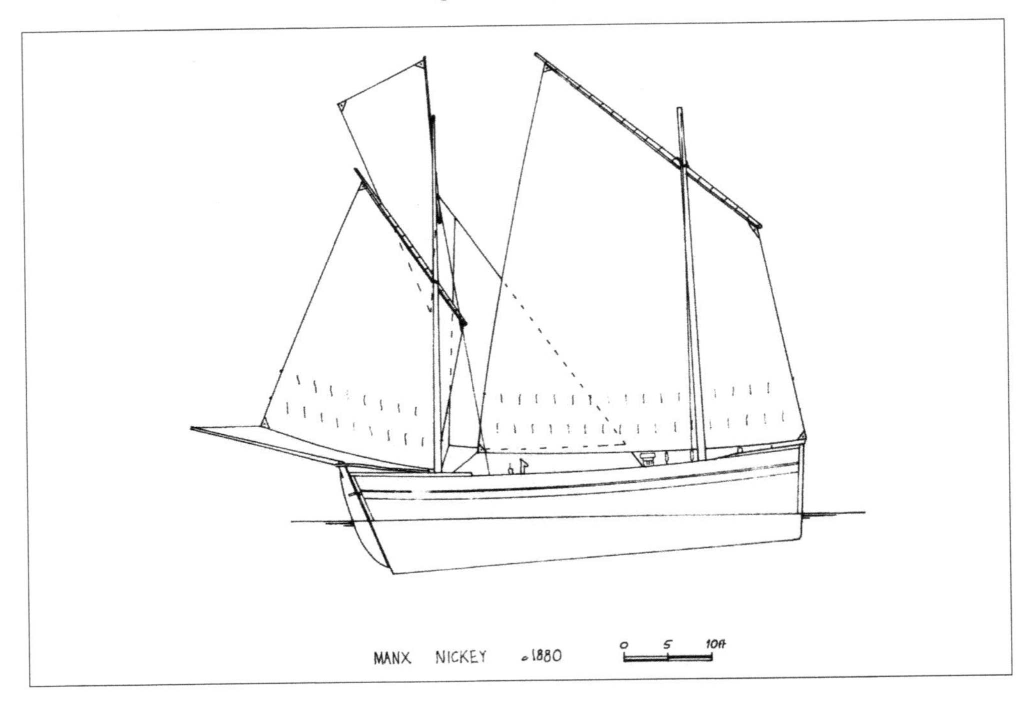

overall, 47 feet on the keel and had a 15-foot 1-inch beam. She was built on a 10-inch-thick elm keel, with white pine planking on doubled oak frames.

The Manx boats tended to be bigger than the Cornish equivalents, and they soon became renowned for their speed. Indeed, some of the boats were rigged with mizzen topsails and mizzen staysails to sail even faster. Their hulls were large with plumb stems and sternposts, not unlike the Scottish east coast fifies. Although generally double-enders, a few were built with counter-sterns, similar to some of the south Cornish boats.

There was accommodation for the seven men and one boy just abaft the midships fish hold, again similar to the St Ives drivers. This accommodation consisted of a space 13–14 feet long with berths along either side, with a stove against the fish hold bulkhead. Access was from a companionway just forward of the mizzen mast. Forward of the fish hold was the net room and bosun's locker.

The nickeys often sailed south for the annual mackerel fishery off Kinsale during March and April, to join all the other Scottish, English, French and Irish boats. Many Manx boats would make the passage in thirty-six hours, a fast time even by today's standards. They would return to the island for the summer herring fishing, and some of the fleet would sail north afterwards for the autumn fishing off the Shetlands.

In 1879, it is said that up to 1,000 boats from Scotland, Ireland and Cornwall as well as from the island operated during the summer in the waters between the Isle of Man and the Irish coast, and as the fishing grew larger, the fishermen grew steadily richer. By 1881, there were 334 Manx boats recorded as taking part in the herring fishing, with the largest numbers registered at Peel and Port St Mary, both ports being favourably positioned for the fleets. Part of the reason for William Paynter moving to Kilkeel,

A Manx nickey off Peel, *c.* 1900. (*Pauline Oliver*)

Scrubbing the hull of a Peel fleet boat. (*Pauline Oliver*)

The Castletown fleet of nickeys lying off the island, *c.* 1880. (*Pauline Oliver*)

as we've already seen, was to satisfy this ever-growing demand. The first nickey he built there for delivery to Peel was the *Zenith*, costing about £700 including sails and nets.

It was around this time that the Manxmen began to follow the Clyde fishermen using ring-nets for the herring. This initially caused local anger, because of the supposedly destructive disturbance of the herring. The trawling was being blamed for the disappearance of the herring spawning grounds off Douglas and Laxey, although this is in no way definite. From available evidence, it seems that the herring was already on the decline, which soon became more apparent and cannot necessarily be blamed on the antics of the trawlermen, although some will disagree.

Generally the nickeys were of a very strong build, and a good, well-proven design. It is hard to see how they were superseded by another design, but their only drawback was their size and need for a large crew. With all the nickeys built between 1860 and the end of the century, it is surprising that not even one has survived.

MANX NOBBIES

The final stage of the evolution of the sailing boats of the Isle of Man fisheries ended with the Manx nobby. It was in 1884 that the first nabby appeared in Peel from Girvan, just two years after the first Lochfyne skiff arrived in Campbeltown from Girvan. Whether this first boat to the island was in fact a Lochfyne skiff or an older nabby is uncertain (the difference between the two not being great), although it was probably of the new, bigger design with accommodation, the smaller boats not tending to sail so far from home.

The Manxmen were so impressed by these smaller double-enders from the Clyde, with their single standing lugsails, that they began to copy them. The first Manx nobby, as they became called, was built in Peel in 1889, and was the *Bonnie Jane*, built for Johnny Hall. The first boats were very similar to their Scottish counterparts in hull shape, but the Manxmen soon altered the design to suit their local conditions and habits. The waters around the Isle of Man are not as sheltered as those of the Clyde.

The boats were fully decked, unlike the Scottish half-deckers. The accommodation was behind the fish hold, like the nickeys, not forward as on the Clyde boats. The mast was raked aft, but not as much as on the Lochfyne skiffs, and set a staysail as well as a jib on the bowsprit. They also set a mizzen mast with a standing lugsail. Keel lengths varied from 30–35 feet for the largest boats to 24–28 feet for the smaller. One or two had a transom stern, but the majority were double-enders with a more rounded stern than the Scottish boats. Early boats had a heavily raked sternpost and long sloping keel, which had proved so effective for manoeuvrability on the confined waters of Loch Fyne.

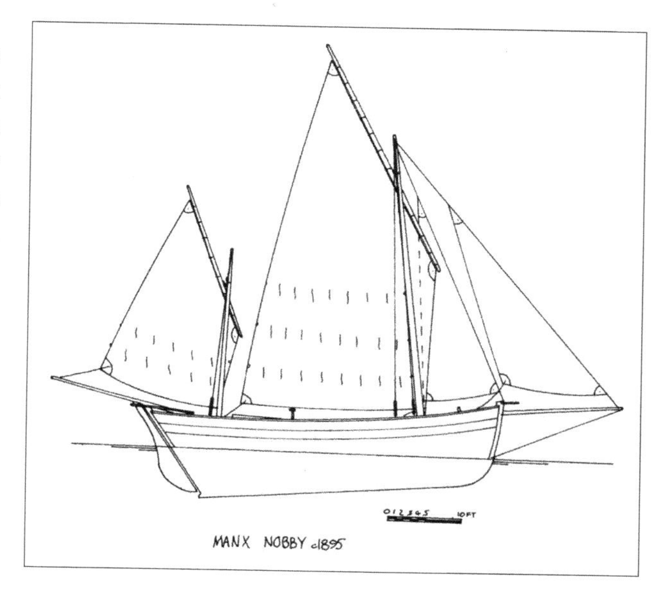
MANX NOBBY c1895

MANX NOBBIES

The first nobbies sailed to the Kinsale fishery alongside the nickeys. The design became favoured because of the obvious advantage of needing a far smaller crew, and the nickey men soon began to convert to the nobby rig. Although many boats made this change, the performance of the old boats was worsened, and so some fishermen did not make the change. As we have seen, the Irish adopted the same design for their fisheries.

Today, only four nobbies have survived. The oldest is *Gladys*, PL61, built in 1901 for Thomas of Peel. She came from the Peel yard of Neakle & Watterson, and measures 41 feet overall, 30 feet on the keel and 11 feet 6 inches on the beam. She fished successfully until the 1930s decline, and is mentioned as being one nobby that had a licence from the Irish Government to fish the south and west coasts. Her shape is traditional, with plumb stem and steeply raking sternpost, although her keel does not slope to the same extent as in earlier nobbies. *Gladys* is currently in Cardigan in need of some restoration. Although she has travelled to Barrow-in-Furness, to Plymouth by land, and to Brest in 1992, then over to the west coast of Ireland where she stayed for a number years, it was not until 2009 that she returned to the Isle of Man for the first time since leaving in 1936.

The next oldest is Mike Clark's *White Heather*, PL5. She was also built by Neakle & Watterson in 1904 for G. Gaskell, and at 15 tons, was one of only seven first-class boats. Her keel length is only 27 feet, although her overall length is 43 feet 8 inches, on a load waterline of 37 feet. The shorter keel is accounted for by the cutaway forefoot, an influence said to have come over from Lancashire, where their nobbies have a similar reduced forefoot.

The Manx nobby *Gladys* in 1909. She remains as the oldest surviving nobby, and sails from Cardigan nowadays. (*Tom Cashin*)

White Heather fished for herring out of Peel until being sold to Ireland in 1913, from where she did not return until 1989. Mike Clark bought her and returned home with her, and in the 1990s rebuilt a considerable amount of her to return her to her former glory. More recently he has converted her back to lug rig.

The remaining two nobbies are both canoe-sterned boats. The elder of these is *Vervine Blossom*, PL65, originally launched as the *Blossom* in 1910 again from Neakle & Watterson's yard. Although having a 29-foot keel, she is smaller overall than the *White Heather*. Mystery surrounds her canoe-stern because when she was launched she had no engine, and did not have one fitted for another year. Perhaps they had planned to have one later because they couldn't afford it at the time. It was in 1911

that the first three boats in the Peel fleet were equipped with engines, the other two being *Cushag* and the *Annie*. All three had Dan engines from the Clifford yard in Ipswich, supplied through James Watterson. By the next year, new boats were being fitted with engines as standard.

The canoe-sterned vessels soon caught on, as the old raked sternposts were unsuitable for receiving stern tubes and props. Some of the older vessels were adapted, the *Gladys* having her engine fitted on the port side, while the *White Heather* had her Kelvin installed on the starboard side (in 1918).

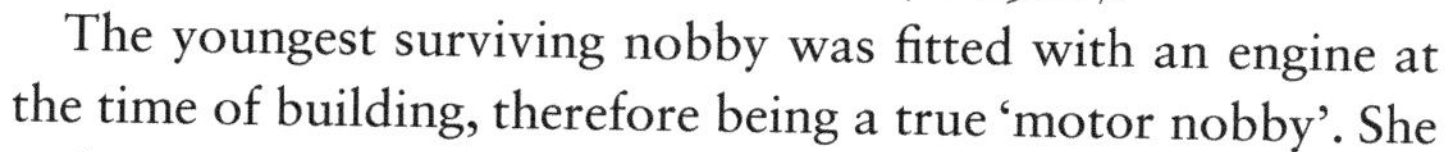

The youngest surviving nobby was fitted with an engine at the time of building, therefore being a true 'motor nobby'. She is the *Aigh Vie*, PL113, built also by Neakle & Watterson and launched in 1917 with a 26 hp Kelvin. Although there have been unconfirmed reports that she sank, the news is that she is now under restoration. All the latterly built boats, although having engines, were nonetheless fully rigged because the engines were so unreliable.

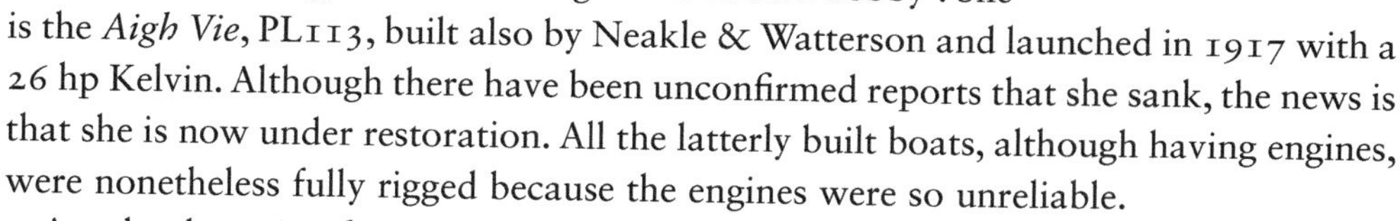

Another boat, *Harbinger*, last seen at Shoreham, is reported as being a Manx nobby, although this has not been confirmed by the fishing registers. More recently it has been reported that the nobby *Sunshine* was built by Neakle & Watterson, Peel, in 1936 for Morley Fox, Sale, Cheshire, and apparently she is at present based in France. Rumour has it that there is also a sister ship to her around somewhere.

The last of the boats, in fact the last sailing boat built for the fleets on the island, was the *Parrot* in 1921. By this time, because of the herring decline, the only boats built were for the Irish fisheries, and by 1930 shipbuilding on the island had ceased, and a tradition of hundreds of years was ended.

Out of 300 nickeys built none remain, and out of some 200 nobbies only four or five have survived. Today, as fishing boats are burned as part of a ridiculous decommissioning policy, we can only look back and dream of the wonderment of seeing the vast fleets of the herring boats. It's also very hard for us to understand the hardships and the great skill that these fishermen inherited from generation to generation. The only sure thing that we can say is that the graceful sailing boats of our fishing fleets were as beautiful as any other boats on the sea that we will ever see, even in the history books. It is hoped that those boats that do remain can keep this spirit alive. Their presence does seem to encourage others to undertake restoration or to build replicas. As present fleets are depleted, it is hard to see how else fishing boats can

Manx baulk yawl, *c.* 1840 (engraving by W. H. Lizais from a drawing by W. Banks).

survive. And as we consider today's fishermen it is not easy to see them obeying a law that was passed in the Isle of Man in 1738 and which required any Manx fisherman who had found a shoal of herring to inform his nearest boat.

THE LONGLINERS

THE BAULK YAWLS

Records also tell us that the practice of long-lining has existed for centuries. A vessel that worked these longlines inshore in the sixteenth and seventeenth centuries around the Isle of Man was the baulk yawl. These appeared as small-scale versions of the scowtes, except that they set the square sail on a smaller yard on a mast much further forward.

Their name originates from *balc* meaning longline (from the Shetland *baak*), and yawl, as we've seen, is from *yol*. Again they were open, clinker-built boats, although timber availability meant that the planks were not as wide as they would have been on true Norwegian vessels. They had four thwarts and four oars, and were, more often than not, rowed. The oars were set into notches cut into the gunwale for the purpose. The lugsail was adopted in the early eighteenth century, as it was on the scowtes.

Sizes varied, but an average baulk yawl was 18 feet overall with a 5-foot 9-inch beam. Larger versions were said to have a crew of up to seven men, but four or five was more realistic. It has been suggested that the baulk yawls were only crewed by old men, the younger men crewing the scowtes!

The yawls had rounded, Viking-like stems and sloping sternposts which made their beaching and launching stern first from the shore easier. The fishing grounds were hardly more than 2 miles offshore during the low season (September to December). After that, the larger boats would go to the deep-sea fishing (January to May), which involved rowing or sailing up to 15 miles offshore. Some boats even sailed over to the Cumbrian ports, and there is mention of fishing in Irish waters in 1577. It is worth remembering that the fishermen of these times were loath to travel too far from home, especially in such small boats, as the sea was an unknown quantity.

When at the fishing, each crew member had his own line of 300 hooks. With no ballast, the boats were extremely light and fast in comparison to the large boats. One crew member was also responsible for emptying the boat of seawater using the 'spoocher' or wooden ladle in the same way as the Shetland men did, before the invention of the bilge pump in the nineteenth century.

These larger yawls survived the transition of the scowtes into decked vessels. They did adopt the transom stern in the nineteenth century, although some retained the double end with two dipping lugs right up to the 1880s.

The evolution of these baulk yawls seems to have ended in what was termed the cod yawls. The *King Orry* was a Ramsey-based boat that had a single standing lugsail and a foresail set on a small bowsprit. The hull was clinker, open and had a transom-stern. The lines were fine and the keel long and straight. The oars worked in swivelling rowlocks, and the main boom sheet block ran on a traveller above the short tiller. What a change these boats were from the original baulk yawls!

In 1872 there were eighty-two second-class boats registered in Ramsey with some of these probably being Ramsey mackerel boats that worked from the harbour as well. This mackerel fishery was entirely different from any other, such as that at Cornwall or Kinsale, in that they used longlines instead of the drift-nets. These mackerel boats were similar to the cod boats, yet slightly longer at 21 feet overall, 3 tons burthen. Although often the same boat would be used for both fisheries, those built specifically

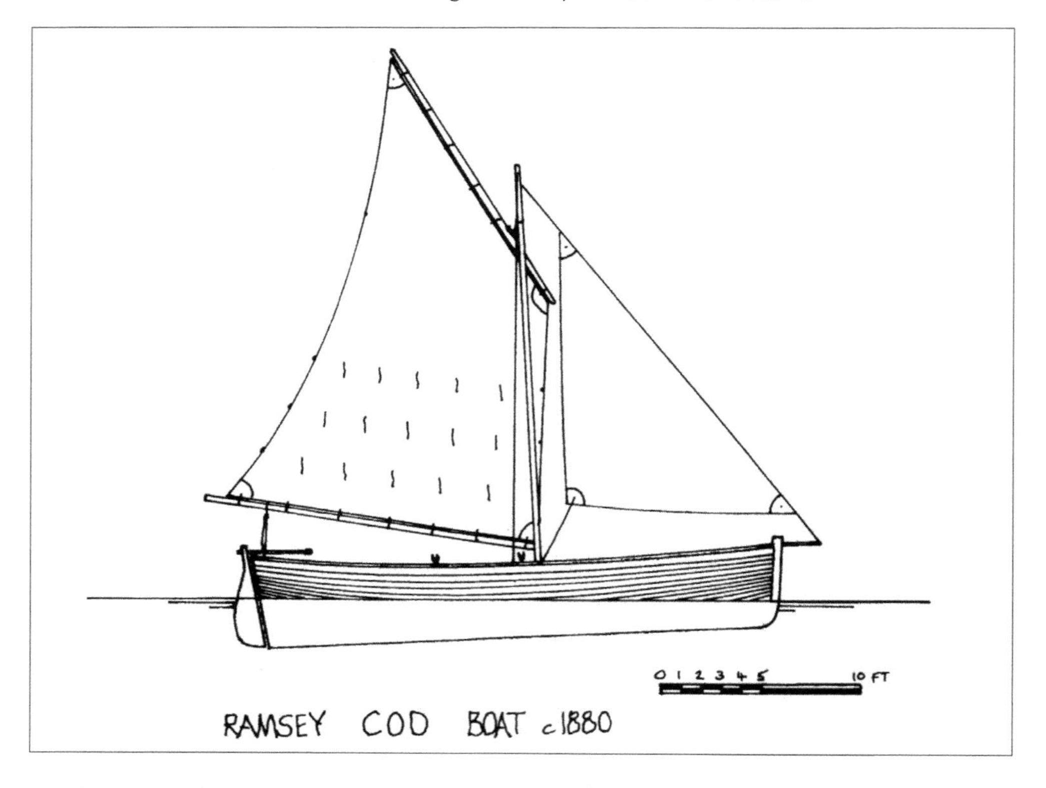

Above left: A Manx baulk yawl, *c.* 1890.

Above right: Ramsey harbour 1938 with some converted Ramsey boats. By this time the majority of these were either being used for pleasure or for taking trippers out.

for the mackerel fishery had a loose-footed gaff mainsail, and often set a topsail. With direct influence from the Morecambe Bay Prawners, these boats adopted the same shape, becoming known as the Ramsey half-deckers. They were up to 30 feet overall, and, although having a straighter stem, they resembled the Lancashire types closely. These boats were built for speed and coping with the short, steep seas often encountered on the fishing grounds over the King William and other sandbanks to the north-east of Ramsey. One boat, the *Master Frank*, was built in Ramsey in 1895, and was relaunched in 1996 after a rebuild and refit by her owner, Joe Pennington.

With the adoption of engines by the 1920s, their shape was such that they were indistinguishable from the Lancashire boats. One particular feature they did have was a hole in the side deck where a man could stand to bait the lines. Today, a few other boats remain apart from the *Master Frank*, among them the *Gien Mie*, PL83, which only recently ceased fishing out of Fleetwood. However, *Master Frank* is generally regarded as being the only genuine surviving Manx half-decker.

One small yawl is the 1926-built *Fisher Lass* hailing from the yard of Arthur Miller of Port St Mary. Although launched as a fishing boat and registered as DO54, RY51 and subsequently as PL9, she worked both longlines and took trippers out. Today, she is still sailing from the island as a pleasure boat and is owned by Michael and Lynn Craine.

The Ramsey boat *Master Frank* soon after her relaunching following a complete rebuild. (*Michael Craine*)

CHAPTER 8

The North West of England: The Solway Firth to the River Dee

Cockles and mussels, and pink shrimps too.

—An old saying

South of the Solway Firth the coast remains indented with muddy creeks, rich with fish that are mostly worked by the Annan boats. The Cumbrian coast does not open up to the sea until near Silloth, a village that became a port because of the railway and the construction of the viaduct across the Solway to Annan. The dock was built in 1855 and flourished briefly after Port Carlisle, established in 1819 and linked to Carlisle itself by canal, silted up. Silloth, though, traded more in cattle, grain and fertiliser than in fish.

The other Cumbrian ports are Maryport, Workington, Harrington, Parton, Whitehaven, Ravenglass, Millom, Barrow and Ulverston, and these grew up mainly on the basis of the Cumbrian mines. Fishing became ancillary, although Whitehaven became an established trawling port in the mid-nineteenth century and Maryport a herring and shrimping port a little later. Smaller communities such as Flookburgh survived through fishing, and boatbuilding flourished in the tiniest of villages, away from the centres of population.

Morecambe Bay has for centuries been the home of a thriving fishery, best known for its shrimps. The fishing was, however, merely at a subsistence level until tourism boomed in the nineteenth century. Historically fish baulks (traps) had been used since the Roman times, but mussels and cockles also turned into an important fishery.

Lancashire, as it used to be, has four important rivers: the Lune, Wyre, Ribble and Mersey. Each has had its own part to play in the development of the region, and each has had its own fishery. At the southernmost extremity, the River Dee had been for centuries the lifeline of Chester, until the river silted up in the seventeenth century. The river has its own fleet of salmon, or draft, boats that fish today in exactly the same way as they did one hundred years ago.

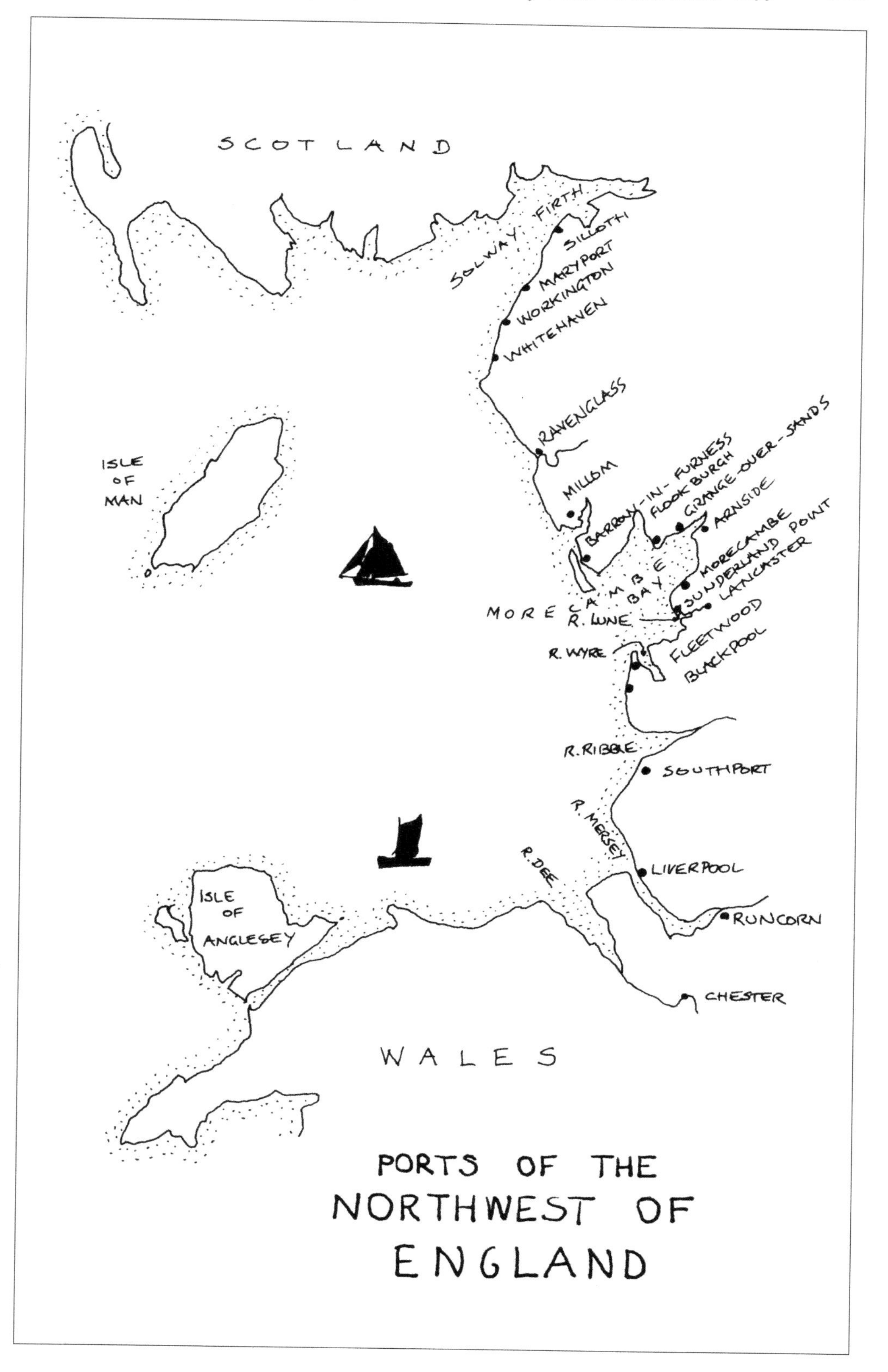
SCOTLAND
SOLWAY FIRTH
SILLOTH
MARYPORT
WORKINGTON
WHITEHAVEN
RAVENGLASS
MILLOM
BARROW-IN-FURNESS
FLOOKBURGH
GRANGE-OVER-SANDS
ARNSIDE
MORECAMBE
SUNDERLAND POINT
LANCASTER
MORECAMBE BAY
R. LUNE
R. WYRE
FLEETWOOD
BLACKPOOL
R. RIBBLE
SOUTHPORT
R. MERSEY
LIVERPOOL
RUNCORN
R. DEE
CHESTER
ISLE OF MAN
ISLE OF ANGLESEY
WALES
PORTS OF THE NORTHWEST OF ENGLAND

THE ORIGINS OF DESIGN

Although the earliest fishing boats along this coast were undoubtedly propelled solely by oars, square-sailed vessels presumably were later used. There is no doubt at all about the heavy Viking influence here, as it remains apparent in place names. Irish Vikings arrived after being chased out of Ireland, and, given the proximity of the Isle of Man, it seems likely that some level of contact was made.

Lug-rigged vessels worked off the Lancashire coast in the eighteenth century, and a 'View near Lancaster Sands' of 1787 shows a single-masted, clinker-built, bluff-bowed vessel with a transom-stern, oars and two thwarts. Boats like these presumably worked in the river estuaries, and along the coast.

Further south, sprit-rigged boats worked in the Mersey, and possibly the Ribble. Freckleton, on the Ribble, became an important port for Preston after silting in the river caused problems for boats wanting to gain the town. There was talk of building a canal from Lytham to Preston, but it never materialised. Remains of a stone quay can still be seen, and boats were built here at one time. In the nineteenth century, even Fleetwood smacks were built here, and across the river boats have a long tradition of being built at Hesketh Bank, so it is presumed that a certain amount of fishing occurred here and across at Banks.

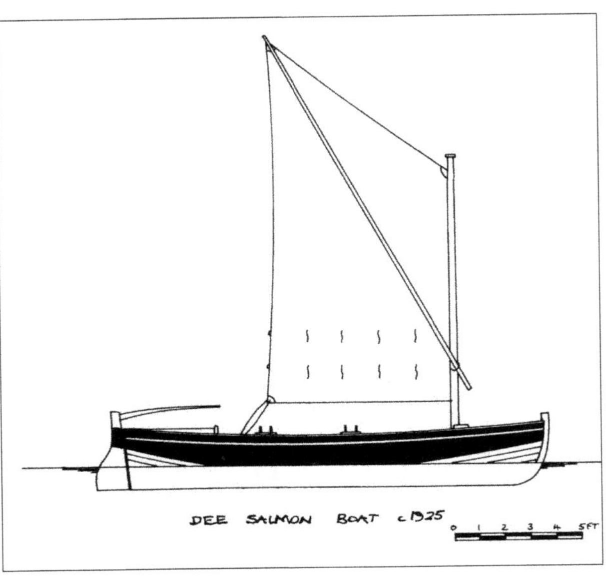

The watermen of the Mersey favoured their gigs, these being mostly three-masted and sprit-rigged, although some only had two masts. The mizzen sprit was always much smaller. These boats were surprisingly seaworthy, considering their size, and they ventured out as far as the coast of Anglesey in search of incoming boats. These survived into the present century, although Daniell shows some of them as having gaff sails with a sprit mizzen towards the middle of the nineteenth century.

The Dee salmon boats were developed at the beginning of this century by a narrow boatbuilder from Walsall, Joseph Harry Taylor, who first moved to Moreton on the Wirral, and built local boats, mostly sprit-rigged. In 1913 he moved to Chester and began building the fishing punts. These were 17 feet overall, clinker-built, transom-sterned and rigged with a single sprit. They draft-netted for salmon using a net that was rowed across the tide and hauled from the shore. The most notable feature of these boats was the chute in the stern – a platform on which the net was kept and from where it was easily shot over the stern. A typical Taylor boat cost £23 in 1923, and Taylor's accounted for most of those built, although 'Old Boaty' built a couple by the quay at Connah's Quay from which several worked. These had smaller transoms than the Chester boats, which were generally kept at Handbridge, on the opposite side of the river. When motorisation took over after the Second World War, the shape was altered to suit power, becoming more beamy and flatter. Three have subsequently been built by David Jones, who took over the Taylor's yard in 1974. These all remain in the ownership of museums, but several other boats have survived from the 1930s and continue to be fished in 1996.

The River Dee was also home to a particular 'jigger' based at Heswall, Parkgate and Neston. These jiggers – clinker-built, square-sterned, vertical-stemmed two-masters with gaff main and standing lug mizzen sheeted to a bumpkin – were specifically used within the estuary for shrimping, cockling, musselling and trawling. They are generally older than the nobbies which superseded them and became known locally as Parkgate nobbies. Influence for their rig has been said to have come from the Manx dandies, although this seems uncertain.

Dee salmon boats on the River Dee at Chester.

MORECAMBE BAY AND THE DEVELOPMENT OF THE FISHERIES

Morecambe Bay gave rise to the variously called Morecambe Bay prawners, Lancashire nobbies or shrimpers. Many writers have devoted whole chapters to these boats, and many others have written short articles, but few it seems have done more than a sketchy amount of research into the origins of these craft. For myself, I am more than grateful to both Len Lloyd of Southport and Nick Miller of Barrow-in-Furness for their help in this matter.

Poulton-le-Sands was a tiny hamlet of fisherfolk prior to 1800, as were most of the present-day towns along this part of the coast. The fishermen were eking out a meagre living catching fish and working the land. Herring, even, was being trapped in the baulks, and in fixed nets, so they didn't have to move far away from the coast.

The shrimpers (as I shall call them for now, as they mostly caught shrimps and not prawns) probably evolved from the early lug-rigged vessels that worked inshore. But the shrimpers were cutter-rigged, and it is most probable that this rig was influenced by the emerging smacks around the waters of the Irish Sea that first appeared around about 1770. Indeed Peel, on the Isle of Man, had a certain number of cutter-rigged vessels in 1775, and the fish-buying boat *Maria* is depicted in 1797 on a Manx plate as a clinker-built, cutter-rigged craft. Spritsailed craft are said to have been seen in Ramsey in 1760, but there appears to be no further mention of this.

The earliest Morecambe Bay boats, therefore, were clinker-built, sloop or cutter-rigged smacks with a raking transom stern similar to the estuary boats. The difference between the sloop and cutter rigs is that the sloops have only one jib hanked to a forestay, while cutters have all manner of jibs and foresails. The cutter rig produced the desired handling in the tricky, confined waters of the bay with its many sandbanks and fast tides, and was later favoured. With the exposed waters of the Irish Sea nearby, the boats also needed to have a good turn of speed to reach safety in the event of the wind whipping up unexpectedly. Many of these early craft were built at Greenodd, where Richard Ashburner was building schooners and fishing boats between about 1830 and 1850.

NOBBY RACE

Top: A drawing by N. P. Miller of a lute-sterned nobby.

Bottom: A small Lancaster-registered nobby. (*Lancaster Maritime Museum*)

Up to the mid-nineteenth century, Poulton, Bare and Heysham were entirely separate and were the main fishing stations in the Bay. Flookburgh, on the north coast, became renowned for its cockles and fluke (flounders), the cockles being gathered by hand and then transported ashore over the sands at low tide by horse and cart. After about 1950 horses disappeared and tractors took over.

Documented evidence of the shrimpers does not actually surface until about 1840, among the origins of the town of Fleetwood. Prior to this, the area was purely one of sand dunes that formed part of the estate of Peter Hesketh Fleetwood. Fleetwood erected his first building in 1836, and so began the growth of the town. His family, like many of those living on the edge of the sea, had owned small boats that fished off the coast and in the Wyre estuary, mainly to supply the needs of the estate. In 1841 four small smacks were hired by the Fleetwood Fishing Co., whose leader was Robert Roskell. These smacks had been brought in from the Leadbetter family at Banks, near Southport, for the fishing season only, and they proved successful. After they were returned, five boats were purchased for the following year, and again these proved worthwhile. Although called smacks, they were in fact probably the earliest of the shrimpers that had been built at Marshside, a part of the North Meols coast.

In 1846, Roskell was joined by a Kirkcudbright fisherman, John Wright, together with his smack. Four years later, the Leadbetters themselves moved up to Fleetwood and the town became, for the first time, established as a fishing station.

It was only a couple of years later that the first 'nobby', as we know them now, appeared from the village of Arnside, a small port nestling on the River Kent. Here Francis John Crossfield and his family began building cutter-rigged craft in 1848, a tradition that was to remain until the closure of the business in 1940. During this time they developed the emerging smack into the Morecambe Bay prawner.

These shrimpers trawled for the brown shrimps that were found on the sand-banks of the Bay, and also for the pink shrimps, locally called sprawns, found in the deeper water, especially around the mouth of the rivers. It is assumed the name 'prawner' comes from this fact, as the only prawns they ever caught were found off the Welsh coast, and then only occasionally.

Crossfield's originally built two classes of shrimper – a boat 31 feet overall by 9 feet beam with 3 feet 6 inches draught and a slightly larger 34 x 10 x 4.5-foot boat – the sprawner. They were normally of larch on oak, with an oak centre-line, although elm was sometimes used for the keel. Likewise pitch-pine was sometimes used for the planking, mostly after carvel construction was introduced around the 1870s. The deck was spruce and the mast and spars Baltic spruce. The Southport shrimpers tended to be slightly larger, up to about 40 feet, as did the Fleetwood boats in later years. By about 1920, Crossfield's new boats were all about the same size.

It was around this time – 1854 – that, as we saw in Chapter 5, the Morecambe fishermen Holmes, Baxter, Woodman and Woodhouse first sailed to Annan in their 30- to 35-foot clinker, half-decked, square-sterned, cutter-rigged trawl boats, thereby taking what were presumably Crossfield boats further afield. Whether the word nobby was a local usage to differentiate between the half-deckers and the bigger deep-sea smacks, or one from Scotland – where nabbies were common by this time – or the Isle of Man in later years (the Manx nobby not appearing until the 1880s) is unknown and various debates still rage.

Although Richard Ayton, in his *Voyage around Great Britain* in 1813, had observed that Blackpool and Poulton 'were overflowing with people', the arrival of the railway in Poulton in 1840 heralded the boom in tourism as the hordes flooded in from the Lancashire mill towns, desperate to get away to the seaside. This, in turn, led to the growth of not only Morecambe and Blackpool, but also Lytham and Southport. At the head of Morecambe Bay, Arnside and Grange-over-Sands became popularised by train-loads of visitors. In thirty years the region was transformed from sleepy fishing hamlets and clusters of several houses into bustling seaside resorts. This produced a need for tripper boats, and these were owned by the fishermen who organised themselves into sailing companies. Morecambe, for example, had six companies with licences for thirty-two sailboats and 158 rowing boats at the end of the nineteenth century. These sailing boats were built on shrimper lines but of shallower draught, many by Crossfield's.

It was during the closing decades of the nineteenth century that the fishing developed quickly alongside the booming tourism. Hundreds of people were employed, both at sea and ashore. What was previously a subsistence activity turned into a large industry. Fleetwood grew rapidly, having some sixty fishing vessels in 1860 and reaching its peak in the 1890s with ninety-five boats, sixty of these being the large smacks and thirty-five being half-deckers, or nobbies. The large smacks were 60–70 feet overall – 35–40 tons – and fished with huge trawls up to 50 feet wide. They sailed as far as southern Ireland, and to Islay and the Clyde, where they would remain for weeks at a

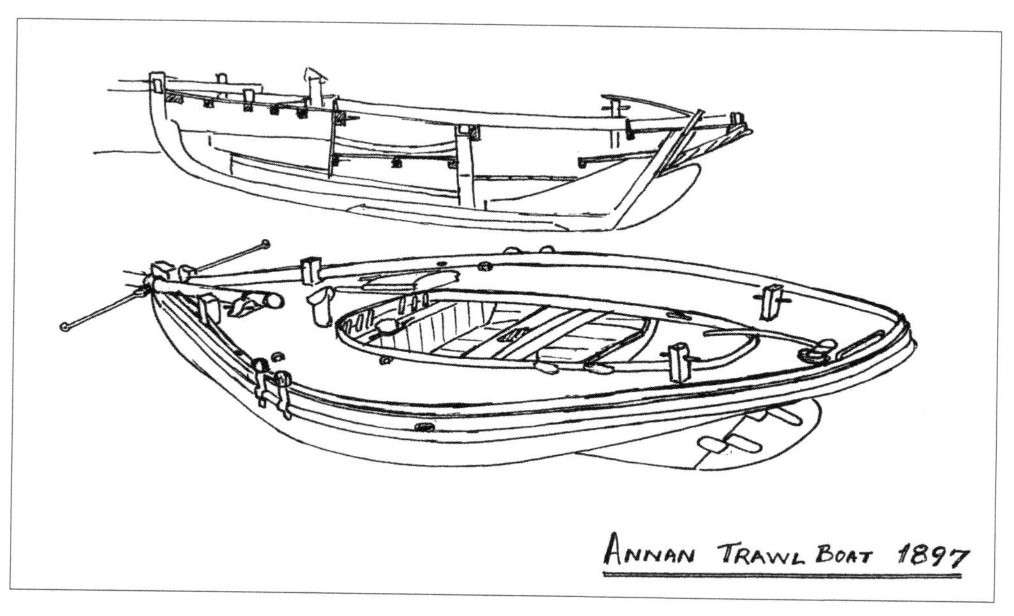

Drawing by N. P. Miller of Annan trawl boat.

A Morecambe Bay nobby off Arnside. By the look of the two ladies aboard, and the hats some of the crew are wearing, this nobby is not about to go fishing. (*Lancaster Maritime Museum*)

Old- and new-style nobbies at Lathom's yard, Crossen Sluice, near Southport, 1890s. (*Henry Lloyd*)

time. They were crewed by four men and a boy and survived until the advent of steam. But steam trawlers were already gaining preference in the 1890s, so the large smacks and ketches were short-lived.

Around this time, the fishermen began to alter the shape of the hull of the shrimpers to improve speed and handling. As in other areas of the country, fishermen often crewed aboard racing yachts and therefore gained experience in competitive sailing. They reduced the forefoot and gave the stem a more rounded profile. The sternpost was raked a little more, the depth increased and the bilges filled. However, this was not wholly at the expense of being able to beach the boats, as those that couldn't be kept afloat at all times were shallower than those that could. Rudder design altered drastically about the turn of the century, and also, by this time the counter-stern had developed from the early square-sterns. Some boats had lute sterns to improve performance, possibly at first fixed onto existing transoms but later embodied in their original construction. The counter stern easily evolved from the lute.

Other builders, apart from those mentioned previously, were Gibsons and Armours, both of Fleetwood. Richard Ashburner's brother, William, set up in Barrow in 1847. William Anderson established a yard at Millom in 1868. A year later, William Stoba was apprenticed to Gibsons, and he remained with the company for thirty-six years, progressing from shipwright to designer. In 1905 he moved across to James Armour, and immediately put into practice ideas he had for improving the design of the shrimpers. He suggested cutting away more of the forefoot and rockering the keel to produce a faster and handier boat. The resultant boat certainly outperformed the older boats, but at the cost of being unsuitable for beaching: a couple failed to lift

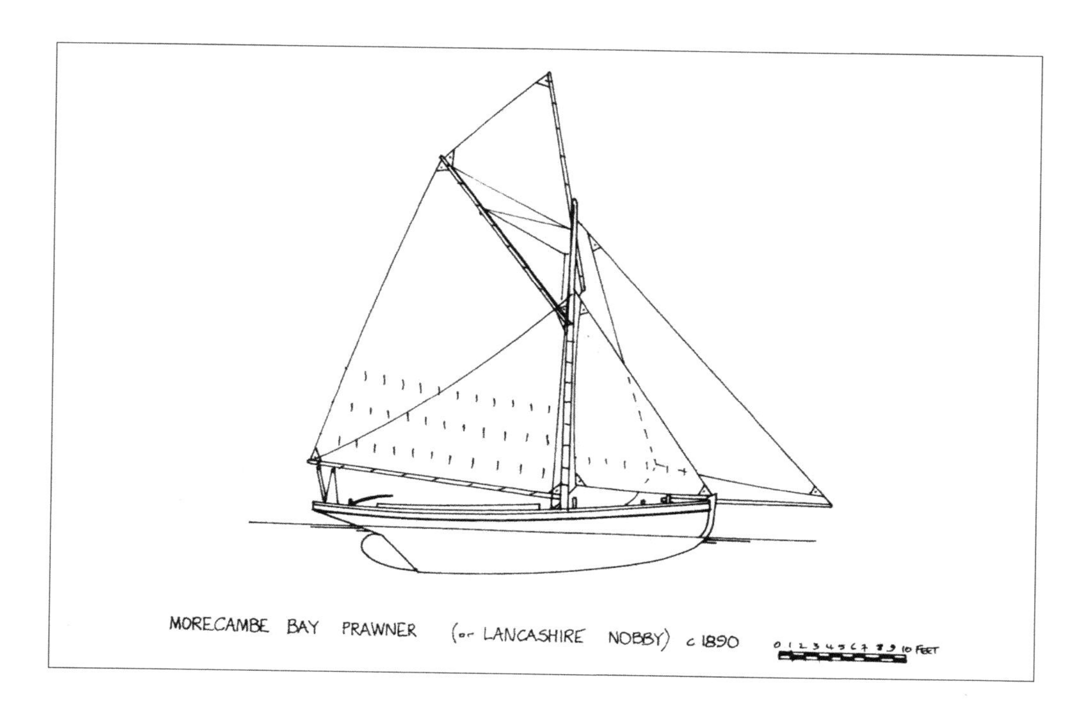
MORECAMBE BAY PRAWNER (or LANCASHIRE NOBBY) c 1890

Boatbuilding at Robert Lathom's yard, again in the 1890s. (*Len Lloyd*)

The launching at the Sluice from Wright's Yard in Sheffield Road, Marshside, *c.* 1895. (*Len Lloyd*)

when the tide rose and were inundated. The problem was resolved by the new boats being kept at the deep harbours of Maryport, Barrow, Fleetwood and Liverpool while the older types remained at Morecambe and Southport.

Crossfield's expanded with one son of Francis John, another John, opening a yard in Conwy in 1905. Ten years later, the two sons of George Crossfield, another of FJ's sons, opened a yard in Hoylake. This enterprise, however, only survived for five years.

The layout of the shrimpers was fairly typical throughout the region. The forecastle of a Southport nobby was some 14 feet in length with 4 feet of headroom and contained two bunks and a coal stove. The smaller nobbies had about 11 feet of forecastle with 3 feet of headroom. The shrimp boiler was situated amidships and the helmsman stood aft. Shrimping thrived up to the First World War, and became highly competitive. The shrimps were boiled as the boats raced home after being out for some ten hours, and the catch was landed quickly to be sold on the beach or from the stalls that attracted the visitors. Huge amounts of shrimp were sent by rail to the city markets as well, this market being regulated by the Morecambe Bay fishermen's co-operative.

Motorisation came to the area in the early 1920s, a few years after the end of hostilities. The first nobby to be designed with an engine was purpose built in 1925. Trawling and shrimping had resumed almost immediately after the war, and very briefly flourished. But it was short-lived, and boats were soon sold off for conversion. Those with engines nevertheless largely retained the rig. As late as 1932,

A prawner at Morecambe, sorting and preparing the shrimps prior to boiling them up. (*Lancaster Maritime Museum*)

Nobbies at Southport Pier 15 August 1897 – time off from fishing for 'quality sailing' or trips around the bay over the Bank Holiday weekend. (*Len Lloyd*)

however, Crossfield's built the *Maud Raby* at Arnside, to complement the seventy boats fishing there.

The Lancashire fleets steadily declined over the next forty years. But because of the number of boats previously sold off for conversion to power – the shape being perfect for this – many survived intact. Even those that continued fishing remained original because the fishermen kept the rig. Today some hundred true nobbies are said still to exist, mainly grouped around Liverpool, Conwy and other parts of the North West. Others have travelled far and wide and are to be found miles away from the dangerous waters of Morecambe Bay, yet their hearts must still cry out for that race against the others as they headed home to Morecambe, boilers all a-smoking and the pungent smell of shrimps mingling with the salty sweetness of the sea air.

HERRING FISHING

Liverpool, it has been said, had up to 1,000 fishing boats in the mid-nineteenth century, both decked and undecked, and a million herring were landed in a good morning during the season between June and September. Although these numbers are, in all probability, an exaggeration, and suggestions of a fleet of 100 boats would seem more reasonable, it does indicate that considerable amounts of herring were caught in Liverpool Bay in the last century. In 1869, when fishing boats were first registered,

Cockle boats arriving at Parkgate *c.* 1900.

records show that there were thirty-nine first-class boats, 146 second-class and ten third-class boats. Many probably didn't register to begin with, so these numbers could be low, but the figure of 1,000 still seems much too high.

Poulton, doubtlessly, had a herring trade until the herring deserted Morecambe Bay in the 1840s. On the Cumbrian coast, Maryport had its own herring fishery that survived into this century. Many Morecambe boats sailed up to Maryport for the herring, considering the trip as a sort of holiday. All these herring boats had a foredeck hatch which was used to allow going below while lying to the drift-nets. The last boat to undertake this 'holiday' was the *Star of Hope* in the late 1920s. Considering the closeness of the Isle of Man grounds, and those off the east coast of Anglesey, it is hardly surprising that herring was caught off the Lancashire coast. However, it is generally assumed that the much larger shrimp fishery, and that of the cod, salmon, mussels and cockles, outweighed the herring in importance.

THE BEACH BOATS OF MORECAMBE AND THE LUNE

Mussel boats worked in Morecambe Bay, especially around Morecambe and Heysham. The latter did not have its harbour completed until the first decade of this century; before that it was just a small shore-side village with a small fishing community and some visitors in the season. The mussel boats worked around the banks, and they were flat-bottomed to enable them to sit on the ground and be filled. They were clinker-built, mostly by Crossfield's of Arnside, up to the end of the nineteenth century. Because of their shallow draught they mostly had centreboards to make them more directionally stable when sailing unloaded. They were also used for drift-netting for herring, stowboating for sprats and whitebait, and for whammel-netting in the River Lune. This usually involved removing the centreboard, and often proved unsuccessful. Original mussel boats were some 15 feet overall, and these were the type that were taken to Annan in 1854 with the shrimpers. Mention has been made of these boats being sprit-rigged, yet the general assumption is that they were in fact rigged with standing lugs like the boats of the Clyde and nearby parts of the South Scottish coast. By the end of the century, the mussel boats were mostly 18–20 feet overall.

In the Lune itself, the Sunderland Point fishermen developed their own type of craft for the whammel-netting of salmon. Whammel-netting involves the shooting of a gill- or hang-net that is staked at one end close to the shore, deployed across the main stream and back to the shore in an arc, before being hauled in from the shore. The method is similar to the draft-net of the River Dee. The 20-foot whammel boats were adapted from the Crossfield-built mussel boats, with a sloping transom and standing lug. A foresail was sometimes set on a very short bowsprit, although most fishermen tended not to bother with this. By the beginning of the twentieth century these whammel boats became more identifiable in their own right, with some sixteen boats working from Sunderland Point and Glasson Dock, where an unofficial co-operative existed enabling both groups of men to co-exist without problem.

Most of these boats were built by Jack Woodhouse, at his yard at Overton, a couple of miles away. The Woodhouse family had been building boats here since 1660, and the last one they built was the *Mary*, LR53, in 1937. She was built for Tom Smith's father, and named after his mother. Tom still owns her, and has restored her for occasional fishing. He has just completed the restoration of the sister boat *Sirius*, LR33, built by Woodhouse in 1923 for James 'Shirley' Gardner and Tommy Spencer. *Daisy*, LR94, is the oldest surviving whammel boat, although she survives in a sorry state, semi-submerged at high water. She was built by Crossfield's in 1894, and has a small stern chute, like the Dee salmon boats, to sit the net on. The others that survive don't have this feature. However, the Woodhouse boats far exceed the Crossfield boats, according to the fishermen. Today, unfortunately, a new breed of whammel boat exists in the form of a double-ended fibreglass hull, the mould being taken from a Yorkshire keelboat that still sits on the marsh. Tom's other boat, *William Arnold*, LR173, is one of these. He hopes, though, to use *Sirius* for whammel-netting before long.

Tank boats evolved at the end of the last century for fishing the disturbed waters of the Lune bar. These had built-in buoyancy tanks, and the fishermen often lashed themselves in when fishing in these confused waters. Like the whammel and mussel boats, these were lug-rigged. Three of these tank boats – one built at Annan and two at Overton – have been reported as having been converted into yachts and are still sailing today.

Pilot boats worked from Sunderland Point, and Harry Gardner uses one of these, *Vera*, for whammel-netting. The Gardner family, too, produced many whammel boats

A River Lune whammel boat with small lugsail. (*Lancaster Maritime Museum*)

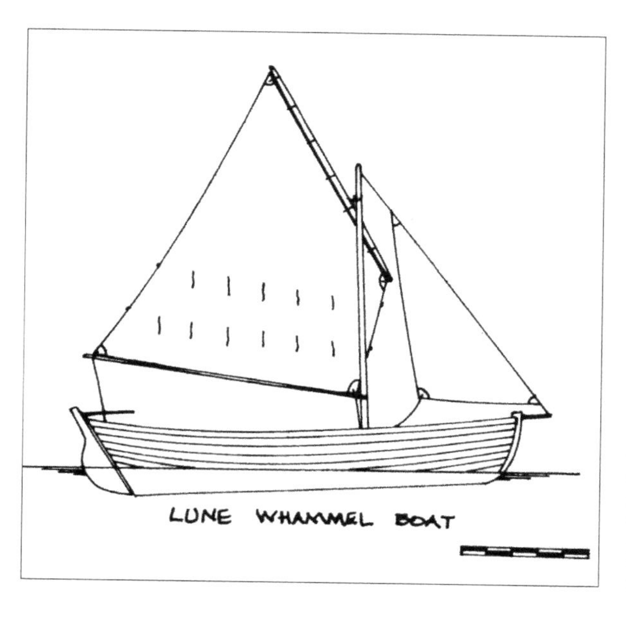

The Morecambe Bay nobby *Arthur Alexander* under full sail at the Conwy nobby race of 1997.

at Overton. The pilot boats are very similar to the fishing ones, and today one sits on the front, outside the Gardner house (1997).

Ivanhoe, LR86, was a double-ended whammel boat built by Jack Woodhouse in 1906 for James Gardner. Only six of these were built, and a few years later three of them were taken down to Chepstow by train, to show the Wye fishermen how to use a whammel-net. *Ivanhoe* returned, but nothing is known about the other five boats. She worked until the 1980s before being retired. Other than *Vera*, *Sirius* and *Mary*, the beach is devoid of these lovely boats. Most have gone for good to be replaced by the newer breed, and as Tom Smith tells of those boats of old – *Agnes*, *Mary*, *Sirius*, *Walrus* – I drift off into a dream trying to imagine the river with these small boats in full sail as they do battle with the tide. There's surely something to be said for history.

CHAPTER 9

The Welsh Coast

Penwaig Nefyn, penwaig Nefyn,
Bolia fel tafamwyr, cefna fel ffarmwrs.
('Nefyn herrings, Nefyn herrings,
Bellies like innkeepers, backs like farmers.')
—The cry of a North Wales fish salesman

The Welsh coast has three distinct topographical regions: the north coast with its comparatively sheltered waters and sandy beaches, the exposed waters from Caernarfon Bay to Pembrokeshire with its inhospitable rocky shoreline interspersed with river inlets, and the south coast, part of the Bristol Channel, which, although relatively exposed to the prevailing Atlantic seas, does not hold any particular fishery of consequence except along its southern edge. The herring fishery, although never developing into an industry as it did on the east coast, was, nevertheless, a very important feature of life along the coast of Cardigan Bay, Pembrokeshire and the Lleyn Peninsula.

Unlike the east coast, the fishery in Wales occurred in autumn and winter. Those who worked within it were part-time fishermen who realised the benefit of adding herring to their diet of home-cured bacon and local produce. When the herring came, agriculture fell to a standstill. Pockets received a healthy lining!

THE GROWTH OF THE HERRING FISHERIES

There is no doubt that the herring was the most important fishery in medieval Wales. In 1206 it was reported that 'in that year there came to the estuary of the Ystwyth such an abundance of fish that their like was never heard of'. Huge amounts were landed at Aberystwyth from where, in the mid-fourteenth century, there were over twenty boats operating. Records of 1302 tell of fishermen being fined for selling herring below the high-water mark to escape paying market tolls. Others were fined for fishing without the requisite licences for their boats. The same fishermen paid a mease of herring – the 'prisemes' – for each herring boat in the harbour.

Although Aberystwyth accounts for its catches of the time, Beaumaris, Barmouth and Aberdaron were the chief herring stations in North Wales, and Tenby, Pembroke and Haverfordwest in the South. Welsh fishermen are said to have gone to the Baltic fishery about this time.

Tenby – in Welsh, *Dinbych-y-pysgod*, which literally means 'Little Fort of the Fishes' – obtained its first quay in 1328, funded by Edward III, and it was one of the first quays built in Wales. The fishermen paid tithes of herrings and oysters for mass to be said on their behalf at the small St Julian's chapel on the harbour.

In 1553, Swansea Customs charged one penny on the import of one barrel of herring. Here, the herring never came, and fishing along the south coast of the country relied upon oysters, cockles and mussels. The Cardigan Bay fisheries developed in all the tiny creeks and bays and was the dominant maritime occupation. When writing about his travels of about 1540 in *Itinerary in Wales*, John Leland wrote, 'Solvach – or Solverach – a smaul creke for ballingars and fischar botes.' The same can be said of the north coast where, at the same time, he reports that Wallasey in Cheshire is a place 'wher men use much to salten hering taken at the se by the mouth of Mersey'. Yet, after the dissolution of the monasteries in 1540, fish consumption decreased as the Church of Rome was no longer there to decree fast-days when only fish could be eaten.

From *The Welsh Port Books of 1550–1603* (Dr E. A. Lewis, 1927) the level of imports of cured herring during the second half of the sixteenth century can be seen. Pembrokeshire seems to have been self-sufficient in its needs, with the fishermen obviously supplying the local demand. Imports of Scots herring were made to Milford in 1554 when there was a scarcity for a few years, but by the end of the century most of the herring traded was for export to France, Ireland, Bristol, Carmarthen, Chester and Liverpool. Carmarthen also imported herrings from Clovelly and Barnstaple, where the principal Bristol Channel fishery was, and from Wexford in Ireland. In the north, Beaumaris was the main export port for salted herring, with the salt coming in from Cheshire and Lancashire. Mostly this herring went to Chester and Liverpool, with Nefyn and Pwllheli sending loads there as well. Irish herrings were selling at 5*s* a barrel both for red and for white herrings, with local shot herring selling for the same price whereas a barrel of full herrings reached 10*s*.

At the dawn of the seventeenth century Aberdyfi was the centre of the fishery when the season started at Michaelmas. At the time Devye, as it was called, was described in *The Ports and Creeks of the County of Merioneth* in 1565 as 'being a Haven and havinge no habitacion, but only three houses whereunto there is no resorte: save only in the tyme of hearinge fishinge at which tyme of fishing there is a wonderfull greate resorte of ffyshers assembled from all places within this Realme with Shippes Boottes and Vessells, and during which abidinge there, there is of the said cumpanye there assembled one chosen among themsellfes to be their Admirall. And there is nother Shippe nor vessell that elongeth to the same Haven otherwise then aforesaid, whereof we have deputed David ap Thomas ap Rutherche, and Thomas ap Humffrey beinge the substanciallest and nerest to the same haven.'

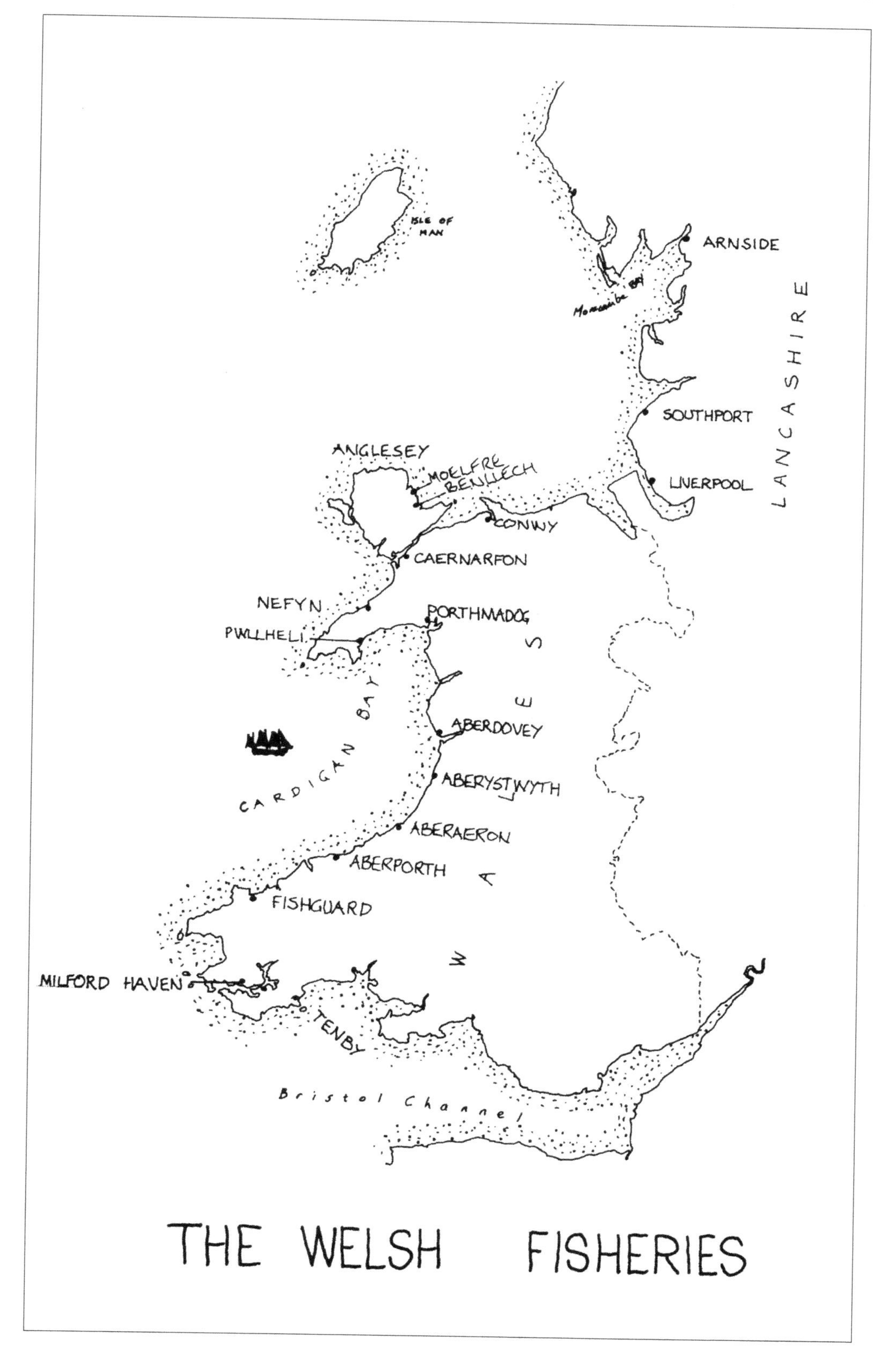
ISLE OF MAN
ARNSIDE
LANCASHIRE
SOUTHPORT
LIVERPOOL
ANGLESEY
MOELFRE
BENLLECH
CONWY
CAERNARFON
NEFYN
PORTHMADOG
PWLLHELI
CARDIGAN BAY
ABERDOVEY
ABERYSTWYTH
ABERAERON
ABERPORTH
FISHGUARD
WALES
MILFORD HAVEN
TENBY
Bristol Channel
THE WELSH FISHERIES

In his book *Description of Pembrokeshire* George Owens writes that the coast of Pembrokeshire was 'enclosed in with a hedge of herring'. He also says of the herring that 'these kinde of fishe is taken on the shores of this Countrey in great abondance … the places of their takeing in this shire most usuallie was in Fishguard, Newport and Dinas, where for manie yeares, and even from the beginninge there hath some quantite beene yearly taken, of later yeares they have resorted to Broade havon, Galtop roade, Martin havon, Hopgain, and St Brides, and have beene plentifullie taken to the great Comoditie of the Countrye, and nowe in the yeare 1602 they have been taken within Milford havon, and in the Roades of Tenby and Caldey, and neere St Davids and generallie in everye parte of the sea shoare about this shire from the fall of Tyvy to Earewere'.

Cardigan developed between 1678 and 1709 into an important herring station. Empty barrels were imported from Wexford and then filled with salted herring and exported back to Dublin. In 1701 Cardigan exported 1,072 barrels, and in 1782, 1,734 barrels. Many of these were sent by boat to the Pickle Herring quay at Southwark in London. Others went to Bristol, Exeter, Falmouth, Dartmouth, Chester and Liverpool. One cargo even went as far as the Canary Islands. Considerable amounts of herring from Aberystwyth were exported through Cardigan. St Dogmaels, across the river from Cardigan, became one of the principal herring ports and soon outgrew Cardigan.

All along the coast, settlements thrived on the herring. Aberarth, where one of the earliest settlements in Wales of the sixth century fished using fish traps, had its own catches of herring. Llangrannog established itself as a religious settlement upon the herring fishing. Newquay was reported by the Admiralty in 1748 as being a hamlet with small fishing sloops. The herring here was 'of very superior quality'. Between the Llangrannog and Newquay the coast is extremely remote, and consequently was a popular smuggling area. Because of the difference in the cost between Irish and English salt, due to the Salt Tax, huge amounts of salt were illegally brought in. Cwmtydu thrived on fishing and smuggling. Ceibach (Little Quay), on the other hand, was a popular little cove used only by fishermen, until shipbuilding developed on the beach in the early 1800s. Aberaeron, an early fishing settlement, had its harbour completed in 1811 and became base to a major herring fleet that numbered over thirty boats in the 1830s. This later rose to over sixty boats. The beaches at Penbryn and Tresaith exported cured herring. During the seventeenth century, mining developed rapidly, until it eventually overshadowed the fishing. When bounties were introduced in 1750, there was no impact on the Welsh fishery. But once the small boats were included in the scheme in 1787, there was immediately a marked increase in the number of boats. Aberporth, whose herring – Sgadan Aberporth (literally Aberporth herring) – were regarded as having a special flavour and were popular in the towns of South Wales, grew rapidly into one of Wales' premier herring ports. Off the north coast of the Lleyn Peninsula, Thomas Pennant reports that in 1,771 herrings were taken in abundance 'from Porth Ysgadan, or the Port of Herrings, to Bardsey island. The value of this catch was normally some £4,000. The catch was either salted ashore or taken direct to Dublin by Irish wherries'. Nefyn soon grew into the other principal Welsh

Tenby beach with some of the fleet of Tenby luggers. St Julian's chapel is behind on the quay, with the fishermen's huts in the arches.

herring port where Penwaig Nefyn, or Nefyn herrings, became renowned throughout the Principality.

When Daniel Defoe visited Tenby in the early 1720s, he found it 'the most agreeable town on all the seacoast of South Wales, except for Pembroke, being a very good road for shipping and well frequented; here is a great fishing for herring in its season, a great colliery or rather export of coal, and they also drive a very considerable trade to Ireland'. The fisheries never really developed here though, because the poor fishermen were said to have never been able to afford bigger boats and hence exploit the resource to its full. Huge catches, however, were made in the seventeenth century and sold locally, as we've already seen.

Aberystwyth still managed to retain its importance as a herring station up to the end of the eighteenth century. In 1745, forty-seven fishing boats caught over 1.25 million herring in one night. This was an enormous catch, even in terms of those caught on the east coast. The Liverpool merchants had gained a footing in the town in the sixteenth century and they continued to exercise control over the fishing throughout the seventeenth and eighteenth centuries. The town would most likely have become one of Britain's chief herring ports in the next century but for the actions of these merchants whose only concern was to exploit any potential there. Most of the herring was shipped back to Liverpool and the town's fishery declined; a situation not helped by the very negative attitude of the town council to fishing in general and the lack of a decent harbour. However, the only influence that remained in the town by the end of the nineteenth century from these Lancashire people was to be seen in the design of their boats.

Milford Haven Docks with steam trawlers and a couple of smacks. (*Tenby Museum & Art Gallery*)

During the nineteenth century, the small coastal villages saw an upsurge in the herring fishery and such places as Porthclais near St Davids, developed small fleets. Similar fleets were to be found at Angle, Dale, St Brides, Little Haven, Abercastle and Cwm-yr-Eglwys. Porthgain, which after about 1830 developed into a substantial quarrying concern, was initially home to a tiny fleet of fishing boats. Goodwick and Lower Fishguard both had small fleets that rivalled each other until the pier at Goodwick was demolished to make way for the new commercial harbour at the turn of this century.

The herring continued to pour into Cardigan Bay right through the nineteenth century, with unsurpassed numbers being taken at Aberystwyth in 1831. Similarly, huge shoals visited the Flintshire coast in 1883. Caernarfon had 300 boats at this time, only seventeen of which were first-class boats over 15 tons. In 1884 huge shoals were seen off the west coast of Anglesey, where Liverpool and Hoylake boats established a fishery in Aberffraw Bay. Meanwhile, the villages of Moelfre, Benllech and Bull Bay were benefiting from vast shoals that appeared from Conwy to Bangor in 1884. The catch on the east coast was at first hawked locally until the railway arrived and enabled it to be sent by horse and cart to the railhead at Benllech and thence direct to the Liverpool market fresh. However, Moelfre herring continued to be hawked throughout Anglesey where fisherwives walked in the season. The cry 'Moelfre herring' is remembered by several people today.

The railway, in fact, brought about the demise of most of the small herring fisheries, and, indeed, most of the sea trading that had gone on around the coasts up till this time. Neyland, a herring station since 1850, soon lost out when Milford Haven opened its docks in 1888, whereupon there was a rush for space in the harbour. With the emerging steam trawlers gaining preference over the older sailing craft, and the speed with which the fish could be moved about the country, Milford soon commanded the fisheries of the Irish Sea, there being no other deepwater harbour between there and Holyhead. In 1924 there were 124,000 barrels of herring being landed, of which 4,000 were even sent to the east coast. By the following year it had become the principal herring port of England and Wales. Both Cardiff and Swansea established themselves as deepwater fishing ports, although not to the same extent as Milford. The same year saw Scottish boats begin landing herring at Holyhead.

The numbers of boats in other parts of the coast dropped rapidly after the First World War. Pwllheli, for instance, had twenty-four trawlers before the war and only eleven afterwards. Aberystwyth was down to three rowing boats and six motor boats by 1928. Cardigan, which had had seventy-eight boats at one time, was reduced to only a handful of boats.

THE WELSH FISHING BOATS

Although the early Celts probably used skin boats to traverse St George's Channel, the earliest of the documented Welsh fishing boats were small rowing boats. In 1565, the creeks and ports of Cardigan Bay were described as possessing 'no shippe belonging to eny of the sayed havons crikes or landinge places or eny other bottes or vessels other then smale shyppinge bottes conteyinge 4 or 5 tonnes apice wich use to fyshe apon the cost of the sayed shire and do use non other trad and that chifly heringe physshinge after Michelmas; in everi of the sayed bottes duringe the fysshinge tyme are continually 6 or seven persons all fysshermen and no mariners'. The same report gives two 'fysshing boates' as being 'two tonnes a peece' used for 'fysshing nere the shore' as working out of Marross in Carmarthenshire. These two boats were called *Sondaye* and *Marye* and were owned by David Bealt and John Hoddyn.

THE WELSH NOBBIES

In Aberystwyth, the oldest pictorial record of a boat is from a drawing by E. Prys Owen in 1820, now in the National Library of Wales. It depicts a clinker-built boat about 25 feet in length overall, with bluff bows and a transom stern. She had a small foredeck and a transverse beam divided her into two different working areas. She was cutter-rigged with a mainsail and a jib and staysail set on a long bowsprit. The main boom extended well aft over the transom, while the mast was stepped well forward against the foredeck. She carried a beam trawl on the starboard side. This was similar to many boats around the Irish Sea during that period, and the shape of the boat reflects influence from the Viking period.

By the time the railway reached Aberystwyth in 1864, the influence of the Lancashire curers was imposing itself on the design of the boats, as previously mentioned. The Lancashire boats of 25–40 feet – carvel-built with steeply raking sternposts, little sheer and cutter-rigged – were common around the north-west of England, and had evolved

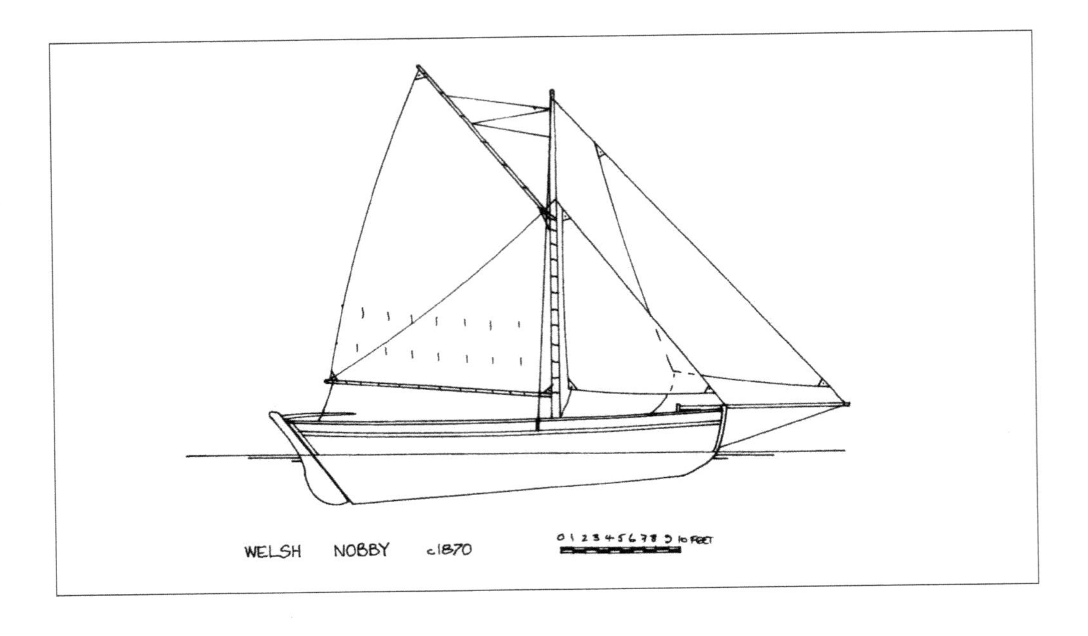

through influences from the Isle of Man and southern parts of Scotland. They travelled to Aberystwyth to work the season in Cardigan Bay, some in fact coming straight from the Isle of Man. This is shown by the fishing register where, in 1872, Aberystwyth only had six of its own clinker-built second-class cutters, whereas there were sixty-eight third-class rowing boats, the remaining boats coming from away.

From about this time, these boats became known as the Welsh nobbies, probably as a result of what the Lancashire boats were called. It wasn't long before similar boats were being built locally, drawing upon experience from these Lancashire boats and the earlier smacks.

During the 1890s, after William Stoba produced the more favoured counter-stern shrimpers, some of them sailed over to Cardigan Bay, but were found not to be suited to the deeper waters there, although some remained working out of Pwllheli and Abersoch well into the present century. Many of these boats worked out of harbours along the North Wales coast such as Conwy, where the Crossfield family opened a yard in 1905. A century earlier, double-ended, two-masted boats worked out of the town.

Both Pwllheli and Abersoch had protected harbours, although they dried out. Some of the nobbies working from these ports had direct similarities to Scottish boats with their deeply raking sternpost, rounded stern and cutaway forefoot. Low in the water because of little sheer, they had a resemblance to the Lancashire boats, although their underwater profile was very different, all the Welsh boats retaining the long straight keel. The Welsh nobbies could be said to be a vague range of boats as few were identical, being the products of various builders and influences. Along with the raked sternpost and straight keel as mentioned, their only other common trait was the cutter rig without topsail. Although most were 25–40 feet long, a few were built up to 50 feet and these were fully decked. They do not seem to have survived much past the 1920s.

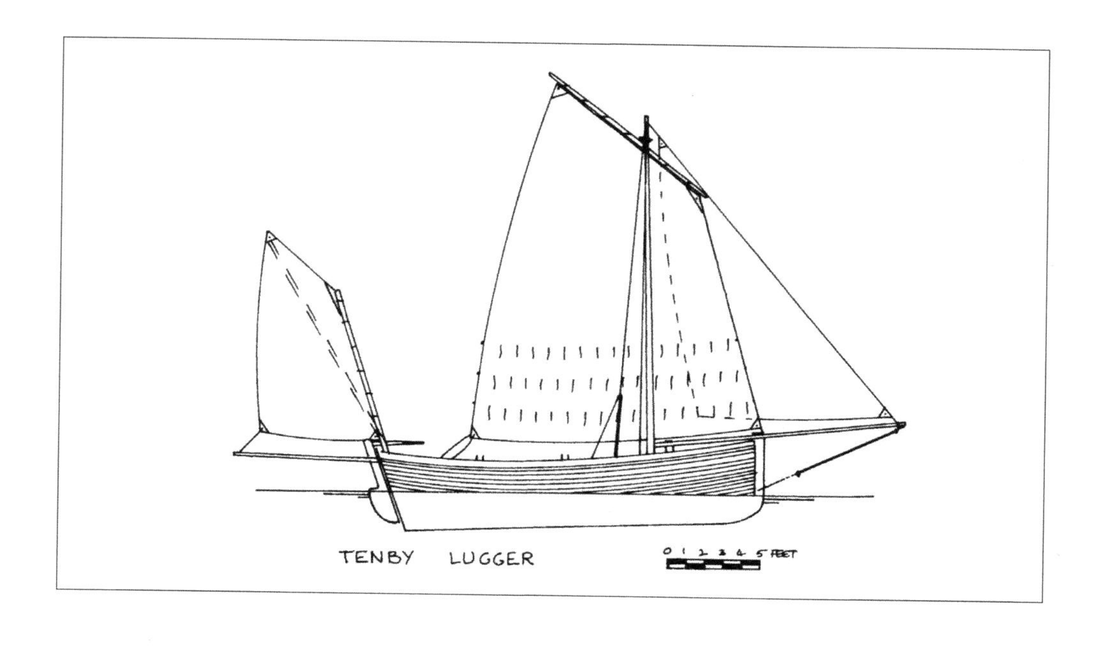

THE TENBY LUGGERS

We've already seen how the fishery developed in Tenby. Contemporary drawings of the early nineteenth century show small single- or two-masted vessels which must have developed from the small open boat type of before. An 1834 print shows a small boat with two square sails that displays particular Viking influence, which, considering their terrorising of the coast between AD 844 and 1091, is hardly surprising. By the end of the nineteenth century these had evolved further into half-decked luggers. The largest boat was built in 1891, the *Eileen*, which had a keel length of 27 feet, though most were a little shorter, at around 25 feet overall.

The earliest boats were clinker-built but carvel construction took over during the latter part of the century. They were straight stemmed, and had long keels and raking transoms. The foredeck reached to the mainmast, providing a limited amount of shelter. There were three thwarts for rowing, and further seating in the stern sheets for the helmsman. The mainsail was a dipping lug with three sets of reefing points while the small mizzen-spritsail was rigged on a raking mast and sheeted to an outrigger. The foresail was set on a bowsprit.

The boats were heavily-built, especially so for their size. The keels could be 18 inches deep and heavy oak floors were fitted. This was possibly because the deep water in the harbour enabled them to be moored afloat at all states of the tide. Some were kept at nearby Saundersfoot which was a sleepy little fishing village that also had a small coal trade until the end of the eighteenth century.

These luggers were used for all types of line fishing, drift-netting and oyster dredging. Drift-netting for herring, though, became unpopular around the turn of the century as the huge shoals of Cardigan Bay never seemed to reach round into Carmarthen Bay as they had done before. However, the town had developed into a well-visited watering hole and the fishermen were able to operate bathing machines and take trippers out to Caldey Island in the summer. This practice survived until the 1950s.

These boats did not exceed about fifty in number at any one time. Unfortunately for the local fishermen, the fishery here was sought after by Devon men, with Brixham trawlers working offshore and only coming into harbour on a Saturday night and staying until Monday morning. Many photographs show them dwarfing the local boats. These foreigners fished both for herring and for oysters, until they had cleared the oyster grounds to the north of Caldey and off Stackpole. Much of the oyster catch was shipped to Liverpool and Bristol, or pickled for the London market. Although a few of the larger boats were brought into Tenby from both Brixham and nearby Mumbles, becoming known as the Tenby cutters or smacks, the local luggers soon disappeared almost completely.

The only one that has survived is thought to be the *Seahorse*, M170, built by James Newt around 1886 as the *Three Sisters* and which has an overall length of 24 feet. Today she is on loan to the West Wales Maritime Heritage Society from the Welsh Industrial and Maritime Museum and is awaiting restoration. A replica vessel is being built by youngsters in Milford Haven.

Tenby luggers and Brixham smacks anchored in Tenby Bay. (*National Library of Wales*)

A clinker-built Tenby lugger similar to *Seahorse*, *c.* 1890. (*Tenby Museum and Art Gallery*)

THE BEACH BOATS

RHYL

Rhyl, or to be more exact Foryd, the ancient port of Rhuddlan, was for centuries a tiny fishing hamlet. In 1800, small fishing boats worked from here and the fishermen lived in 'tiny miserable thatched cottages with turf or cobbled-stone walls' which were in constant threat from high tides and the sand-dunes. Tai un nos – one-night cottages – were small dwellings that could be erected in one day so that smoke came out of the chimney by sunset. The builder of such a cottage would automatically own it, and if that owner threw an axe from his front door, then all the land encompassed within the radius of that throw would be his as well. Or so he believed! This belief existed in many other parts of rural Wales. These early fishing boats appear to have been rigged with one square sail, and they caught all manner of fish. Thomas Pennant, who lived at Mostyn nearby, recorded in 1796 the fishermen as catching 'flounders, plaice, small sole, ray, dab, cod, weaver, and even anchovy' as well as mackerel and herring of which he stated, 'Herring in this sea are extremely desultory. At times they appear in vast shoals, even as high as Chester. They arrive in the month of November, continue until February, and are followed by multitudes of small vessels which enliven the channel. Great quantities are taken and salted but are generally shotten and meagre. The last time they appeared here in quantities was in the years 1766 and 1767.'

With the passing steamer trade and the coming of the railway in 1848, tourism developed rapidly in Rhyl, and the fishing community declined. Some Morecambe Bay nobbies worked out of the little harbour in the early part of this century and came to be called the Rhyl nobbies! They went shrimping during the spring and summer, boiling the catch in the same way as the Lancashire fishermen.

ANGLESEY

On beaches such as those at Benllech and Moelfre the favoured craft were small, transom-sterned, open boats normally about 20 feet overall, although towards the end of the nineteenth century 30-foot boats were to be found. These were all lug-rigged, though oars were normally used for the herring fishing. Although the boats were undecked, they were often sailed over to the Isle of Man, a distance of 40 miles or more, for the summer herring there. Most of these boats were built by Matthew Owen of Menai Bridge, were clinker-built and constantly beached, so did not have a long lifespan. The Anglesey herring fishing survived into the 1930s, but after that the small beaches fell quiet, and the beach boats that had cluttered these tiny villages for generations gave way to the tourists.

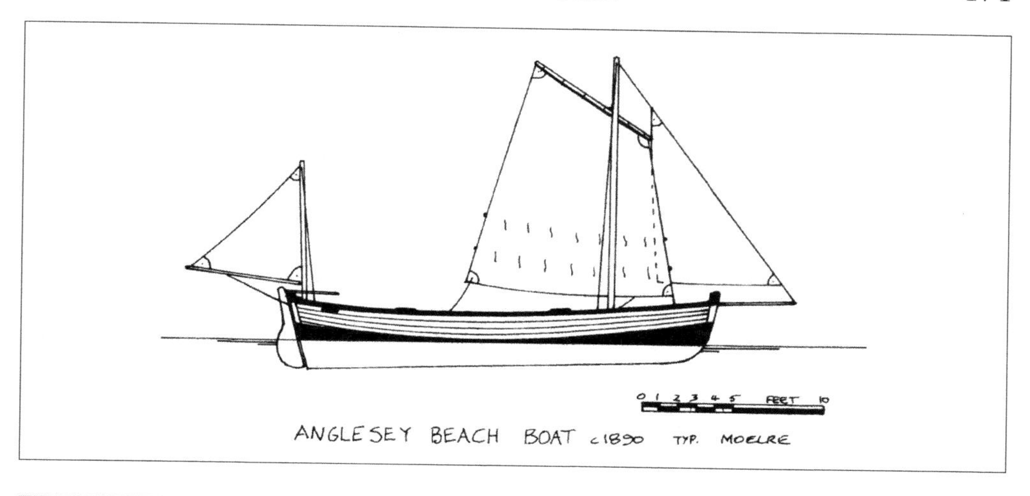

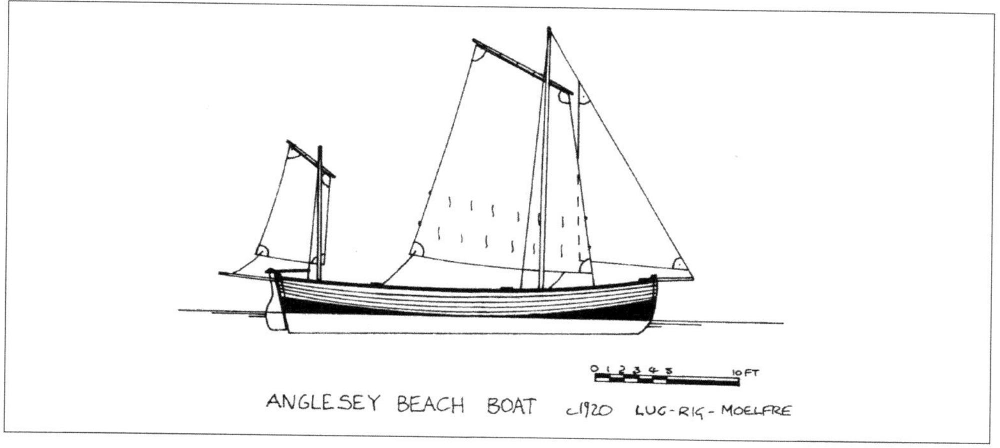

The beach at Moelfre on Anglesey with both double-ended and transom-sterned herring boats, *c.* 1920. Note the smallness of the transom.

NEFYN

Nefyn, and its neighbour Porth Dinllaen, are small bays protected by arms of land pointing north-east, so giving shelter in all but easterly winds. Most of the fishermen there were part time, being seamen or quarrymen according to season, and were mostly centred on Nefyn. The latter was more renowned for its shipbuilding. The herring season lasted from September to January, and during that period many visiting fishermen would lodge at Nefyn, a tiny village a mile or so away. On the beach a considerable number of ships were built for the slate trade and general shipping, and the most common type of fishing boats were small, open double-ended craft about 18 feet overall that cost about £20 to build. These beach boats were also built by Matthew Owen of Menai Bridge and were similar to the Anglesey boats. Nefyn was a busy herring station, yet it appears that none of the nobbies originated here, although the majority of the bigger boats were of this type. Records show that in 1910 forty Nefyn boats sailed to fish out of Ireland. These would most likely have been the larger nobby type, similar to the Manx boats, which were, at that time, beginning to be fitted with engines. A later photograph of the beach, taken around 1930, shows fifie-type motor boats that must have come in from the east coast of Scotland.

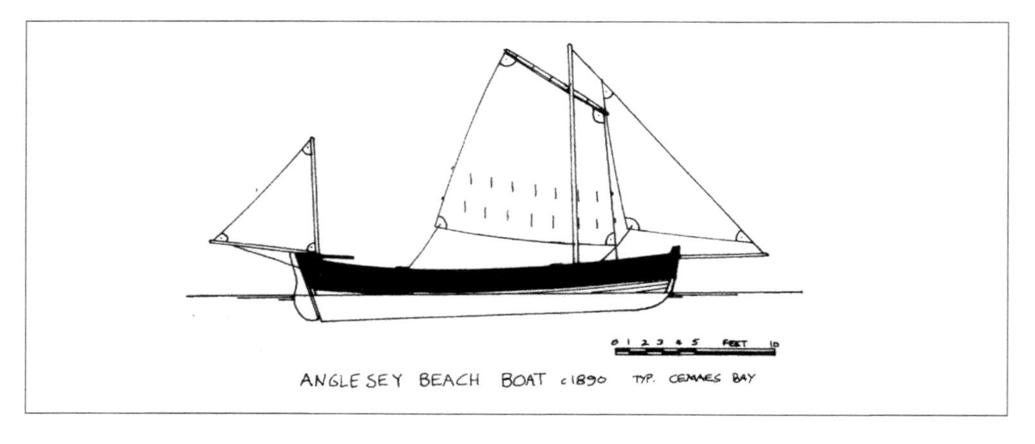

ANGLESEY BEACH BOAT c1890 TYP. CEMAES BAY

HERRING NOBBY AT PORTH DINLLAEN, NEFYN AT AROUND THE TURN OF THE CENTURY – POSSIBLY ISLE-OF-MAN NOBBY

A typical double-ended herring boat as used on Bardsey Island. Here it is seen on Aberdaron beach, complete with crew, obviously for the camera! (*National Library of Wales*)

Porth Ysgaden (the herring port) is a tiny sandy landing situated south-west along the coast from Porth Dinllaen. Here a fleet of small herring boats worked towards the end of the nineteenth century. These 25-foot boats were similar to those from along the coast but the accessibility of the beach just shows that the fishermen would base themselves at whatever place they could find. Every cove, every hamlet would have at least a couple of boats that were used primarily to search out the herring shoals in season. Further along the coast in the same direction, another inlet can be found at Porth Iago, where even today a couple of boats can be found.

Just as in Scotland and Cornwall, these fishermen, whether part-time or not, had their own methods of finding the fish. Whether watching the seabirds diving, glimpsing the 'fire in the water' or just netting in a successful spot, the fishermen were adept at finding shoals. And so that there was no disagreement about fishing limits, there was an unwritten law that prevented fishermen from one place impinging on another's ground. As elsewhere, they were a community on their own. As the old saying goes, 'A fisherman is born, not made.'

ABERDARON

The tiny village of Aberdaron sits peacefully at the extremity of the Lleyn, and is one of the most inaccessible places on the Welsh coast. Here, one still finds a couple of

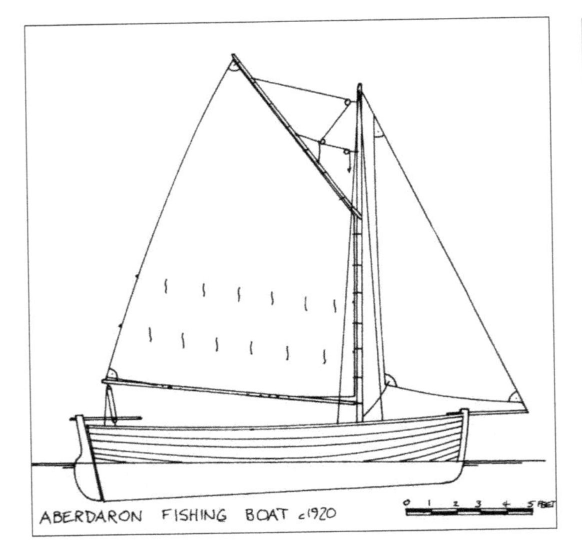

Another double-ender lying on a Gwynedd beach *c.* 1890. (*National Library of Wales*)

boats beached in the little cove of Porth Meudy or Fishermen's Cove, yet a hundred years ago it was home to a large fleet of small, 12- to 16-foot clinker-built, open boats. Originally these were double-enders from being herring boats, but, as the fishing declined, they fished for crabs and lobsters, and adopted a transom stern and a skeg so that they could be rowed backwards onto a pot. Previously, many boats had capsized when the pots were being hauled in over the pointed stern.

Herring were abundant in the sixteenth century and were landed and shipped to Ireland and Chester, although a lot of it went to local consumption. Over the next two centuries the fish was taken to London and Liverpool in small vessels that were loaded on the beach. Salt was smuggled in from Ireland as it cost only 1*d* a pound there, but 4*d* a pound in Britain, so smuggled-in salt at 2*d* a pound was worthwhile.

As the fishermen found they had to sail further for the herring, they found readier markets for their shellfish. By the 1880s nearly all the boats were potting. They were also being built locally, whereas previously some had come from David Williams of Aberystwyth.

One local family of builders was the Thomas family. John Thomas was born on Bardsey in 1880, where his father and grandfather had long been building ships on the beach. But after being at sea until 1918, John began building what is now termed the Aberdaron fishing boat and, indeed, he built over 100 of these vessels, a few of which remain. His father also built many; the oldest surviving boat, the *Annie*, having been built by him in about 1865.

These boats were originally lug-rigged, but adopted the gaff rig around the turn of the century, and today, the dozen or so that have survived are gunter-rigged. They race every weekend during the summer, weather permitting, and present a reminder of the vast numbers of beach boats that once worked off these coasts.

ABERYSTWYTH THREE-MASTED BOATS

Perhaps unique in their design and popular for herring fishing, the Aberystwyth three-masted boats were numerous in the harbour in the nineteenth century. Although technically speaking not strictly a beach boat, they spent most of their time working off the beach and were moored in the harbour where they dried out. The first vessel of this type is shown in a drawing from 1844, but it is assumed that they were introduced earlier. It has been suggested that the design originated from the village of Borth, an ancient fishing community above the beach, a few miles north of Aberystwyth.

These boats were undecked and originally were lightly built. Lengths of keel were 23–25 feet and perhaps a foot or so more in overall length as they had straight stems and upright pointed sterns. They were built locally from yellow pine with larch beneath the waterline. They had four thwarts, and local beach shingle was used as ballast, to be thrown overboard to compensate for the weight of fish when a catch was to be taken on board.

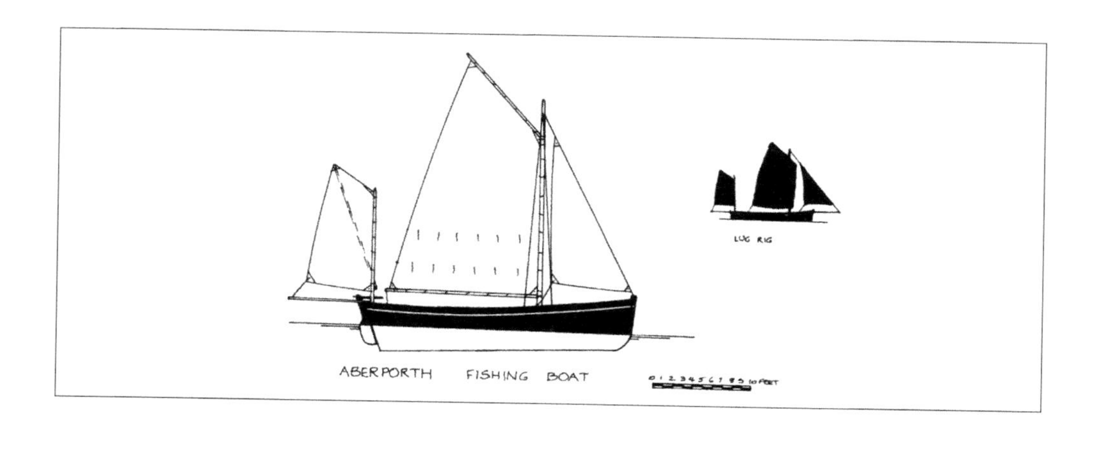

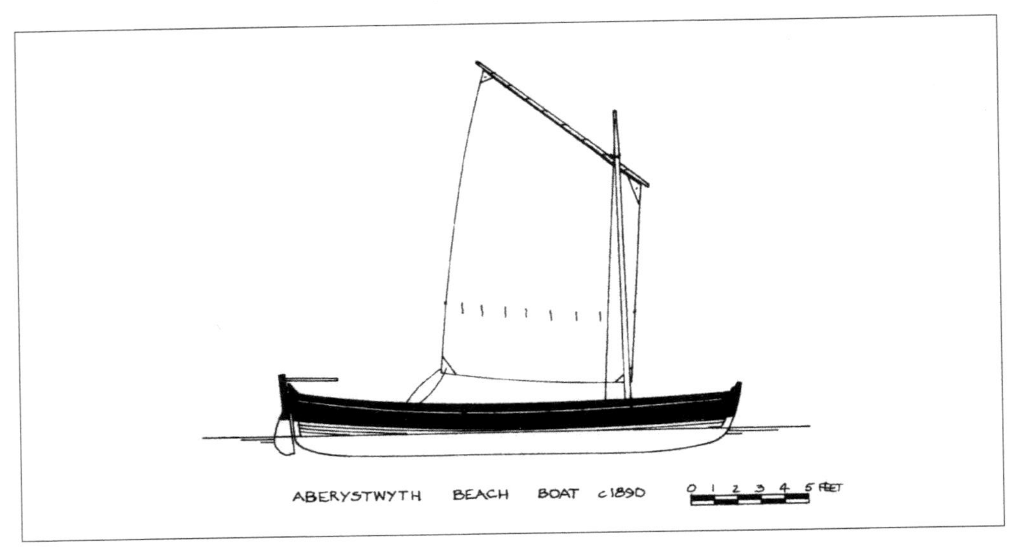
ABERYSTWYTH BEACH BOAT c1890
0 1 2 3 4 5 FEET

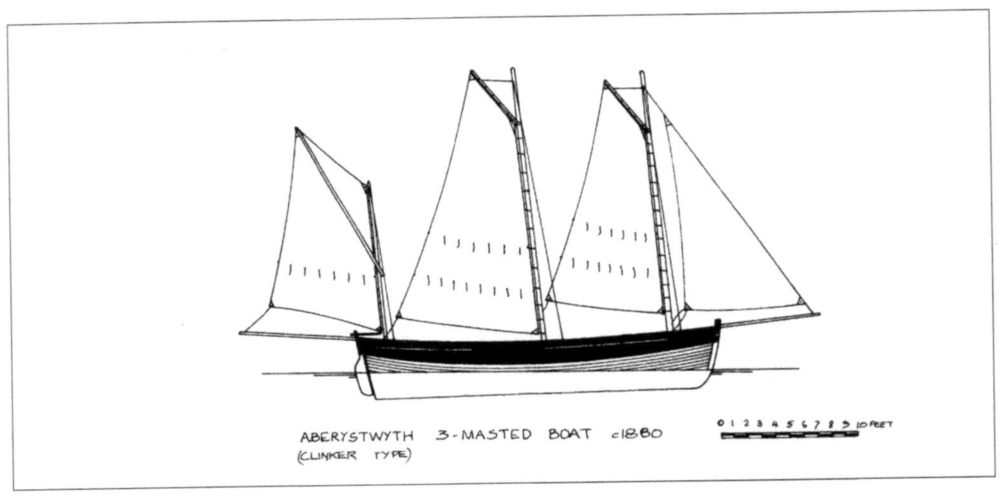
ABERYSTWYTH 3-MASTED BOAT c1880
(CLINKER TYPE)
0 1 2 3 4 5 6 7 8 9 10 FEET

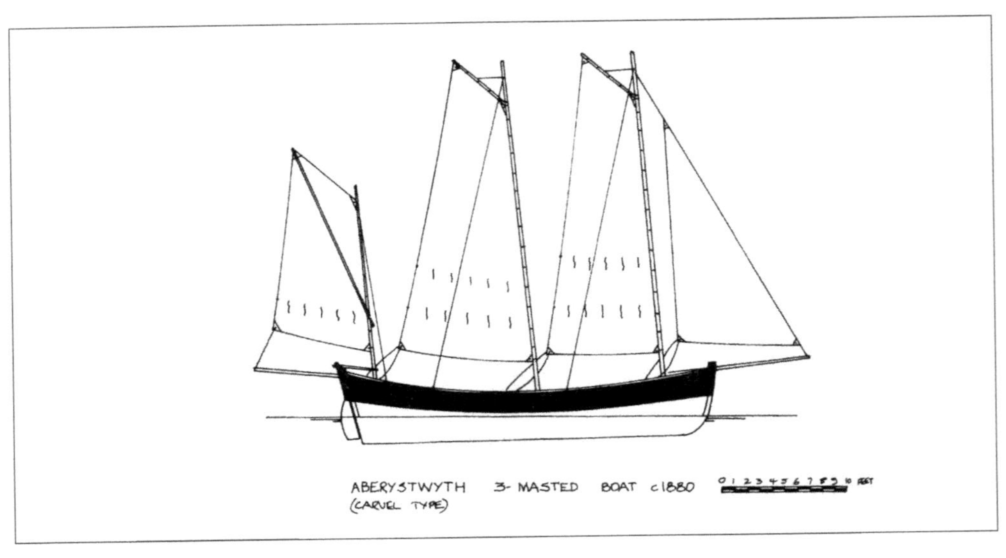
ABERYSTWYTH 3-MASTED BOAT c1880
(CARVEL TYPE)
0 1 2 3 4 5 6 7 8 9 10 FEET

A typical small salmon boat on the Dyfi estuary, *c.* 1890. (*National Library of Wales*)

With three masts, they were sailed some distance offshore, although when drifting for herring, the mainmast was sometimes left ashore. The main and foremasts were of a similar length, and carried a boomless gaff sail on each. Earlier boats were either square sailed or lug-rigged. The mizzen mast was much shorter and carried a spritsail which was sheeted to a long outrigger or bumpkin. Some carried a foresail set on a bowsprit, while others did not. No topsails were ever carried. Originally, these boats were clinker-built but around 1875 they underwent a change in design and habit.

Firstly, some boats introduced a gaff sail to the mizzen. Others, locally built by David Williams, were double-ended. Williams also built a smaller version, about 18 feet overall in a clinker construction, which proved popular upon the beach. However, when carvel-built boats arrived, lengths increased and some bigger boats appeared. The first of these was said to have been a lifeboat, but soon the others copied. These were developed for the lucrative tourist season: taking holidaymakers for trips around the bay made more money than fishing, as became true in many other parts of the country. The last boat was the *Lizzie*, built in Porthmadog around 1890 and a little over 40 feet overall. By the turn of the century, again, the decline in the herring meant that the summer business was the only business. The merchant seaman remained at sea; the rabbit hunters stayed with the burrows.

ABERPORTH HERRING BOATS

In Cardigan and St Dogmaels the boats were reported about 1830 as being 8–20 tons burthen with masts and sails, mostly open and manned by six or eight men. Similar boats were used at Aberporth, from where they took their name. These herring boats were 25–30 feet overall. Like both Cardigan and St Dogmaels, Aberporth has no harbour, so the boats were beached. They were generally of a heavy construction, transom-sterned and carvel-built. Mostly used for drift-netting, locally called 'drifio', they did work another type of fishing, known as 'tranio' or 'setin', which involved a fixed net weighted with stones and set with anchors at specific places. These herring boats had two masts, with a gaff mainsail, a spritsail mizzen and with a jib set on a bowsprit. As in many other parts, the fishermen did not use the sails when fishing for herring and relied solely on the oars. Years earlier, they had used a lug mainsail, and were similar to the Tenby luggers. Some carried a topsail while few were double-ended. Generally, however, sails were not used for drift-netting. Similar craft worked out of Aberaeron, New Quay, Newport, Llangrannog and even as far as Dinas and Fishguard. When the herring declined at the turn of the century, the numbers of boats fell, so that by 1905 this type of boat had disappeared and, unfortunately, no examples remain.

HERRING GIGS, CORACLES, COCKLE & MUSSEL BOATS

These small clinker-built boats with their narrow transoms and lines were similar to the Aberystwyth three-masted boats and were mostly found in that area, although others were to be found at places such as Aberporth. Generally they were rowing boats, although the larger ones carried a mainsail and even occasionally a mizzen spritsail. They, too, were used for rowing tourists around during the summer season when the fishing was slack.

Similar small clinker-built rowing boats, sometimes with a sail, can be found in many other small fishing communities, of which there are many along the Pembrokeshire coast, as we've seen. Such places as Abercastle, Porthgain, Solva and Little Haven which are all tiny coves on the rocky south-west coast, Dale and Angle in the Milford Haven and Ferryside on the River Towy, renowned for its cockles, all have their own fleets. Cockles were mostly taken off the beach by horse and cart; there were also thriving cockle gatherings at Laugharne and Penclawdd.

Dale was home to one particular boat which was built in 1906 for local fisherman Frederick Knights, who was only fourteen years old at the time. This Dale gig, as it became known, was 17 feet overall, had an extremely fine line with flat floors, and still survives, albeit in need of restoration. *Agnes*, named after Frederick's mother, was used for 'long-lines, nets and pots for shellfish within the confines of the Haven'. Although she was a perfect rowing boat, she was rigged with a dipping lug main and standing lug mizzen, the latter being changed to a tri-sail at a later date, as was common about the waterway.

No book about fishing craft would fail to mention that much-written-about champion of river craft, the coracle. Salmon and sewin (sea trout) fishing was the most popular of river fishings. The coracles themselves are one of the oldest of all fishing craft in Britain, but obviously, due to their peculiar shape and size, they didn't venture too far out to sea. Yet today they still survive, whereas their seaborne counterparts have not, so that many of them can be found in various corners of the Principality in

A drawing of an original Mumbles oyster skiff, *c.* 1855, by E. Duncan.

one form or other. In addition, many coracles continue to be built by purists, some of whom even run courses in the art of their construction. One tradition that is alive and kicking!

THE RIVER CLEDDAU

Talking of salmon, the upper reaches of the River Cleddau was the home of the compass-net boats. This method of fishing involves a bag-net suspended on two long poles that are stuck into the muddy riverbed around the time of low water, when the water is 12 feet or so deep. The poles form a 'V', and are set and hauled using 14-foot beamy transom-sterned boats. The most notable feature of these boats was the tarring inside and out, said to originate from the need to not be seen at night while poaching, although this isn't necessarily the reason! It's just as likely a cheap way to preserve the timbers. The boats were carvel-built with substantial oak frames every 12 inches or so, larch-planked, had three thwarts each with rowing points and two holes aft through the heavy gunwale for the mainsheet of the single lugsail. Their heavy scantlings are said to have come from their building by fishermen who worked in the nearby Pembroke Naval Dockyard out of the fishing season. Some boats were built in the yards when the work was slack, so it is said.

The Llangwm fishermen were also particularly renowned for their herring fishing, once the salmon season was over at the end of each August. This was really when the sails of the small boats were used to get further downstream and back. Most other times, they were solely rowed. The herring they caught was a freshwater sub-species that entered the sheltered waterway to spawn between December and March. For this the compass-net boats were also used. At one time there were supposedly a hundred such boats working the herring, but, as only eight salmon licences were handed out annually, I doubt it. Today the herring have long gone, and so, it seems, have the salmon, considering that only one was caught in 1997. Only two wooden boats keep

Mumbles oyster skiffs *Snake* and *Hawk*, after the adoption of the East Coast type. Although many were imported from the East Coast others, including these two, were built at Appledore, just across the Bristol channel.

this traditional way of the compass-net alive, even though the eight licences are still fished, some using plastic boats! Four or five other boats still remain lying around the riverbank, although, judging by their state now, this is more through good luck than attention.

Across the waterway, similar boats worked from Lawrenny and even as far up as Haverfordwest, where the quality of the fish was unsurpassed in the eighteenth century. These were all built locally by boatbuilders upon the riverbank, two of which were at Coheston until relatively recently.

THE MUMBLES OYSTER SKIFFS

A healthy oyster trade grew up at Oystermouth, close to the Mumbles, in Swansea Bay. In 1580 oysters from here were shipped to Tenby, and later on Mumbles oysters were renowned as being the best in Britain. They were dredged between December and

February, when, up to about 1850, heavy, open boats were being used. These boats had a shallop rig similar to the Swansea Bay pilot boats – Swansea Bay being one area where the shallop or two-masted boat, developed by the addition of a headsail into a simple schooner, persisted longest in Britain. The pilot boats were about 30 feet long overall, three-quarter-decked and had a round bilge with strong runners to enable them to be beached. The oyster boats, it is assumed, were similar, albeit probably smaller in length. At about this time there were some thirty boats working the oyster grounds, each fisherman having his own 'perch' where he kept his landings just above the low-water mark, until such time as they were sold.

Around the middle of the century, there was an influx of dredgers from other parts of Britain, most notably the east coast. The local boats found that they could not compete with these bigger boats, and consequently a group of local fishermen travelled to Colchester and ordered a number of these smacks. The first boat to arrive was the *Seven Sisters*, and the type became known as the Mumbles dredge boats in contrast to the previous Swansea fishing skiffs, although reference has also been made to them as Oystermouth lug boats. By 1863 there were some ninety locally owned cutters, and 188 in 1871, the majority of these being built at Appledore but also in other parts of Devon and Cornwall, at a cost of about £300. None, it seems, were built locally, although many of the boats were fitted out, repaired and annually overhauled at the Mundick yard that existed between 1860 and 1892 near the beach. The fishery peaked in 1873 with the landing of 6,600,000 oysters, but thereafter declined through overfishing and pollution. As well as the local boats, there were still outsiders fishing – a dozen or so from Colchester in 1870 – and the fitting of steam capstans in the 1870s, further aggravating the decline. With no harbour facilities and a blossoming tourist trade, many boats gave up dredging yet continued to trawl out of the oyster season in places as far away as the Solway Firth, but by 1910 there were only fourteen oyster skiffs remaining. The *Emmeline*, 14SA, was the last boat to dredge, and she had been built in 1865 by William Paynter in St Ives. She was 40 feet overall, decked and had a low freeboard aft to facilitate the towing of the two dredges.

Further west, several similar oyster dredgers worked from Port Eynon, where particularly tasty oysters were said to be taken. In 1864 there were twenty-two boats based there – together with sixty-six at Mumbles and eight at Swansea. A small pier had been built at Port Eynon in the late seventeenth century, but the place generally suffered from a lack of proper facilities and the boats were terribly exposed to southern winds. Today, the remains of a salt works here is said to be a relic from the times that herring were landed on the beach. Other than that, no signs of any activity of a fishing nature are obvious today and the beach is more often than not a colour of windsurfers and the like.

Other oyster grounds fished by these boats were at Porthcawl, Tenby and off Stackpole. Yet from 188 skiffs working in 1871 to forty-seven seven years later, the decline of this peculiar oyster fishery was as sudden and final as any that occurred around our British coasts.

A Milford smack, M99, which, although its name is not known, was probably built on the East Coast of England. (*Tenby Museum & Art Gallery*)

POSTWAR DEVELOPMENTS

We have already seen how the arrival of the railways and the later advent of steam trawlers caused the demise of the herring fishing around the coast of Wales. White fish were landed in quantities, and oysters, cockles, mussels, crabs and lobsters continued to support several local boats in all the aforementioned creeks and harbours. But once the 127-ton, 98-foot steam trawler *Sybil*, from Lowestoft, had sailed into Milford Haven Docks as the first boat there in 1888, there was no hope remaining for the traditional fishing boats of Wales. This pattern has been constantly apparent in other parts of the country. Only Milford Haven, with sixty-six trawlers and 150 smacks in 1904 and 200 boats (mainly steam and motor boats) in 1924, along with Cardiff, Swansea and Holyhead, continued to have any reasonable fishery. But, as motor took over from steam, increasing numbers of ex-Scots boats, both fifies and zulus, found their way into the Welsh fleets. These were generally already fitted with engines, and survived up to the Second World War, and even beyond. However, the cruiser-sterned MFV became an increasingly common sight around these ports as the developments of the twentieth century took hold. Fishing became an intense industry, and Wales, from this time, supported only the smallest of fleets.

CHAPTER 10

The Somerset and North Devon Coasts

Sole, fresh 'ake and Clovelly herrings.
—The cry of the Bideford, Instow and Appledore fish sellers

Much in the same way as the Irish considered Dungarvan cod to be superior to Newfoundland cod, so the fishermen of Clovelly thought their cod tastier than the Canadian variety. But it wasn't their cod that gave them their identity, but their herring. In 1908, Charles Harper reported that the 'Clovelly fishermen are famed for their endurance and Clovelly herrings for their flavour' yet this was eighty years after the peak of a trade that was already well into decline.

Clovelly sits on the side of a hill leading down towards the sea, with steps leading down the last hundred feet or so of cliff giving access to the beach. It has been a centre of tourism for many years, day-trippers flocking to walk the angled streets, keen to reach this beach with its picturesque harbour – Pool Quay – that is always dotted with boats. It is hard, then, to imagine that this small shelter was at one time considered an important fishing station – in the time of Henry VIII it was rated seventh in importance in Devon, while in the nineteenth century it is mentioned as being one of England's chief fishing ports. Likewise, when one considers the herring industry, one does not immediately think of the Bristol Channel, yet huge shoals of herring were to be found here at one time.

The Welsh Port Books (see Chapter 9) tell us that herrings were sent by boat from both Bideford and Clovelly to Carmarthen in 1602. Undoubtedly there was a continuous trade between either side of the Bristol Channel, and cured herrings were just one such commodity.

In 1630, Thomas Westcote noted that herring – 'the king of fishes' – was being caught at Clovelly and Lynmouth, the latter being 'notable for the marvellous plenty of herrings there taken'. Not as much is known about the early Lynmouth fishery except that many of its smokehouses were washed into the sea during a storm and resulting flood in 1607.

Clovelly had its quay built originally in the fourteenth century, and it was then reconstructed by an eminent local lawyer, one George Cary, at a cost of £2,000 in the

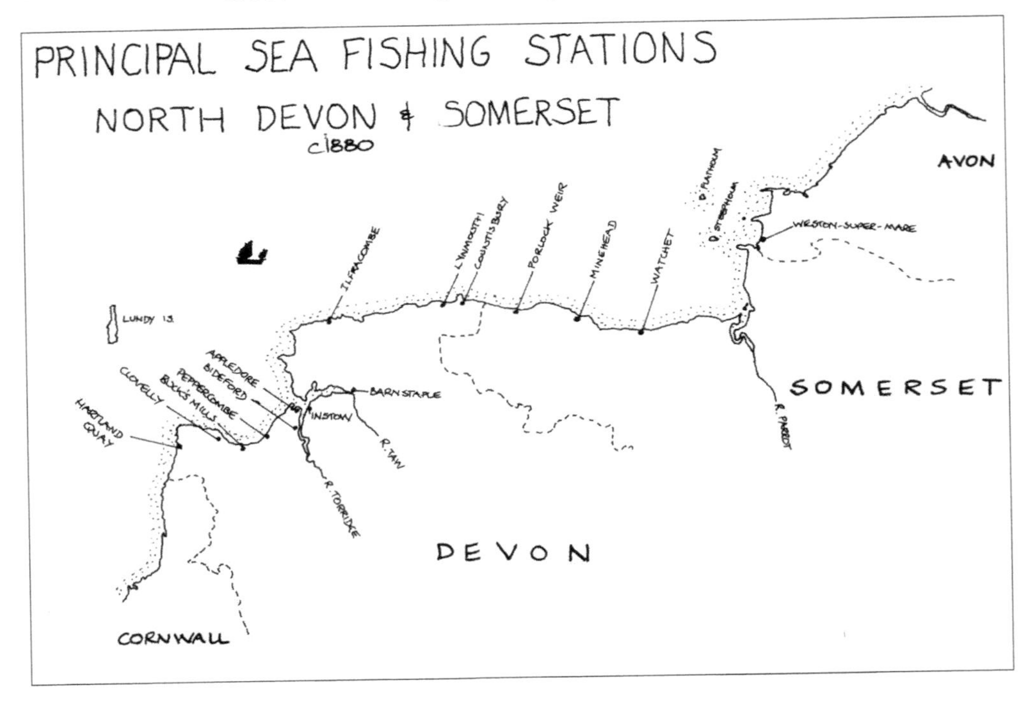

Above left: General view over Clovelly harbour in the 1920s. The smaller boats are the picarooners.

Above right: Landing a catch of herrings at Clovelly. Note that these boats have a Clovelly registration. This photograph dates from the early 1890s.

late sixteenth century. Although this was mainly to promote the import of coal and limestone (there was a limestone kiln there as at many spots all along the coast), the harbour was always full of fishing boats. In 1749 there were about 100 boats, and these appear to have been about 20 feet overall, of 5–6 tons, and had two small sails, probably a square main and sprit mizzen, and, as often as not, were propelled with oars. It seems that they were clinker boats, which suggests that some influence may have come from the Vikings who were continually raiding this coast in the ninth and tenth centuries. Indeed their shape at that time resembles many other craft of the west coast, whereas it has always been assumed that influence along this coast came from the south coast. It is probable that later developments came from there around the beginning of the nineteenth century.

From the parish records of the Reverend John Robbins, vicar of Clovelly in 1730–1777, we know that the herring fishing was intermittent: 'In this year 1740, God was pleased to send his blessing of a great Fishery among us after a failure of many years. This thro' his mercy continued in 1741. In this year 1742 the fish was small & poor & in less quantities. In this year 1743 but an indifferent fishing. In this year 1744 worse than in the preceding. In this year 1745 still worse. In the year 1746 much worse.'

We do know that the herring improved after this seemingly disastrous time. Around 1800 there were about sixty fishing boats in the harbour, and in 1839 the harbourmaster recorded '72 boats in all, 6 large-boats, 50 small boats and in addition, 16 boats that are employed in the taking of large fish', the latter being a reference to the smack-rigged trawlers, or 'long-boomers', as they became called, that had only just realised the possibility of trawling this area for skate, sole, plaice and, of course, cod. One such locally owned of these was *Teazer*, 218BD, which was 30 feet in length. Skiffs were similar boats that dredged for oysters, not unlike the Mumbles oyster skiffs from Oystermouth in Swansea Bay or indeed those working the oyster dredges off Porlock Weir.

The herring fishing lasted from September to December. Mackerel was caught in summer, and at one time pilchards were seined from this part of Devon. In the spring, the fishermen longlined for cod and hand-lined for congers, and the catching and taking of lobsters supplemented their fishing through the late spring and summer. And when the fishing was poor in summer, there were the day-trippers who would take a trip around the bay at 2*s* an hour in 1890, or maybe a trip up to Mouth Mill for 5*s* all in. During the winter the boats were laid up, and the men would disappear into the woods to find hazel and withies (willow) to make lobster pots. When the herring was at its peak, even the seamen would head home from wherever they were to spend time at the herring just as in Wales as a two-man boat could make as much as £100 by Christmas in the early 1880s with up to 10,000 fish being caught in one day.

This catch would have been auctioned on the beach originally for local consumption and carried up the village on sledges pulled by donkeys, but some of it did find its way to other parts of the county. In 1832 the *Exeter Gazette* reported that the first catches had arrived in Clovelly in September, and were immediately dispatched to Bideford by horse and cart where the fish were sold at 2*d* each. The next day the market at Exeter 'was well supplied with them at 10 for *6d*'. Obviously this was not a business to be in for profit. When the railway arrived at Bideford around 1850, the markets opened up, the fish travelled further afield and less was retained locally.

By the middle of the nineteenth century the type of boat being used was peculiar to Clovelly, perhaps because of the influences mentioned earlier. These Clovelly herring boats, as they became known, were between 19 and 24 feet generally, although one or two have been reported as being 30 feet. They were a development of the earlier small boats, yet were carvel-built as were most of the other boats in the West Country. They had deep keels aft, upright transoms and were very bluff in their appearance, and were recognised for their seaworthiness. Furthermore they were half-decked with wide waterways extending right to the transom and surrounding the comparatively small well for the fishermen to sit in. Unlike the smaller boats they were crewed by three men, and had two square-headed lugsails, a dipping lug as the main and a standing lug as the small mizzen. The tack of the main was secured to a hook on the end of a short iron bowsprit that extended a few inches forward of the stem. The top strakes aft were noted for having a pronounced tumblehome to meet the beautiful heart-shaped transom. All the boats were registered at Bideford, where there were seventy-six second-class boats in 1872, although some photos show boats with a Clovelly registration.

In 1880, a smaller type of boat was introduced into the Clovelly fleet. These two-man craft were mostly about 20 feet overall, yet were rigged in a similar manner. Being smaller and much lighter in construction, they were easily beached, and therefore were also capable of being launched at a moment's notice at all states of the tide, whereas the bigger, older boats could only leave harbour at or around high water. These smaller boats were consequently able not only to reach the herring shoals quicker, but to net the fish and return home to land the catch more quickly too, thereby often obtaining a higher price. They soon became known as the 'picarooners' from the Spanish for sea-robber or pirate, a name obviously applied to them by the owners of the larger herring boats which sometimes had not even left harbour by the time the small boats had returned back with a catch! The picarooner's rig was far handier than that of the bigger boats, and as they were also simple to maintain, it was not long before the bigger boats were sold and all the fishermen turning to the new type.

This coast was extremely exposed with little shelter, and several disasters occurred, especially before the picarooners when the big boats were unable to enter harbour if the tide was down. In 1821, sixty boats were caught out in a sudden gale, and thirty-one fishermen drowned, eleven of whom were from Clovelly, twenty-four boats were destroyed and another sixteen damaged. In 1838 another gale wrecked fourteen boats, taking with it twenty-one lives. In Buck's Mills eleven fishermen drowned during a gale on a summer's day.

Clovelly supported three bark houses for the treating of the fishing nets with preservative. Various boatbuilders worked around the beach area where they built the picarooners using local timber from the woods (just as they built the herring boats in earlier years). Peter Mills and Tom Waters both built at Clovelly, the latter moving to East Bideford around 1855 because of the railway there. Captain John Mills was Peter

The harbour at Clovelly *c.* 1890 with two large and a few smaller herring boats on the right. Note the lovely wine-glass transoms. (*Author*)

Clovelly harbour with various boats on the beach. Crazy Kate's Cottage is behind. Kate Lyall lived here in one of the oldest houses in the village, and each day she watched her husband fishing out in the bay, until one day he was drowned in front of her eyes. She became demented and remained so until her death in 1736, hence the name Crazy Kate.

son, a seaman who built several boats in the 1880s. Jim Whitefield was there at the end of the century. J. Hinks & Son and H. Ford both had yards in Appledore that survived well into the twentieth century. Much earlier, between 1811 and 1818, a boatbuilder called Morris is reported as working from Clovelly.

The picarooners survived through the first and second world wars, yet, as the herring declined, so did their numbers. The bigger boats disappeared altogether although the last, the *Pearl*, was measured up before finally rotting away in Fremington Pill. Because of the nature of the fishing, it was able to survive only by fishermen working singly. Clovelly Ledge boats were one-man boats, and were about 11 feet overall. These ledge boats got their name from a submarine 'ledge' that lay offshore, 2.5 miles further east at Buck's Mills or Bucksh. Here there is a small beach landing between rocks at the foot of the cliff. The beach is steeper than at Clovelly and more exposed, yet there is no harbour. A quay was built by Richard Cole, who owned large estates nearby and died in 1614, 'to shelter ships and boats' and for the import of limestone and coal. Today the limestone kiln is still there, yet the

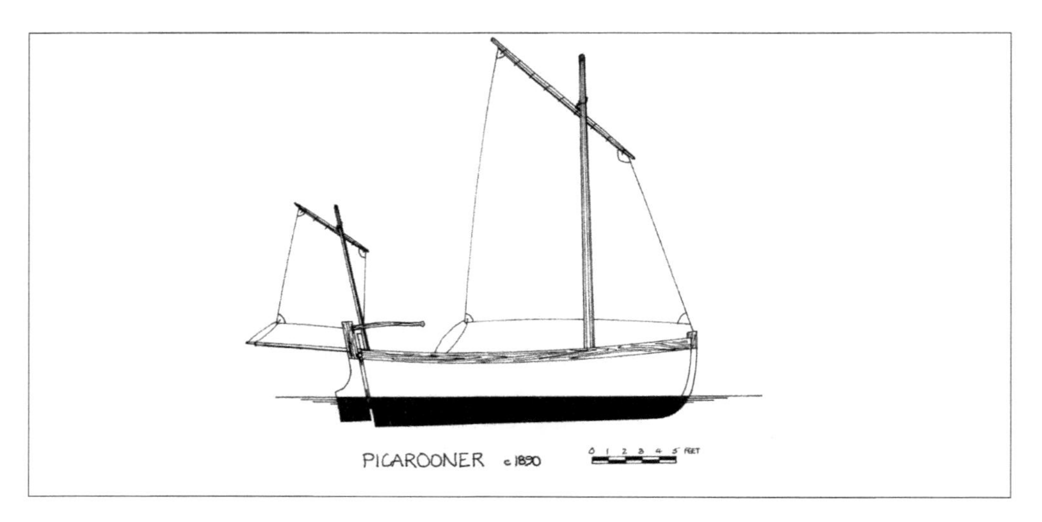

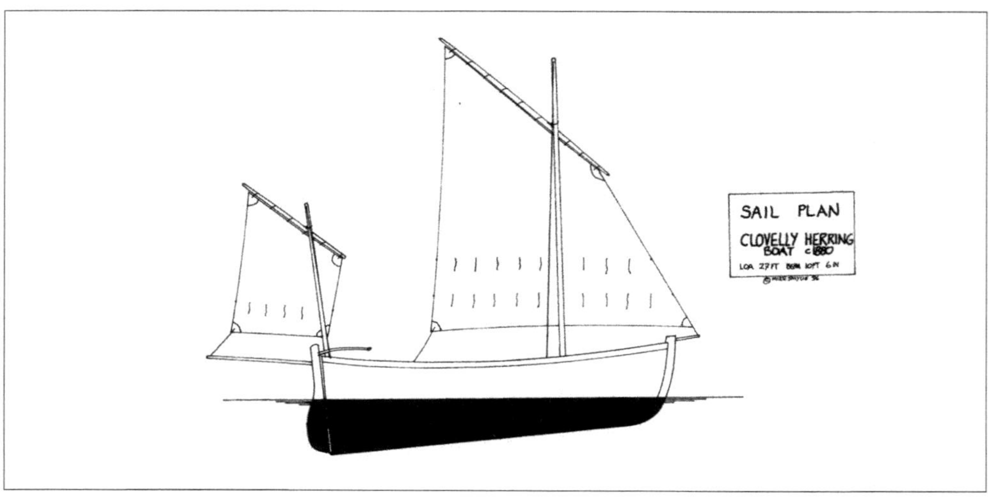

quay did not survive for long, although signs of it are still visible at low water. The fishing boats, had therefore, to be hauled right up the beach, and so were much smaller for this reason, although some carvel-built, two-man luggers of up to 17 feet overall were reported as working out from the beach around 1900. These Buck's Mills boats were similar to the Clovelly boats, in being open craft with two lugsails and heart-shaped transoms.

A good example of one of these Buck's Mills Ledge boats built in the 1880s survives in the ISCA collection at Eyemouth. The Clovelly Ledge boat *Wave*, built in 1888, is at North Devon Maritime Museum at Appledore. She has a purple line painted around her, which is the West Country 'mourning line' for her dead owner. Normally boats are taken out and filled with stones to sink them on the owner's death, but, luckily, *Wave* escaped this fate. Other examples have survived though are not currently in a seaworthy state. However, one replica picarooner, *Little Lily*, was built by students from the Falmouth Marine College and was brought to Clovelly where she is used to fish for herring by the village's harbourmaster and fisherman Stephen Perham.

Along with Clovelly and Buck's Mills, Peppercombe – 2 miles east of Buck's Mills – is the only other significant place on this stretch of coast where boats were based. The boats there were presumably the same as the ledge boats.

Hartland Quay, built around the same time as that at Buck's Mills by William Abbot, has been reported as not being 'unconcerned in that gainful fishing trade', although the harbour was primarily built to further 'quicken the pulse of seaborne trade'.

By 1900 there were only fifteen fishing boats in Clovelly, and today there are only two fishermen there catching herring, Stephen and his brother Tom. Catches can sometimes be good – eighteen mease being caught in the autumn of 1995 in one catch, this being

A Clovelly herring boat with two masts on the beach. (*National Maritime Museum*)

A small rowing boat in Clovelly in 1995, as is used nowadays to drift for herring a mile or so off the shore.

the largest single catch for nearly twenty years. (The mease is the local measure for herring and normally represents 600 fish, but in Clovelly it is five long hundreds and the tally, making 612 fish).

Today gone are all the herring boats and ledge boats, and most of the picarooners. *Little Lily* still fishes and occasionally so does one of the 10-foot wooden punts, both setting their 300-yard-long nets half a mile offshore and 2 miles in either direction along the shore. More common in the harbour are the motor boats that still take trippers out around the bay in the summer. Though Clovelly is a very different village than it was a century ago, there still remains a sense of awareness in some quarters of these traditions.

ALONG THE DEVON COAST

Appledore, Bideford and Barnstaple were bases to salmon fishing, and consequently were never centres of the herring fishery. The craft here were similar to the watermen's gigs that were kept busy attending shipping out of Appledore. They were rigged with two spritsails, although, after about 1900, the fishermen adopted the lugsail on the mainmast, retained the sprit mizzen and sometimes set a small foresail. These continued salmon fishing until the 1940s when the motor took over. Further north, Ilfracombe is the next harbour of any significance where herring were landed in quantity. The

A Buck's Mill ledge boat, very similar to the Clovelly boats, above the beach. (*Author*)

Reverend John Swete, in 1789, referred to the 'capture and seasoning of Herrings which at periodical times are here taken in vast quantities, and being cured are afterwards exported to the Continent'. The bounty payments of 1787 no doubt encouraged the herring fishery and the fish was either cured 'in a common pickle of salt' or 'by salting and smoking'. Combe smoked herring was deemed a delicacy in parts of Europe.

The boats working out of Ilfracombe were probably similar to those at Clovelly. Some were simply rowing boats, others having square sails, and possibly a sprit mizzen. The herring were caught close inshore, and, because of the tendency for fishing to complement an occupation at sea or in shipbuilding, the boats never did develop into anything particularly individual to the town. Indeed it has been said that one reason for the Bristol Channel boats never being much over 25 feet overall was because the Viking raids made people wary of venturing far out to sea in small boats. How this relates to the West Country fishermen who sailed out to the Newfoundland cod fishery, I do not know. My guess is that the reasons are more likely to be cost, the erratic nature of the herring

Fishermen with their typical boats, almost identical to the Clovelly boats, at Lynmouth in the 1890s. Lynmouth also had its own lifeboat, which was crewed largely by fishermen.

and the fact that, east of Ilfracombe, the Channel becomes quite narrow and shallow. Furthermore, the vast amount of trade between this part of the country and South Wales meant that work at sea was never in short supply aboard the smacks.

Lynmouth, as mentioned before, was the centre of a large herring fishery in the sixteenth century with vast amounts of herring being shipped to Bristol. It is said that men from all parts of the Bristol Channel were attracted to this fishery, which lasted from Michaelmas to Christmas. Many of those coming into the town rented cottages for the three-month period. The fishing flourished into the eighteenth century, and it was a good source of local employment. Like Ilfracombe, it was renowned for its red herring.

In 1787, Lynmouth Bay was 'one mass of herring'. So much so, in fact, that there was a glut. The surplus was used as fertiliser and superstitious fishermen were so shocked that they blamed the disappearance of the herring the next year on the 'insult offered to the fish by using them as manure'!

Early prints about this time show small two-masted vessels with a square main and tiny sprit mizzen. This seems to have been the general type right along the Devon coast. These early boats were no doubt clinker-built, and clinker-built boats were still in use in Lynmouth in 1890, although the last herring curing houses had been demolished in the 1870s. With the arrival of the railway in 1898, tourism quickly followed, and within years the fishing disappeared.

SOMERSET

The most westerly harbour in Somerset is at Porlock Weir, centre of a healthy oyster fishery until the nineteenth century when the entrance was improved to take cargo

1. A typical Scottish east coast salmon coble.

2. The fifie *Reaper*, owned by the Scottish Fisheries Museum at Anstruther, under full sail. This boat is the sole survivor of this size of vessel and is crewed by volunteers, who cope easily with its massive rig. (*Scottish Fisheries Museum*)

Left: 3. The Orkney yole *Gremista*, built by Len Lloyd in 1999, under full sail during the annual regatta at Stromness in 2003.

Below: 4. Tommy Isbister, boatbuilder of Shetland, rowing his Ness yoal in 2003.

Above: 5. Inside the Unst Boat Museum, where various traditional open Shetland boats are on display.

Below: 6. The Lochfyne skiff *Clan Gordon* sailing in 2002 on the Firth of Forth. At the time, she was the only one of these boats rigged traditionally.

Above: 7. The Lochfyne skiff *Perseverance*, which belonged to the author in the early 1990s, under sail on the west coast of Scotland after celebrating her eightieth birthday in Campbeltown in 1992.

Below: 8. The Grimsay lobster boat *Lily* on the beach there in 2001.

Right: 9. Cormac Levis's towelsail lobster boat *Saoirse Muireann* under sail during the Baltimore Regatta of 2006.

Below: 10. Achill yawls moored in 2006.

Above: 11. A River Boyne salmon pram in a derelict state in 2006.

Left: 12. The Morecambe Bay nobby *Empress*. She was built by Crossfield's of Arnside in 1928 and restored in the late 1980s and early 1990s.

13. The Ramsay boat *Master Frank*, restored by her current owner Joe Pennington.

14. A Dee salmon boat in the 1980s.

Above: 15. Simon Cooper and the author attempting to demonstrate the use of coracles at the International Festival of the Sea in 2005.

Left: 16. One of the fleet of Aberdaron lobster boats sailing *c.* 1995.

Above: 17. The author sailing the new Clovelly picarooner *Little Lily* without much wind in 2009. (*Ann Cooper*)

Below: 18. A River Severn stop-net boat at anchor off Gatcombe. (*The Environment Agency*)

Above: 19. Cornish luggers sailing in calm conditions at Looe in 2008.

Below: 20. The lugger *Guide Me* off Hout Bay, in the Cape, South Africa, in 1990. (*Judy Brickhill*)

Above: 21. Beer luggers racing during the regatta, *c.* 1999.

Right: 22. Another Beer lugger sailing well in light winds.

Above: 23. A typical transom-sterned Weymouth boat near the crane at the southern tip of Portland, where these boats were hoisted into the water.

Left: 24. Two Portsmouth-based seine-net boats boats.

Above: 25. Motorised Hastings boats in the 1990s, which display some of the same characteristics as their former ancestors. Note the elliptical and lute sterns.

Below: 26. Hauling in a Hastings boat in the time-honoured way. Hastings remains the only beach-based fleet in the UK, although it is under threat from increasing legislation, fish quotas and the effects of over-fishing.

EMMA

Left: 27. The Faversham bawley *Emma*, F22, recognisable from the wide transom-stern, sailing during the Swale Barge Match in 2011. She was built by Haywards of Southend in 1850, though has undergone restoration since.

Right: 28. The small Essex smack *William & Mary* at the Swale Barge Match in 2011.

Below: 29. The Lowestoft trawler *Excelsior*, LT472, leaving the port. Built in 1921 by John Chambers, she remains one of several sailing trawlers surviving. (*Excelsior Trust*)

Above: 30. The Yorkshire coble *Gratitude*, built by Hector Handyside in 1976. (*Dave Wharton*)

Left: 31. A stern view of the Bridlington coble *Three Brothers* in the harbour there in 2010.

Flatners at Weston-super-Mare – the Parrett flatner on the left contrasts with the Weston flatners taking trippers out.

vessels of up to 12 feet draught. Lock gates were added so that boats could load and unload at all states of the tide. Coal, limestone and cement were brought in, while agricultural produce, tan bark, charcoal and pit props were exported. Nearby Culbone (which has thirteenth-century origins) was populated by convicts, lepers and PoWs, who made a living cutting down oak woods to produce these goods as well as timber for shipbuilding.

Minehead and Watchet both had a herring fishery, and an early 1820 engraving shows trading vessels and a 20-foot open fishing vessel again with a square mainsail and mizzen. The masts appear to be the same height, suggesting a type of wherry rig that is found in parts of the Bristol Channel in vessels such as the Swansea Bay Pilot boats. It has been written that the fishermen from these parts spent more time smuggling than actually fishing, and that the Excise men were continually watching their comings and goings, often pouncing just at the wrong time. Herring continues to be landed at Watchet, but only in small amounts, and, like Clovelly, caught in small boats.

Somerset rivers were home to various flat-bottomed hard-chine flatners fishing, most notably the River Parrett. These sprit-rigged craft were suited to these conditions, yet appear somewhat medieval in appearance. To say that these vessels were primarily fishing boats would be incorrect. They were, as were many boats of a similar size and construction, the workhorse of riverbank dwellers and workers. Thus they carried withies and turf around the Somerset levels, reeds from where they had been cut, coal from South Wales, animals to market and trippers out to sea. Fishing, nonetheless,

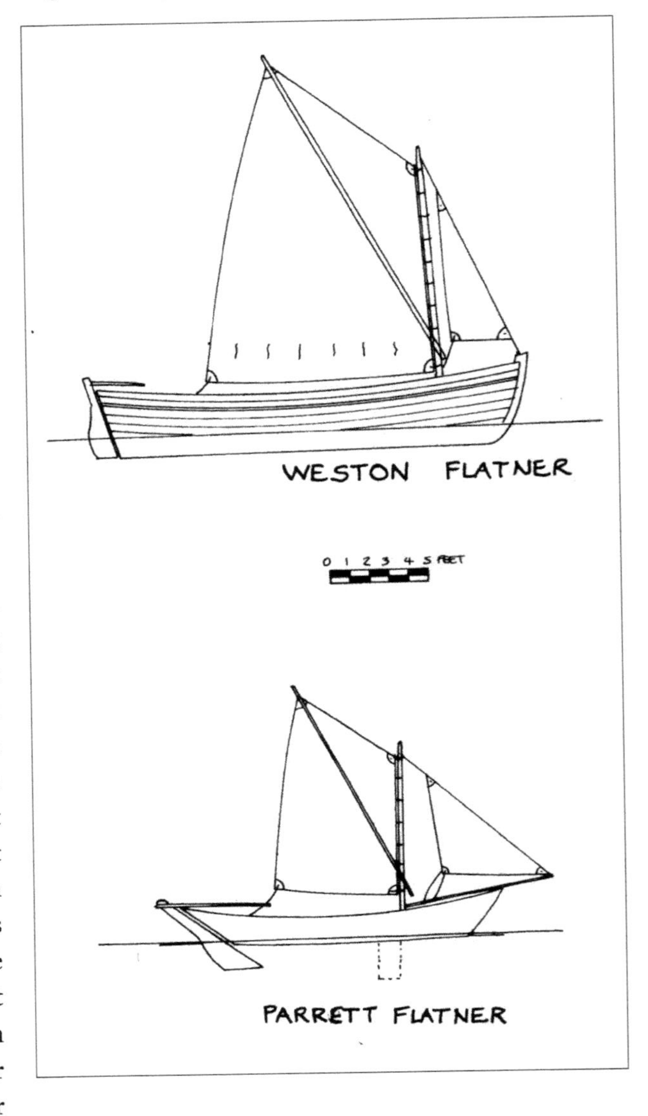

probably had the greatest influence upon them in their development. Whereas the 16-foot-long so-called withy boats and turf boats were small and crudely built with a single one-inch-thick oak bottom simply to work on the shallow waterways of the levels where they were pulled along, the fishing boats worked in more of a constraining environment. The River Parrett has been described as one of Britain's most dangerous rivers because of its high spring tidal range which brings its strong currents and the fact that it is immensely muddy with all the silt it carries down. In the nineteenth century over a hundred people drowned in it and many vessels were caught out and subsequently sunk in its grip. Thus the river boats were a bit longer than the upstream craft – always about 19 feet and 5 feet 6 inches in beam – and they fished under oars upon the river mainly for salmon, using a dip-net or to spear eels. These larger craft fished for sprats, salmon and shrimps, by sailing out into the bay. Some sailed as far away as the Welsh coast. One replica, *Yankee Jack*, was built and launched in 1997. In Watchet, where many such vessels worked from, they were usually referred to as 'flatties'.

Weston-super-Mare, too, was a fishing village until the Victorians transformed it into a flourishing seaside resort town. The boats used in the fishing were the shallow-bottomed boats, also called flatners, yet having a different appearance from the double-ended flatners, and these were suited to fishing around the bay. These craft were clinker-built, normally around 20 feet overall, although one or two up to 30 feet were built, and had a relatively deep transom and a dagger board. They were renowned as being good sailing and pulling vessels, and most had a single spritsail for sailing as far out as Flat Holm and Steep Holm, the latter being to where a group of Viking raiders were driven after an unsuccessful raid on Watchet in 914, and upon which, in 1776, a refuge hut was built for the safety and use of any fisherman. Later in the nineteenth century the bigger flatners had two masts, but these craft were generally used for taking the day-trippers out.

Fishing was mostly confined to the 20-foot boats, winter fishing being the most reliable for sprats and herring, while fishing for salmon and haddock complemented the tourist season. Some fishermen went shrimping between the end of the tourist season and the arrival of the herring in October. The boats were also suited to trammel-netting in shallow water, and some laid swing-nets at other times of the year. The flatners do bear a certain resemblance to the Devon boats, with local adaptations suiting local conditions. By 1900 there were some ten boats or so working out of Weston, but twenty years later these had halved in numbers. A few survived past the last war, albeit fitted with inboard engines, but, by the early 1970s, these were merely used for pleasure. The ubiquitous motor boat had arrived. *Flare*, although a motor flatner, today exists as a reminder of these vessels in the museum at Weston, while the 1936-built *Ann* is under restoration.

THE SALMON BOATS

Finally along this coast, although not actually within the boundaries of Somerset, mention must be made of the small boats used to catch salmon upon the upper reaches of the River Severn. We've already seen how in Wales the coracles and compass-net boats worked in the Welsh rivers. Here flat-bottomed craft with transom sterns are known as stop-net boats. These boats were moored across the stream in the river with their net lowered and thus were heavily built to cope with the stresses that this placed upon the vessels. Similar boats also worked on the River Wye.

Other craft used for the salmon are the flat-bottomed punts – or Severn scows – which are one of the most primitive craft working on British waters. Like the Scottish coble, they are square-sterned, yet the Severn punts have a square bow making them, in fact, rectangular, long and narrow. Their shallowness and shape allowed them to come very close to the riverbank while they were fishing. These punts were traditionally the work of punt builders who lined the river bank. However, only a couple of punts have survived.

On the rivers Taw and Torridge salmon boats were 18 feet in length, clinker- then later carvel-built, and had four or five thwarts for the four fishermen to row, the fifth being used to hold the net at the stern. Each builder built his boat slightly differently to the other as there were no drawings. These 'North Devon salmon boats', as they are known, were mostly owned by Appledore fishermen and used to fish the salmon drafts on both the Taw and the Torridge rivers. It's close to Appledore that both rivers meet before flowing out into Bideford Bay, and Appledore had been a port of substance since the Middle Ages.

CHAPTER 11

The Cornish Coast: Bude to the River Tamar

Heeva!

—The cry of the Cornish pilchard fishers

Cornwall is probably better known these days for its smuggling past rather than its fishing heritage. Yet it was, in fact, the fishing industry that created the circumstances that allowed smuggling to develop.

Cornwall was home to the second biggest fleet in the British Isles at one time. The coastline stretches for over 300 miles from its border with Devon near Bude on the north coast, right around the rocky extremes of Land's End and eastwards to the River Tamar. Nearly every cove, beach or inlet has, in the past, had some form of association with the fishing trade – whether it is from the shore or afloat.

The smuggling lore has, it seems, come about through the many exaggerations of a healthy tourist trade of latter times, and helped by the writings of various well-known novelists. The fact is that smuggling began in the seventeenth century to counter the imposition of the salt duties on imported salt. Continental salt was used extensively in the pilchard industry, as it was in the herring industry all over the country. The fact that it was easier to obtain it from over the Channel rather than across a country with a poor communication network probably ensured this thriving business. After the salt duties were imposed the fishermen continued to ship it, but only into quiet coves under the cover of darkness. It was not long before the boat owners saw the advantage of this type of importation for other goods, especially those that were heavily taxed. So they started to bring in wine, spirits, tobacco and even tea, and a flourishing 'free-trade' began that lasted well over two hundred years.

The sea had long been used for the transport of goods and people around the coast, even before smuggling. Roads hardly existed, and Cornwall was cut off from the rest of the country, as were many other coastal areas in Scotland, Wales and the more remote parts of England. Thus harbours were built for these boats. Copper and tin were exported in vast quantities, and stone was quarried. Harbours were built exclusively for fishing fleets, and consequently the fishing industry grew and prospered. Among this industry evolved boats to suit the local requirements.

MAIN FISHING HARBOURS OF CORNWALL
PRINCIPAL PORTS UNDERLINED - OTHERS ONLY HARBOURS FOR PILCHARDS.
CORNWALL
PORT GAVERNE
PORT ISAAC
PORT QUIN
PADSTOW
NEWQUAY
POLKERRIS
FOWEY
LOOE
PORTWRINKLE
POLPERRO
MEVAGISSEY
PORTREATH
ST. IVES
PORTLOE
PORTSCATHO
ST. MAWES
FALMOUTH
PENZANCE
NEWLYN
MOUSEHOLE
LAMORNA COVE
SENNEN COVE
PENBERTH
MOUNT'S BAY
PORTHLEVEN
MULLION COVE
PORT HOUSTOCK
COVERACK
CADGWITH
CHURCH COVE

1620
1790
1814
1870
THE DEVELOPEMENT OF THE ST. IVES LUGGER.

EARLY BOATS

Not much is known about the early boats, only that they were open boats with single lugsails. The first mention of a type was in 1620 when the 'kok' – a pilchard boat with two masts, the foremast being placed well forward – was depicted in drawings that still exist. Drawings dated 1694 and 1790 also still exist of other similar types of boats, again with two masts. Three-masted vessels appear to have been adopted prior to 1790, as was common among other fleets along the south and east coast of Britain.

The lug rig appears to have been introduced by the French. There is no doubt that the French boats sailed regularly across the Channel to Cornwall to trade legally and illegally. Some of these boats were the large *Chasse-Marée*, powerful three-masted lugsail vessels of great speed often crewed by 'pirates and vagabonds', that sometimes attacked the local fleets. The Cornishmen, being so impressed by these boats, soon copied the rig, and the lugsail quickly grew in popularity.

The lug rig is generally regarded as an intermediate stage in the development of the fore and aft rig from the square sail. It appears to have originated as far back as the days of the Phoenicians. Some have suggested that the influence came directly to the Cornish when the Phoenicians sailed here in search of tin. This could explain the lack of clinker boats built in Cornwall. Nearly all boats were carvel-built like Mediterranean craft, and not clinker like the Viking boats. On the other hand this could just be an influence transferred across France to this country as the Cornish boats also drew more water aft than forward – a Breton influence. What is sure is that the Cornish boats developed on similar lines to many Scottish boats, although through different influences, and this, perhaps, is because all were evolved to suit local conditions that were in fact quite alike. And, like the Scottish boats, the Cornish luggers were developed for drift-netting.

ST. IVES SEINE BOATS c1900

HEEVA! – THE CRY OF THE PILCHARD FISHERMEN

The pilchard industry has been a major part of the economy of Cornwall since the sixteenth century, and probably well before that. St Ives, always the centre of it, had its first pier built in the fifteenth century to protect its fishing vessels.

The fishing began each year around late July, and lasted four months or so, and generally ran in tandem with the harvest, hence the Cornish saying:

Corn up in shock,
fish in rock.

The seine-net was considered the only way to catch pilchards, so that laws were passed to prevent boats breaking up the shoals until they reached the shallower waters off the shore. These shoals were massive, appearing as a stain of red, purple and silver in the sea with excited flocks of seagulls and gannets overhead.

The operation was undertaken by companies of seiners who were licensed to work a particular area. The seine, as it was called, consisted of three boats, two nets and a cellar ashore to store the fish. The main boat was called the stop-seiner and would be an open, low boat of about 35–40 feet in overall length. Double-ended carvel boats were regarded as best, for they were fastest and could carry the heavy pilchard net. They were recognisable for their maximum beam being well abaft of amidships. They were without sail, being propelled solely by six oars.

Aboard this stop-seiner was the main net, or stop-seine. The second net, the tuck-seine, was carried aboard the second boat – known as the folyer (literally 'follower') or tuck boat – which was 2–3 feet shorter but of a similar type. The third boat was called the 'lurker' – much smaller at about 16–18 feet long – and was used to direct operations. Often the lurker was dispensed with as operations were directed by a series of signals by the master-seiner from the cliff top. Stationed here in his hut, too, was the 'huer', whose job it was to spot the shoals (sometimes the master and huer was one and the same person). When the shoal was spotted, the huer traditionally cried 'heeva, heeva, heeva' through a trumpet-like speaking horn. The two main boats then sped out, watching the master for directions. The net was shot by the stop-seiner, and swept around the shoal to encircle it and enclose as many fish as possible. The folyer would help if necessary, using the second net if the catch was huge. The whole lot would then be towed into the shelter of a bay and anchored in very shallow water. Thus the fish could be taken ashore for curing, or balking as it was known. If the capacity for balking was too great, then the pilchards would be left in the net in the sea for up to a few days, and taken ashore as needs demanded. Massive amounts of fish were caught in a single haul – easily up to half a million. The largest single seine was said to have been in the autumn of 1851, when it was estimated that 17,908,800 pilchards were caught. The catch took a week to land and the overall profit was said to have been more than £7,500, probably the equivalent of £500,000 in today's terms.

Meanwhile, the larger drift-net boats would catch the shoals offshore, much to the annoyance of the seiners. Arguments between the two parties raged for years, before the drift-nets finally gained precedence, and seining gradually lost favour from the 1850s onwards, so that by the end of the century it had virtually ceased to exist. Figures for 1870 show this: in that year there were 379 seines, 280 of which were in St Ives alone, and at the same time there were 635 pilchard 'drivers' as the drift-net boats were termed.

All around Cornwall each harbour had its fishing 'palace' or fish cellar – a large underground courtyard for salting, packing and pressing the catch. Communities grew up on the pilchard fishing, a good example being the three ports of Portquin, Port Isaac and Port Gaverne on the extremely exposed north coast. Port Isaac had its own pilchard fishery before the sixteenth century, and there were forty-nine fishing boats registered there in 1850 as well as four palaces. Other than seining, crabbing was a seasonal alternative. Port Gaverne had its pilchard fishery in Tudor times. All the Cornish ports were hives of activity in the nineteenth century: the exporting of the ores from the mines, importing of coal and limestone and other goods, and boatbuilding on many a beach all occurring alongside the fishery. The ancient fishing port of Portreath, on the north coast, was built in the seventeenth century to ship the copper ore from the nearby copper mines to the huge smelter at Swansea. Fourteen vessels – from a 17-ton lugger to a 244-ton barquentine – were built between 1822 and 1884 and when the copper industry slumped in the 1890s, coal and timber continued to be brought in. Alongside this the fishing flourished up to the end of the century. Today the quays lie quiet, with only an occasional fishing boat working from there. Newquay, Perranporth and St Agnes all had their pilchard fisheries, while all along the south coast there were seines operating from any number of coves. Today all that remains is the sole pilchard curing factory in Newlyn where up to 100 tonnes of pilchards are still salted, pressed and packed in wooden barrels and boxes each year for export to the same Italian family that started buying from Cornwall in 1905.

A view of a large seine of pilchards at Porthgwarra, near St Levan. (*Bryan Roberts*)

A pilchard haul aboard a seine boat at Sennen Cove, *c.* 1890. (*Bryan Roberts*)

:ENTURY BOATS – THE LUGGERS

the nineteenth century, operational influences created two quite ornish lugger. In the eastern ports of Looe, Polperro, Fowey and Cornish luggers were to be found; while to the west of Cornwall, erally referred to as 'drivers'. This was because it was considered that, when boats were drifting with the tide with the main mast lowered, the nets were being 'driven' along by this tide.

Whereas the east boats were transom-sterned, the west boats were double-ended, showing some possible Viking traits. Among these western craft, there were again two distinct types. Although double-ended to fit snugly in the crowded harbours, the boats of Mount's Bay were finer than the boats of St Ives on the north coast. At St Ives, where the harbour protection was poor, full-bodied boats were built to sit upright on the beach at low tide. These were strongly built with huge stems and sternposts so that they could continually stand up to the pounding as the tide went in and out, without the need for legs – as these would have been smashed in the surging tide around the harbour, especially in harsh weather. The harbours at Penzance and Newlyn were more sheltered, so the boats could be safely beached with legs. These finer boats were, of course, faster, but the St Ives boats lasted longer because of their stronger build.

The pilchard drivers were the smallest at around 30 feet overall, and were half-deckers. In 1870 these cost about £120 to build. They were developed purely for the pilchard fishing and were rigged with two lugsails only.

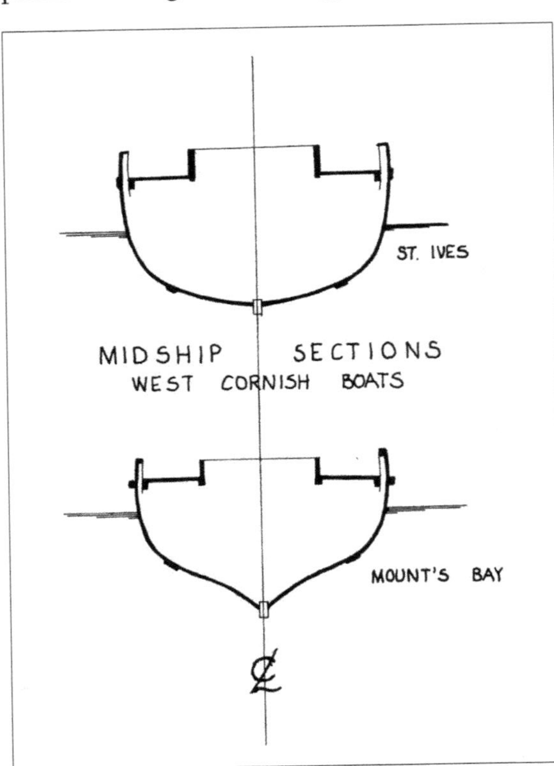

The larger mackerel drivers were fully decked, and much stronger. Lengths varied according to needs, but an average boat was, say, 40 feet overall. Keel length, then, would be 37 feet, beam 12 feet, and depth 6 feet. These boats would cost approximately £200. Some boats towards the end of the century were over 50 feet overall, and would have cost £600 with all the gear, nets and sails. Mount's Bay boats tended always to be bigger than St Ives boats.

Both of these types, the pilchard and mackerel boats, developed from the earlier open boats that had small forecastles. Washington's report refers both to St Ives and Penzance boats, and he produced examples of both.

Deck view of a St Ives pilchard boat in 1906. (*National Maritime Museum*)

One of the most renowned of Cornish boatbuilders was William Paynter of St Ives, whose yard was at the top of the beach (where Woolworths was until recently). Between 1840 and 1880 he built many of these fine luggers, mainly of the fuller St Ives type. Many of these boats found their way to other west coast ports as we've already seen – Campbeltown, Isle of Man and Ireland – and Paynter himself moved to Kilkeel to build from there for a short time (see Chapter 6). His apprentice, John MacKintosh, carried on building there after he returned to Cornwall.

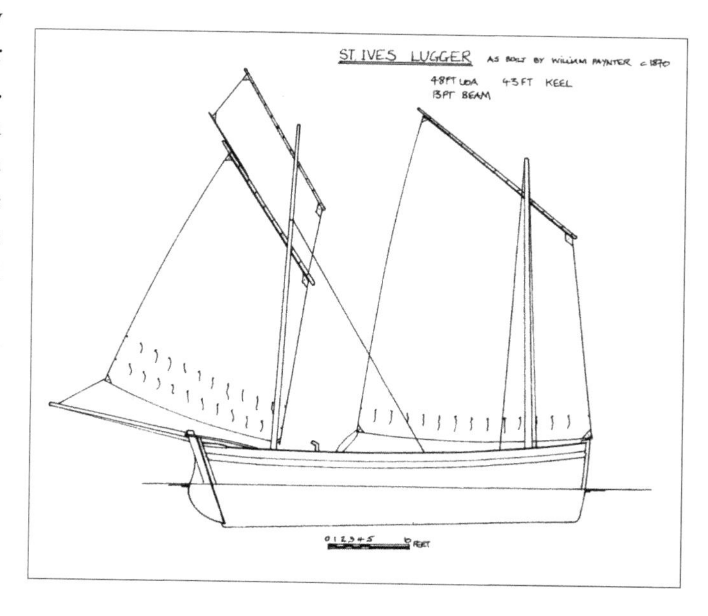

One fine lugger built by another Cornish builder was the *Willing Lass*, SS635. She was 52 feet overall on a 47-foot keel, and had a 14-foot 6-inch beam. She was sold to Kilkeel before the First World War and registered as N385.

Other builders of these luggers were Richard Warren and James Wills, both of Newlyn, and William William and his brother at Mousehole, to mention but a few.

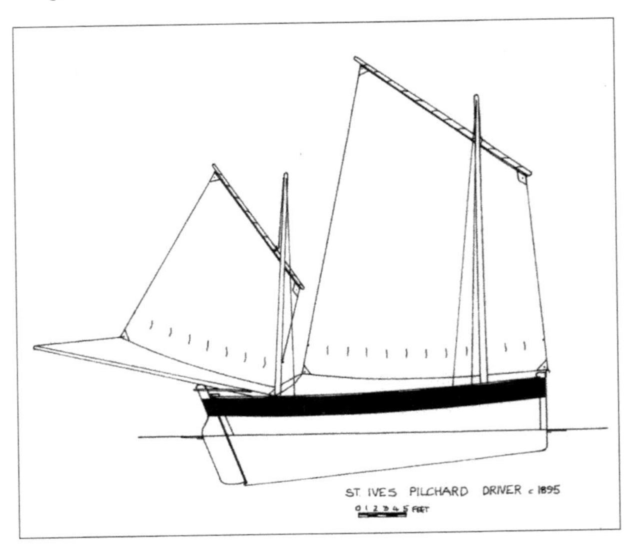

Left: The only surviving St Ives lugger, *Barnabus*, under sail in the early 1990s.

Below: St Ives harbour *c.* 1895, with a misture of West and East Cornish luggers. (*National Maritime Museum*)

Pilchard boats at St Ives *c.* 1908 – note the way they remain upright when dried out. (*National Maritime Museum*)

Henry Trevorrow also built the *Barnabas*, SS634, in 1881, and today she is the only one of these boats left. Owned now by the Maritime Trust, she has been lovingly restored to her former self, and is a common sight among the other boats in and around Falmouth.

Dolly Pentreath, SS679, was built by Norman Laity as a replica of the St Ives mackerel driver *Godrivey* that sailed to Australia, while ocean sailor Pete Goss has recently sailed aboard his lugger *Spirit of Mystery* to Australia. The original *Mystery* was sailed to the Continent via the Southern Ocean in the late nineteenth century by a group of Newlyn fishermen eager to join the gold rush. Goss's venture was to mirror this journey.

CONSTRUCTION DETAILS

No boat was ever built the same as another one. As was the custom in most parts of the country, all the Cornish boats were built by the rule of thumb. No plans were ever drawn – only a few half-models, some of which still exist were used. Boat improvement was a matter of making small changes in the light of new experience.

Around 1860, however, the design improved immensely. Floors rose more sharply, especially in the Mount's Bay boats, and the entry and exit into and from the water became finer. This was to gain speed. Because of the opening of the railway into Cornwall at that time, the markets of the expanding industrial towns created an instant boom in the fishing, and it became a chase to get the fish to the railhead, instead of, as previously, to the local market. Faster and bigger boats were the order of the day.

Materials were fairly similar, and because of the already established coastal trade, there was no difficulty in getting decent timber to build with, unlike in the remote parts of Scotland. Keels were set up in American elm, stem and stern posts were oak, although elm was used occasionally. Frames were oak, home grown to shape normally, although sawing was necessary at times. They were always fixed perpendicular to the keel. Some boats had double frames, while others had scarphed frames. To form the sharp entry into the water, the frames were bolted directly onto the deadwood, and aft they were fixed, again directly onto the deadwood, using chocks inside them. Midship frames were in five sections – the floor and two futtocks or 'footicks' as they were locally called each side. Knees were oak. The hull planking was nominally 2-inch-thick white, yellow or pitch pine. Yellow pine was favoured for the deck because of its impermeability and hard-wearing quality. Spars were spruce, fir or yellow pine.

THE RIG

The West Cornish drivers only had two lugsails and no foresails. The smaller pilchard boats had two dipping lugsails, with the mizzen higher-peaked than the main and rarely used. In fact all of the Mount's Bay boats were recognisable because of their high-peaked sails, a French influence said to have been first adopted by the St Ives fishermen and then taken to Mount's Bay during a regatta.

Meanwhile, the larger mackerel drivers had much larger rigs by comparison. On a typical 50-foot driver, two large masts, the mizzen being the bigger, reached up to more than 50 feet above the deck. The St Ives boats stepped the foremast much further forward than the Mount's Bay boats, to give the foresail a more upright luff. The mainsail hung on a 25-foot foreyard and had an area of 700 square feet. The mizzen supported a 500-square-foot mizzen sail, hanging on a 30-foot yard and was sheeted to a 30-foot bumpkin, 20 feet of which was outboard. Above this was a jackyard mizzen topsail of 120 square feet on its own 21-foot topsail yard. This huge rig allowed the boats to achieve incredible speeds of over 10 knots on their journeys around the coasts. And travel they did.

The mackerel season started in January and lasted until June. Many boats would sail one hundred or more miles due west out into the Channel approaches to catch the early shoals, following them up-Channel as the season progressed. After the mackerel season was over, they would sail to catch the Kinsale and, later, Howth and Isle of Man herring up to August, whence they would sail north, through the Forth and Clyde Canal for the North Sea late summer herring, and down to the autumn herring off East Anglia. Some boats would even go further north, perhaps through the Caledonian Canal, to catch the Shetland fishing in early summer before returning south in the autumn. Often the boats would not return home until November, the

A pilchard driver being rowed out as the sails remain limp in the calm air. (*National Maritime Museum*)

journey home being a race for the fastest passage. The *Lloyd*, SS5, later sold to Ireland, sailed from Scarborough to St Ives in 1902 in fifty hours, a speed that would be difficult to beat today in similar boats when one considers that the distance is over 600 miles. The distance from Peterhead to St Ives – approximately 850 miles – was covered in 100 hours.

After returning to Cornwall, the fishermen would have one month or so to refit their vessels before returning to the mackerel. Boats would be brought as high up the beach as possible, and when all the necessary repair work was done, the scrubbing, scraping and tarring of the hull was tackled, finishing off with the paining above the waterline. Some of the Cornish boats were the best kept of the entire fishing fleet of the British Isles, although I know of no part of the country where fishermen were not noted for the upkeep of their vessels.

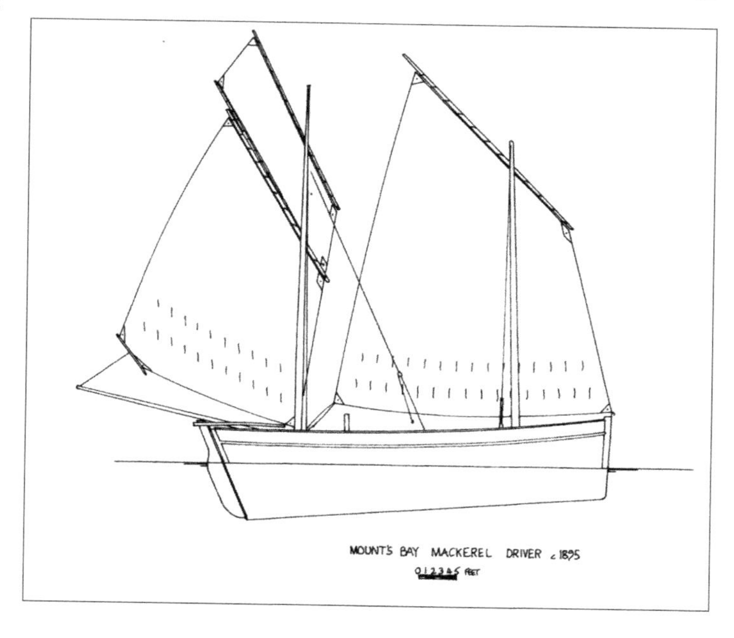

The colouring of the vessels was thus: as already said, the hulls

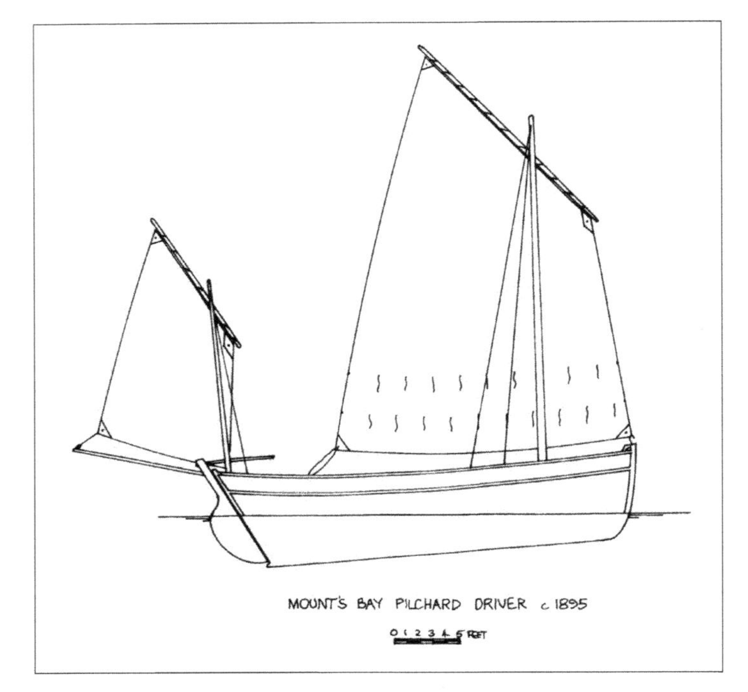

Below: Another view of luggers at Newlyn about 1880. Note one is registered at Plymouth.

The beach at Newlyn about the turn of the last century. This spot is roughly where many decommissioned fishing boats have been smashed up by a JCB, as part of Europe's plan to reduce Britain's fishing fleet and hand over her fishing grounds to other countries within the European Union.

were tarred, usually up to the bulwarks. Above here, the bulwark rail was white. Inboard the bulwarks, the stem head, sternpost head, masthead and mizzen band and bumpkin end were also white. Hatch coamings were either blue or white, and all the spars varnished, or dressed with pilchard oil. The white parts were painted black when the owner died as a sign of mourning, and would be left so for twelve months.

ACCOMMODATION

In the mackerel drivers, where they were fully decked, the accommodation was aft. Here were the bunks, lockers, cooking stove and all provisions, as well as some gear. Headroom would have been about 5 feet between beams normally. Just forward was the net room, then the fish hold and line room over the ballast, and up at the forward end, ahead of the mast, was the forepeak used for sail stowage and general gear. Compared to many fishing boats of the end of the last century, the accommodation was fairly good, but it must be remembered that the crew of six or seven would spend long periods, often months, aboard.

THE EAST COAST

As we've seen, all the eastern Cornish boats were transom-sterned, as were many craft of the south coast. This is said to have originated from French influence. These luggers were some 40 feet overall, later increasing in size to around 50 feet. Their appearance did not alter substantially from port to port, the only difference being the rake of the transom. The Mevagissey luggers were said to have the most prettily raked stern, it being less upright than the Looe boats. Unlike the West luggers, these boats set a foresail on a bowsprit. The larger of the boats were fully decked, and had full accommodation

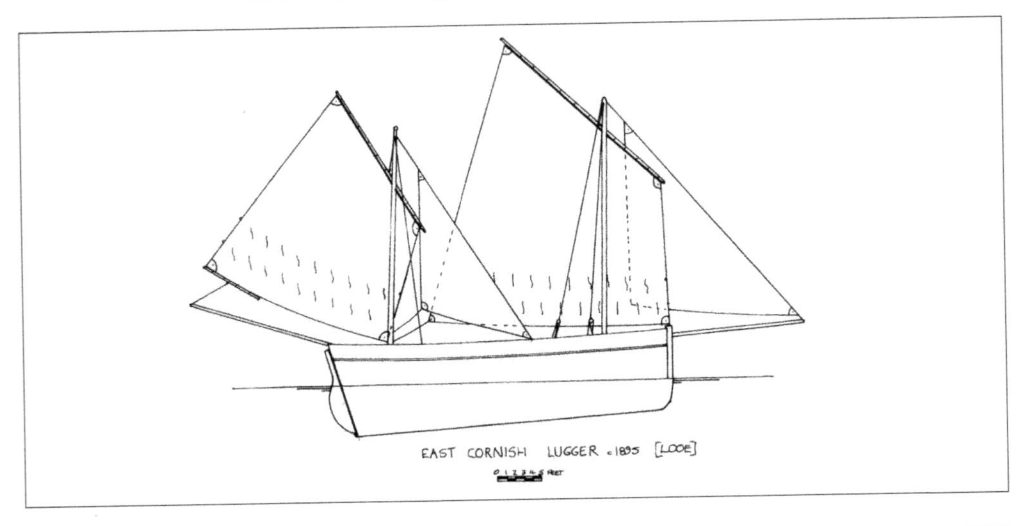

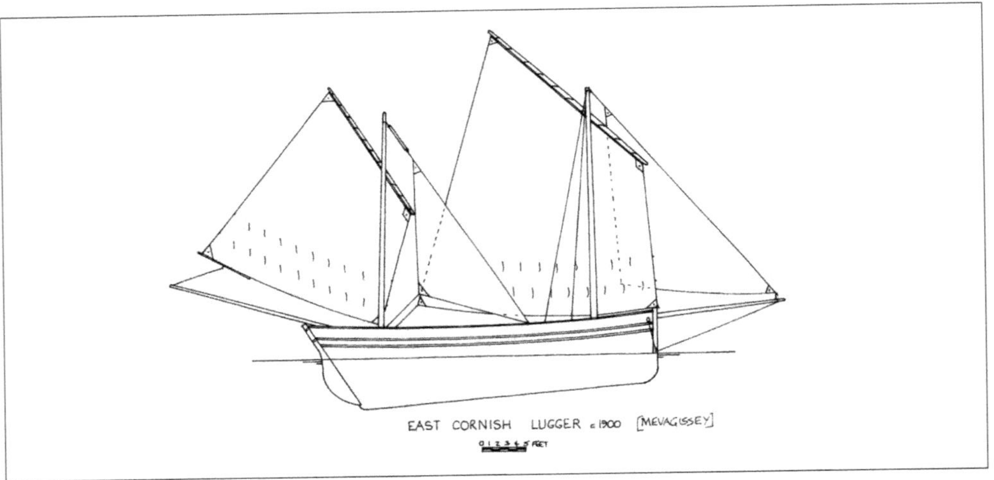

for up to six crew, sail rooms, net room and fish hold. Others had large hatches over the fish hold and net room. All these boats were used for mackerel and herring fishing, the pilchard fishing being more concentrated on the northern coast.

Many of these luggers were built by Oliver of Porthleven, while others were built at Looe and Mevagissey. Generally they worked out of Looe, Polperro and Mevagissey, although they sailed further afield. By the mid-eighteenth century the East Cornish luggers could be seen at the great autumn herring fishery in the North Sea, and afterwards they would fish for herring and hake all winter. In 1889, there were 218 boats employing 765 men and fifty-eight boys in the East Cornish fleet. By the turn of the century these numbers had been severely depleted, and within another decade the steam trawlers and motor boats had virtually seen off the last of the large sailing luggers. Smaller luggers that survived longer were the 'toshers' of Mevagissey. At 19 feet 11 inches, they were developed to avoid paying harbour dues payable for all vessels over 20 feet. Some of these one-man boats were to be seen with Kelvin engines up to the Second World War.

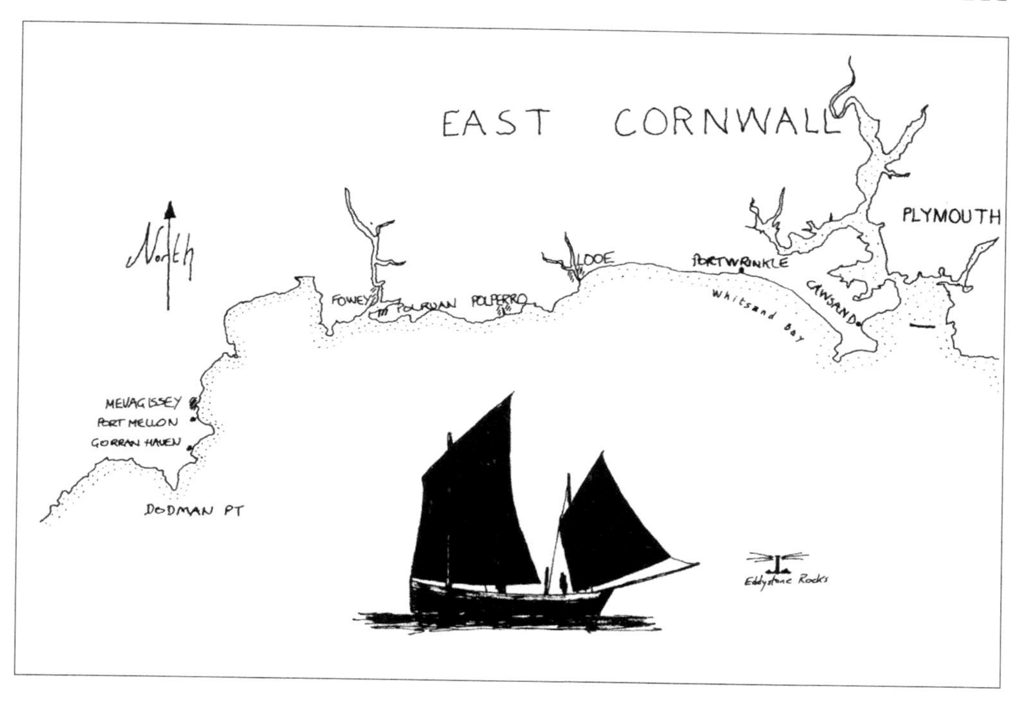

East Cornish luggers in Porthleven harbour, *c.* 1880. (*Author*)

Above: Polperro with the gaffer *Lady Beatrice* leaving the harbour, *c.* 1880.

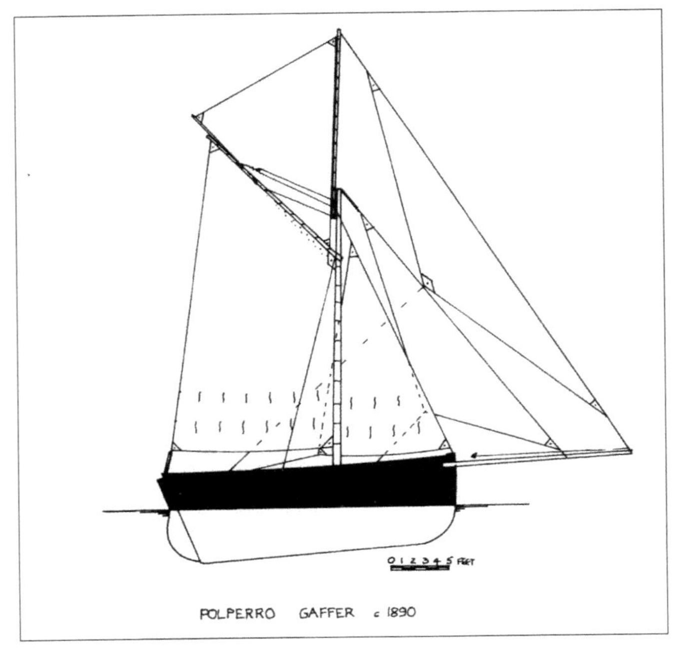

The 'Polperro Gaffers' fished out of the harbour there in large numbers. Up to the end of the eighteenth century, these boats were sprit-rigged and clinker-built. Along with the spritsail, they set a foresail on a short bowsprit, and some set a yard topsail. The larger boats also set a mizzen standing lugsail. In 1891, many of these boats were lost in a gale, and were replaced with carvel-built boats. Developed to suit local requirements, they were 25 feet overall, 20 feet in keel length and 9 feet in the beam. They set a cutter rig with a boomless mainsail, and only had a cuddy in the forepeak under the foredeck with bunks for basic accommodation. As we shall see, the same influences affected the boats from Plymouth, where some of the hookers were indistinguishable from the Polperro and Mevagissey craft. Several still sail including *Vilona May*, FY20, built in Looe in 1896. She is rumoured to have sailed to Australia in the 1950s though there seems to be no evidence to support the theory. However, she has sailed across the Atlantic, up to Boston and as far north as Greenland under her present ownership, before returning home to Cornwall.

THE NORTH COAST

Although there were several harbours along the exposed northern coast, fishing wasn't as important an occupation as was mining. Nevertheless, as mentioned before, boats did work out of ports such as Newquay, Padstow, and the three villages of Port Isaac, Port Gaverne and Portquin. Over time they have been home to small fleets of, mostly, small crabbing and inshore boats. The boats tended to reflect those from the south coast though a few larger luggers did work from here. Portquin had its own pilchard cellars and fleet of small luggers. The whole fleet was, it is said, lost in a storm overnight at some time towards the end of the nineteenth century (much earlier dates have also been suggested) and fishing immediately died although no evidence of this fact survives. Today it is a deserted village owned by the National Trust. Port Isaac, on the other hand, thrives from tourism but it had a pilchard fishery before the sixteenth century and a pier was built sometime around then. In 1850, when forty-nine fishing boats were registered, there were also two boatbuilding yards, the last of which, run by Harry Hills, closed in the 1960s.

THE NEW AGE

Just as has been spelt out all around the country, the coming of motor power forced the demise of the sailing luggers. Some of the Cornish luggers were fitted with engines as the straight sternpost made simple the provision of an aperture for the shaft. Others were rapidly sold off as fishing declined. By the early 1930s, those that had survived were themselves laid up to rot as the new era dawned.

Many east luggers remain sailing today – *Guide Me*, FY233, built in 1911, *Our Boys*, FY221, built by Dick Pearce of Looe – a prolific builder – in 1904, *Britannia*,

The *Boy Willie*, PZ602, at Mousehole. (*National Maritime Museum*)

PZ8, built in Mevagissey 1904, *Guiding Star*, FY363, built by Angear of Looe in 1907, *Girl Sybil*, PZ595, a pilchard driver built in Porthleven in 1917, *Lois*, PZ626, built by Oliver of Porthleven in 1913, *Iris*, *Eileen* and *White Heather* – and there are many more too numerous to mention that are still sailing today or under refit. Yet to the west, as mentioned before, only *Barnabas*, SS634, is a survivor of St Ives. At least two of these luggers, *Guide Me* and *Guiding Star*, and probably more, have crossed the Atlantic within the last few years, which can only be interpreted as an honour in memory of those that built these fine craft.

One of the grandest of all the luggers – the *Ebenezer*, built by William Paynter in 1867 – was one of the last to fall apart on the beach at Lelant, not far from where she was built at St Ives. Today bits of her still remain in the sand, but that is not much of an epitaph for such a fine boat. But the story is the same everywhere. Mankind, through his cleverness and experience, built these mighty craft, yet the same creature, through his greed and stupidity, allowed them to rot and fall apart into the ground from whence they came. But perhaps, after all, that's the best way that they should end their days.

THE FALMOUTH OYSTER DREDGERS

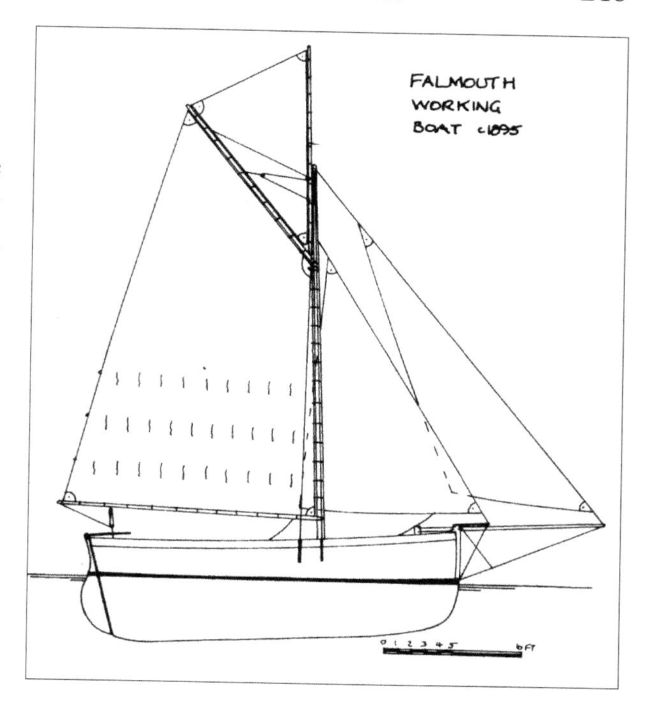

The oyster dredgers of the River Fal are more commonly referred to as the Falmouth Working Boats or, less often, the Truro River Oyster Boats or the Restronguet Creek Oyster Boats. Although these boats are not in fact an actual category of boats, being more a collection of craft from perhaps all around Britain that have worked the Falmouth oyster fishery, they do all fit into the same grouping. They are typically 22–30 feet long, gaff rigged, have a long bowsprit, are carvel-built, transom-sterned half-deckers, and have long, straight keels with a draught of not more than 5 feet. They are indeed very similar in shape to the Falmouth Quay Punts, the watermen's craft of Falmouth, which often sailed out to Wolf Rock in search of an incoming ship. Originally, the oyster fishery was fished by the drivers out of season of the pelagic shoals. These luggers were gradually adapted over the years to suit the oyster fishing, and so developed into the dredger type. Although many were built for the purpose close by, many others were brought in from all parts of Britain, so that many types were used, yet only those that were suitable remained. There were instances of Morecambe Bay nobbies dredging, and east coast smacks, but the more successful boats seem to have been those with transom sterns. One German-built boat was the *Zigeuner*, FH89, rebuilt at Restronguet Creek about 1840, and still fishing in 1940.

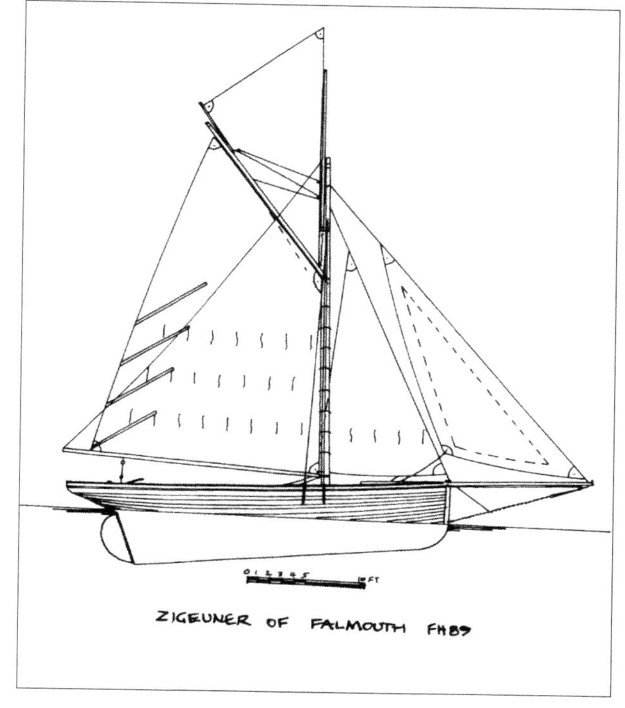

The oyster punts were small boats, around 14 feet and mostly carvel-built, that generally had two standing lugs and were used for all manner of jobs – long-lining, lobster potting, dredging for oysters, pleasure and racing, and even transporting plums from the fruit farms to the market in Falmouth.

As the Falmouth Oyster Fishery survives today, and local byelaws insist that all boats licensed must work under sail alone, the last of Britain's sailing fishing boats can still be seen working the grounds. And these boats today are hardly altered from those of a hundred years ago. Today's fishing legislators could learn a lot from this example of man's ability to react to nature's needs.

THE CRAB BOATS

Many coves of Southern Cornwall still have small, open boats moored, and the majority of these still fish for crabs. Some of these coves had harbours built in the last century: such places as Mullion, Sennen Cove and Coverack being capable of sheltering a fleet of vessels, while Cadgwith, nearby Church Cove, Porthoustock and Porthallow were, and still are to some extent, bases to beach-based fleets. The boats that evolved for the crabbing were similar again to the drivers. The Cadgwith boats are similar to the toshers of Mevagissey, and *Minerva*, FH58, built in 1935 is one of the oldest of these boats and she has just been restored even though she was in

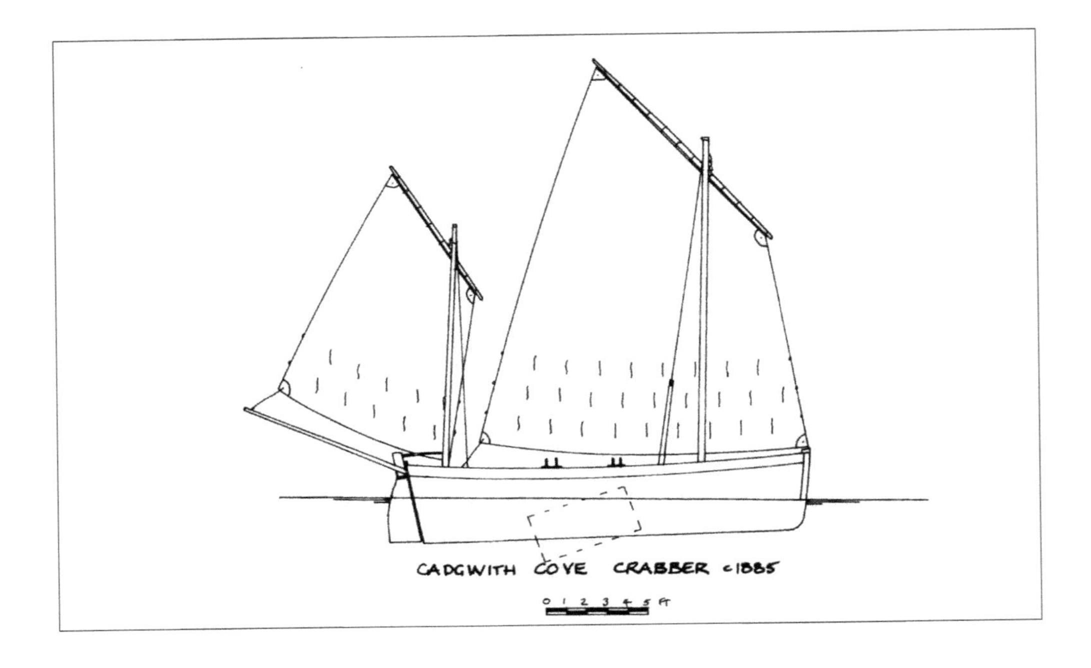

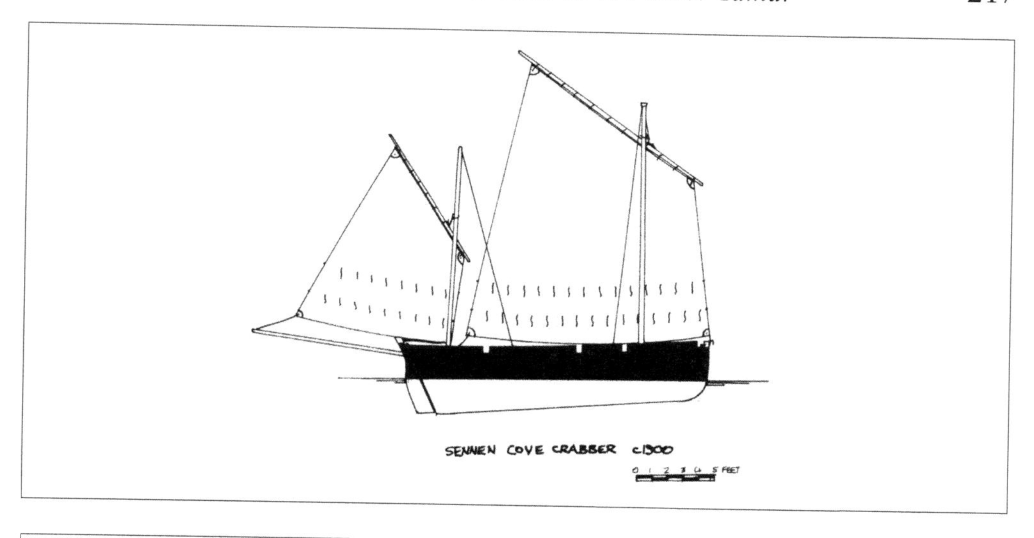

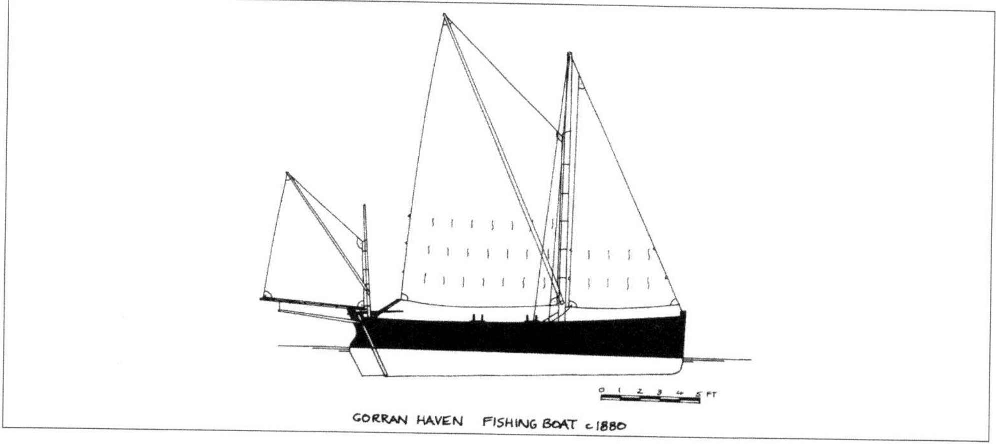

Above: Mullion Cove with several small crabbing boats, *c.* 1910.

a surprisingly good condition when her present owners found her. Further west, at Penberth, one of the last crab boats in commission now sits in the National Fishing Heritage Centre in Grimsby. *Tunny*, PZ145, was built by George Peake in Newlyn in 1936 and is 17 feet overall. She potted and lined all her life while under the ownership of the same family.

Sennen Cove, one of the most exposed shelters in the whole of Britain, developed its own unique crab and lobster boat. In 1850 there were eighteen of these boats that employed eighty fishermen, who either potted or fished for pilchards or red mullet. They were two-masted luggers of about 20 feet overall, transom-sterned and were totally open even though they worked off some of the wildest waters in Britain around Land's End. They had cut-outs in the gunwale for the oars to sit in when rowing; when sailing they were closed off with wooden shutters. These boats would sail all around Land's End and as far as Scilly during the summer crab season, hauling the pots, taking the catch, re-baiting and resetting the pots amid the sharp rocks that the area is renowned for.

Today most of the boats are of fibreglass, and crabbing is still a popular fishing all around the South Cornish coast.

At Gorran Haven, sprit-rigged boats worked out of the small harbour. One typical boat measured in the 1930s, the *Cuckoo*, was built in the village by John Pill in 1881. It appears that he built only a few boats, being a carpenter of all items including coffins. This carvel-built, undecked boat was 16 feet 5 inches overall, had an upright stem and well-raked transom stern, with sections that gave a steep rise of floor. These boats were all built for rowing as well as sailing, and were used for long-lining and lobster-potting, and sometimes for drifting for mackerel and pilchards during the season. By the time they were having motors fitted they had increased in length to about 18 feet, and were typical of the beach boat along this coast. A couple of restored boats still exist.

Although irrelevant to the types of boats, it is perhaps worth mentioning that there are two good examples of fishermen's capstans to be seen on this coast – the big Round House at Sennen Cove and the beach windlass at Penberth.

GIGS, JUMBOS, JOLLIES AND PUNTS

In differing parts of Cornwall, small boats had their own names. In St Ives, a small boat was a punt, and their length could have been anything between about 12 feet and 18 feet. They were generally used as tenders to the bigger luggers, although they were sometimes used for seining the pilchards. Just around the coast, the same boats were called jollies.

Slightly larger than these were the gigs, which were fast, clinker-built boats used for the inshore autumnal fishing. They were smaller versions of the pilchard driver and yet were heavier than the pilot gigs, measuring 26–32 feet. They were either rowed by four oars or sailed with a dipping lug mainsail and a small standing lug or sprit mizzen. Some had centreboards enabling them to sail fast and most were built in St Ives though a few worked out of Newquay to the north. Their main use was being able to get to sea while the heavier boats were still ebbed by the tide thus giving the fishermen a quick shot or two in advance of the bigger boats. With the onset of motorisation in the 1910s, gigs increased in length, up to 40 feet, but they retained their openness although the engine – or engines – had a cambered cover over. Tommy Thomas was a prolific builder of these craft in St Ives, working from his boatshed on the promenade, as was Henry Trevorrow and a fellow called Landers. One or two double-ended seine boats were converted into gigs by having a transom stern built on. Some forty carvel motor gigs were built in total in St Ives mostly in the 1920s. Whereas Newlyn built motor pilchard drivers, St Ives built motor gigs, which were capable of carrying large amounts of herring. They kept the mizzen lugsail for lying to their drift-nets. By the end of the 1920s these gigs had developed with wheelhouses being fitted, short foredecks added and a higher freeboard. Most were painted light blue whereas the sailing gigs were white. By the 1940s they were hardly recognisable, with capstans fitted for crabbing and trawling. Few remain though St Ives still has two working as canopy launches while *Caronia*, SS70, a rare gig built in 1927 by Peakes of Newlyn, now called *Starfish*, is now a pleasure yacht.

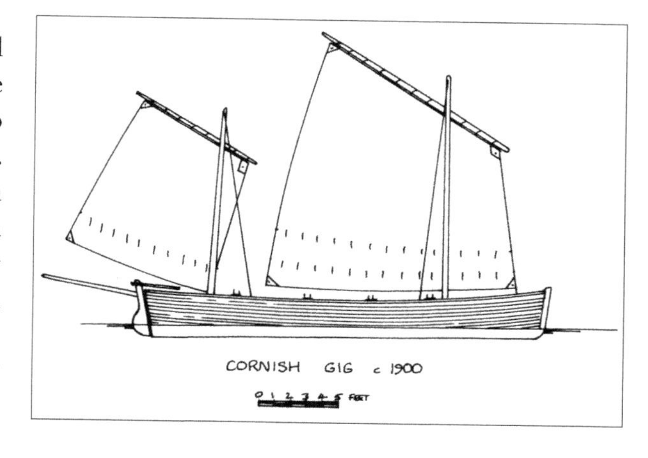

It must also be remembered that gig fishing was not just confined to St Ives and that gigs from other beaches and harbours were used for inshore fishing, especially hand-lining, while others were at one time used for mullet seining. These fishing gigs, however, should not be confused with the pilot gigs which today are numerous, having undergone a renaissance in the last decade or two with weekly rowing regattas during the summer and an annual international championship in the Scilly Isles. Moreover, questions have been asked as to why gig racing is not an Olympic sport! Indeed – why not?

In the 1880s, some St Ives fishermen were not keen on the gigs, perhaps due to the fact that many were wrecked, and when they wanted a smaller vessel than the pilchard driver they opted for what became the 'Jumbo', a vessel unique to the town. The name is perhaps misleading for 'jumbo-ising' referred to the habit of lengthening boats, something that some Cornish owners opted for in the 1920s when new boats cost too much and there was a dearth of the older boats. Jumbos in St Ives were, it is said, named after an African elephant in London Zoo, the most popular attraction, which was sold to the USA in 1882, causing a national outcry.

In St Ives, though, a Jumbo was another baby version of the pilchard driver, a small double-ended clinker or carvel boat, totally open, crewed by some three fishermen and rigged with two lugsails in a similar fashion to the pilchard boats. Length was about 24–29 feet and those below 25 feet were clinker while larger craft tended to be carvel-built. According to the late Eddie Murt who spent a lifetime researching St Ives

A typical Guernsey fishing boat as used mostly for potting and catching mackerel, *c.* 1890s. (*F. W. Guerin*)

fishing boats, there were over twenty Jumbos in the 1880s, two of which were probably converted ships' lifeboats. Small in number, they were also short-lived, often referred to as 'old men's boats' and were generally not favoured. William Paynter of St Ives certainly built one or two for a copy of his draft has survived, now in the National Maritime Museum, from the 1880s, which Paynter describes as being built 'for John Uren & others'. Boatbuilder Jonny Nance, also of St Ives, has recently built two small clinker replicas of these craft, his first, the *Celeste*, being the first to sail in the Bay for many years, and he hopes to build many more so that these Jumbos can race together once again.

The Looe lugger *Guide Me* on the Surinam River while on her way back to the UK from South Africa, 1990. (*Judy Brickhill*)

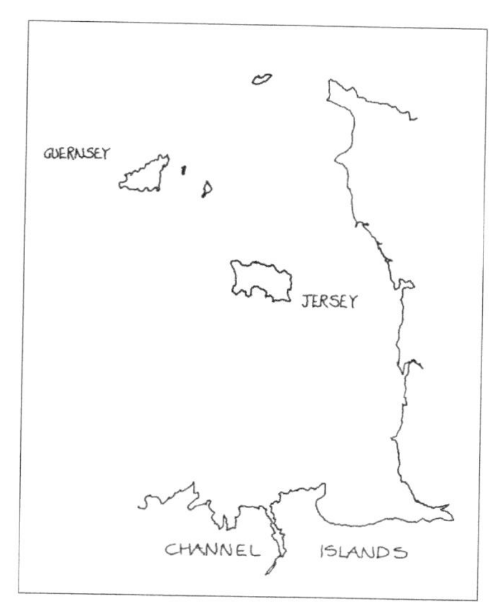

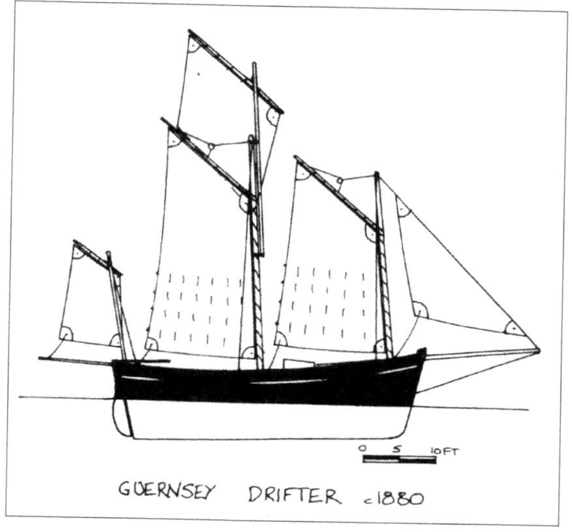

GUERNSEY DRIFTER c1880

THE CHANNEL ISLANDS

Fishing was the mainstay of the Guernsey economy at the beginning of the nineteenth century, and for centuries before that. As far as the other islands were concerned, Guernsey was the main fishing centre – Jersey men tending to join the Newfoundland fishing rather than that at home.

Fairly predictably, the fishing boats of the islands resemble closely those from the adjoining coast of France, and, like the Cornish boats, were influenced by the *Chasse Marée*. All the boats were transom-sterned, and generally they all fell into one of two categories. The largest group was that of the mackerel boats. These were up to about 36 feet overall, and were renowned for their wide beam – 12 feet – and deep draught – 8 feet. This ensured stability. These 12–15 ton boats were originally three-masted luggers. They were renowned for their sea-keeping abilities and were easily manoeuvred, both qualities being essential in the vast tidal range waters of the islands. They were carvel-built of pine planking on oak frames. About the middle of the nineteenth century they adopted the gaff rig on the foremast and mainmast, retaining the mizzen standing lug. To gain more canvas they set a topsail, and a foresail was set on a bowsprit. Around 1880 the mizzen was done away with altogether.

The smaller boats, although similar in shape, were the crabbers, which were anywhere in length between about 12 feet and 30 feet. These were mostly sprit-rigged initially, yet they, too, adopted the gaff rig during the 1870s.

A few of the mackerel drifters survived up to the 1930s, and the crab boats for many more years. But today's boats are as those to be found anywhere else – fibreglass and bristling with aerials. It seems that the only part of fishing tradition that remains from the island is that of another nature, for it was here that the Guernsey sweater, or gansey, originated. The gansey, knitted by a fisherman's wife for her man, became as much part of any fisherman's life as his boat.

CHAPTER 12

South Devon and Dorset Coast: Plymouth Sound to Portland Bill

Grumble you may, but go you must.
—Trawlermen's saying

Crossing the Tamar into Devon, a change in emphasis within the fisheries is immediately apparent. The Devon coast is, at its western end, similar to the Cornish coast in that it has hidden coves and bays, yet once round Start Point the coast is less rocky. By the time the River Exe is crossed, the coast has become a shingle beach, backed by cliffs. On Chesil Beach, Britain's longest at 18 miles, the steep shore is home to a unique fishing boat, the lerret.

Fishing has for centuries been the mainstay of the Devon economy, especially away from the coastal towns. Although pilchards are caught on the western end of the coast, it is with far from the same intensity as that of the Cornish. Plymouth is renowned for its hake, Bigbury and Start Bays for their crabs, and, of course, there's Brixham – the mother of British fishing ports. Plymouth, too, had a long relationship with the Newfoundland fisheries, it being one of the first British ports to send boats out to catch the cod only a few years after John Cabot returned to Bristol after his epic 1497 voyage. Some sixty ships with about forty crew sailed from Plymouth each March during the early part of the seventeenth century. They returned each September full of dried cod and created an important trade for the city. The fishery disappeared for a time during the Civil War, and it never really recovered so that, by 1700, Plymouth was only sending twelve ships. Dartmouth sent ships out – eighty boats in 1631, but these numbers had decreased drastically by 1652.

Plymouth fish landings peaked in 1892 when some 7,352 tons were landed. By that time, however, new ways were changing the industry. Steam trawlers were outdoing the sailing boats, and the landings were being concentrated in a few ports. By the turn of the century the landings were severely reduced. Brixham continued to prosper, and the small beach fisheries survived for a time. Today, a few boats work from harbours such as Lyme Regis and Exmouth, but only Brixham and Plymouth have survived with inshore fleets of any substance.

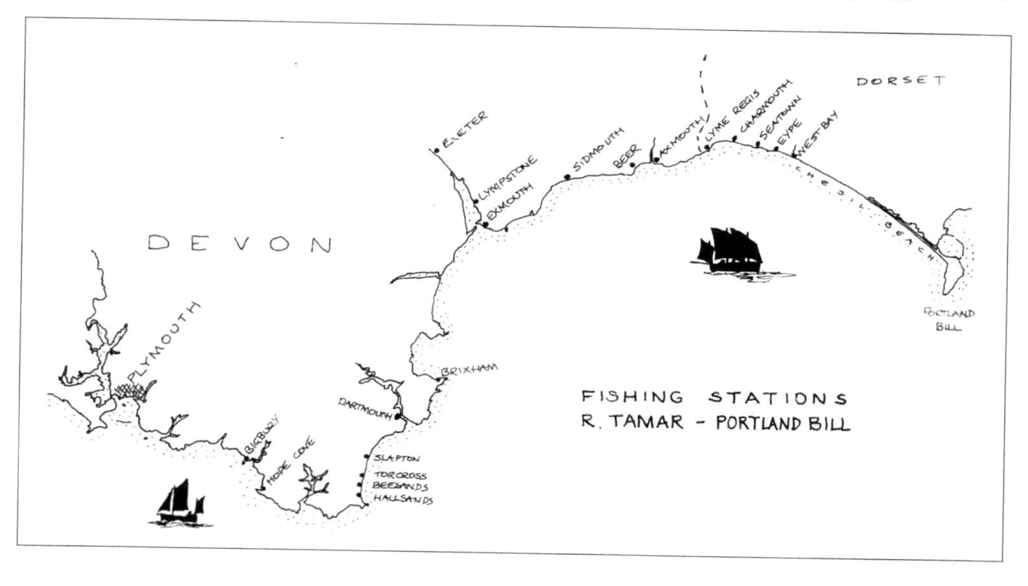

THE PLYMOUTH FISHING BOATS

Sutton Pool was the quayside where, in the nineteenth century, the home fleet of Plymouth was based. This bustling quay had its own market squashed into the narrow Barbican, and it was said to have been the finest fish market in the country. Here, around the middle of the century, following the building of the railway in 1848, came the Plymouth trawlers to land their fish. Although the first trawlers came from Brixham in the 1780s, there were by 1820 thirty trawlers working from Plymouth. By 1850, there were sixty boats. These trawlers were unlike the Brixham boats in that they retained the cutter rig, and the hulls were designed for speed. The fishing grounds lay inside and to the west of the Eddystone lighthouse, so the boats didn't have far to sail out. They mostly left Sutton Pool in the early morning and returned in the afternoon. The only exception was when fishing for soles at night.

Most trawlers were built locally at Cattewater, many coming from the yard of W. H. Shilston. The original sail plan of one of their cutters still exists in the National Maritime Museum. Another one of their boats, the *Erycina*, PH63, was deemed to be the fastest boat in the fleet.

By 1874, these trawlers were fishing in Mounts Bay, and landing in Newlyn. It seems that, while fishing in the Channel, many of these trawlers were run down by shipping. Others were wrecked on the Plymouth Sound breakwater when running into harbour during gales. Others that survived worked up to the First World War, but by the 1920s steam trawlers and the advent of motors meant that none of the sailing trawlers have survived through to today.

Another type of boat working out of Sutton Pool was the Plymouth hooker. These vessels worked out of Dartmouth and Brixham as well. These heavily built cutters were called hookers to denote fishing by hook and line, a term that has been traced back to 1567. When long-lining, they set lines with up to 3,000 hooks, and at other times they

Above: Fishing boats at Sutton Pool, Plymouth, during the heyday of the port at the turn of the century. Many of these steamers are from the East Coast, and landed catches here during the mackerel season.

Left: Plymouth hookers leaving the harbour are having to be rowed out to find some breeze to fill the sails.

hand-lined for whiting and bream. In the season they fished for mackerel with extra lines and for hake at certain times of the year.

There were two classes of hooker, differing in size, but otherwise the same – transom-sterned, deep-heeled and carrying a boomless mainsail. The bigger boats were up to just over forty feet, and were dandy-rigged with a standing lug on a small mizzen mast. They were fully decked, and had accommodation at the after end. The smaller boats were about 31 feet overall and only had a small cuddy under the foredeck and with waterways aft. Both types set a topsail, with a jib and foresail on a long bowsprit. In the late nineteenth century there were some forty hookers working from Sutton Pool.

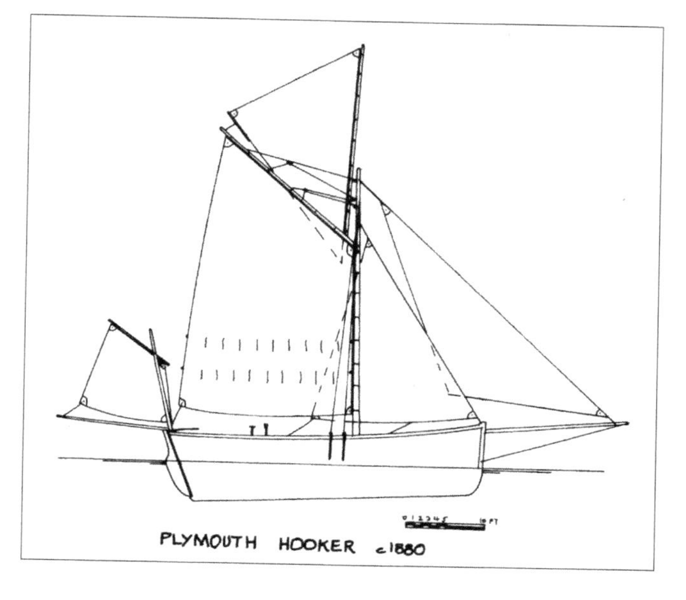

The hookers sailed up to 50 miles outside the Eddystone into deep water away from where the trawlers worked. Others worked close inshore along the coast as far as Looe where the bottom was rocky so that trawling was impossible.

Similar hookers worked out of Brixham, where there were seventy such vessels in 1833. At Dartmouth, where a local byelaw stated that 'the foreshore had to be cleared of all moorings and obstructions between Bayard's Cove and Warfleet by Michaelmas' for the autumnal sprat fishery, there were small 20-foot carvel-built, sprit-rigged, open boats with transom sterns working this fishery up to the 1900s. The town had its own fleet of hookers working from the harbour.

As was seen in the previous chapter, these hookers bore resemblance to the East Cornish boats in their hull shape. As many were built in Porthleven this is hardly surprising. During the 1930s several boats continued fishing, having had motors fitted. These retained the sails, and continued to fish until after the Second World War. As stocks declined, and the modern boats surpassed the older boats, the hookers petered out, so that none were working by the middle of the century. One or two remain in converted form, these being *Little Pearl* and *Princess Marina*, both of which are of unknown vintage. Whether there are more is as yet uncertain, although others are certainly mentioned in various publications. These include *Dayspring* (1893), *Doris* (1880), *Dolphin* (1909) and *Mary* (1861), while *Certa* (Pearce of Looe, 1897) is referred to as a Looe hooker. The lines of the hooker *Water Lily*, built about 1900, were lifted and then *Water Lily II* was built from these by W. Trout & Sons of Topsham and launched in about 1997.

THE BRIXHAM TRAWLERS

Whether Brixham or Barking is home to the modern-day trawl seems immaterial these days when almost the entire fleet of British sailing trawlers has gone. Brixham-built trawlers are in the majority of those that remain today. *Provident*, BM28, and *Leader* are household names these days, yet *Vigilance*, BM76, *Pilgrim*, *Kenya Jacaranda*, BM57 – ex-*Torbay Lass* – and *Deodar* are all still going strong; others are based in Sweden and further afield.

Brixham's association with fish seems to go back to about 1200, when herring, cod and hake were landed in substantial amounts. At the same time, Brixham vessels were joining in at the annual Yarmouth Herring Fair. Most of the fish caught off the Devon coast were dried, and were subsequently taken across the Channel or inland to serve the local population. At the time of William of Orange's arrival in 1688, there was a tiny pier for him to step ashore upon. Facilities were gradually improved until the end of the eighteenth century when the basis of the present harbour was set out.

The Brixham trawler *Leader* was built in 1892 and measures 100 feet and 110 tons. Currently, after a refit, she is chartering in the West Country, Channel Islands, Brittany and Isles of Scilly from her base on the River Dart. (*Struan Cooper*)

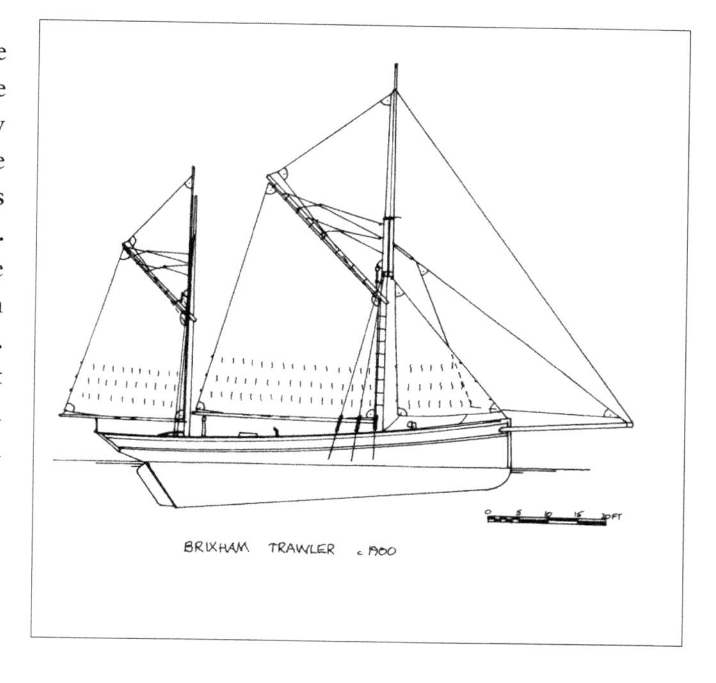

The origins of the trawlers go back to the late eighteenth century when, in 1785, there were seventy-six decked boats working out of Brixham. Then the boats were bluff-bowed sloops with a long, open transom. These were some 46 feet overall, and 40 feet on the keel. The Brixham men had previously drift-netted in the North Sea in similar vessels, but had given that up and taken to trawling. By the mid-century, lengths of keel had increased to 46 feet, when about seventy trawlers were working. Twenty-five years later they were even bigger – 57 feet of keel, and cost around £500, with the biggest being 77 feet overall. Some 120 boats were working from Brixham then, and 136 out of Dartmouth. But, by this time, the boats had adopted the ketch rig for which the Brixham boats remained famous.

Fishing smacks at Brixham in 1868. Nearly all are single-masted as the ketches haven't yet gained popularity.

These smacks were all built locally. Possibly the best-known builder was Upham, yet Jackman's yard was next door. Other builders were Matthews of Fishcombe, Gibbs of Galmpton and Philips of Dartmouth. Many of the Brixham boats were taken to the east coast, where eventually their design was altered to suit the shorter seas of the North Sea – the Brixham boats being suited to the Atlantic conditions.

All the trawlers were similar above water, being fuller in the stern to support the trawl when under tow, unlike the finer-sterned drifters. The draught was relatively shallow, yet the boat was powerful enough to pull a 2-ton trawl through the water. But, as elsewhere, the decline in fishing and the advent of motorisation caused the change in methods that saw the disappearance of these boats. The last two trawlers were built in 1926 – *Vigilance* by Uphams and *Encourage*, BM63, from Jackmans.

The Brixham boats sailed much further afield than the Plymouth boats, and, as we've seen, they were common in the North Sea, as well as Swansea. They often trawled in Cardigan Bay and Dublin Bay, some going even further north. When the fish was landed at Brixham, it was taken by fast cutters to Portsmouth, whence it went by horse to London. After the arrival of the railway, it went direct.

MULES AND 'MUMBLE BEES'

Two variations of the trawlers existed. The mules were smaller versions of the bigger boats, retaining the ketch rig, and generally under 40 tons, the bigger boats being referred to as the 'big sloops'.

The 'Mumble Bees' – sometimes known as 'bumble-bees' – were cutter-rigged vessels of around 50 feet. When the Mumbles oyster trade vanished in the 1880s, the fleet of oyster smacks disappeared. It seems that twelve smacks transferred to Brixham and the boats so impressed the fishermen that they built further types. *Little Mint*, BM355, is sailing again today after a long refit though looks a little sad in Gweek. She is 41 feet overall and 17.43 tons, and was built by Richard Pearce of Looe – who was renowned for his Plymouth hookers, Looe luggers and Cornish drifters – in 1913. *Golden Vanity* is another Mumble Bee, built in 1908 by Sanders & Co. of Galmpton primarily as a yacht, although she now is somewhat altered from the way she was then.

THE BIGBURY AND START BAY FISHING BOATS

Burgh Island, joined to Bigbury-on-Sea by an isthmus of sand that uncovers at half tide, still has the remains of a Huer's Hut atop, and the Pilchard Inn, which dates from the fourteenth century, reflects the seining that the local fishermen used to do. Records suggest that the rights for this fishery go back to medieval times and were given by the priors of Buckfast Abbey. The landed pilchards were cured in the local cellars at either

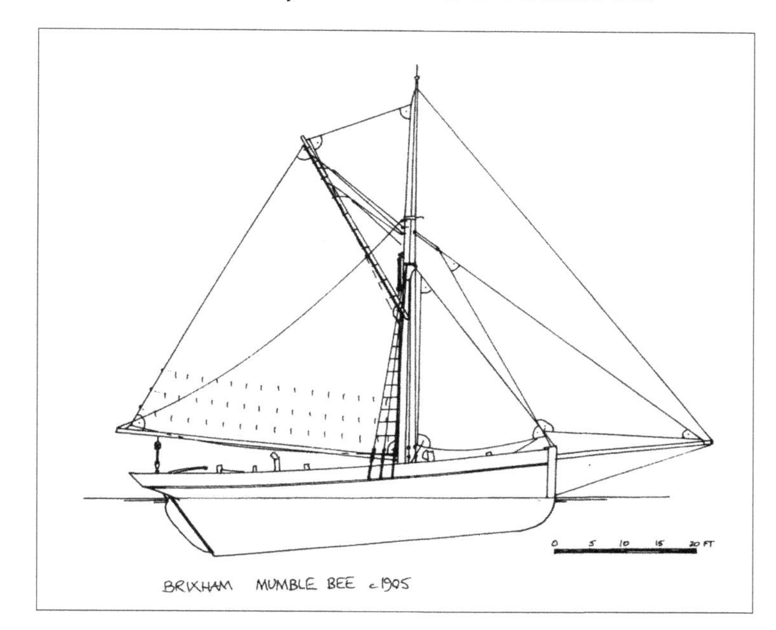

Below: An 18-foot Hope Cove crabber built in the 1920s by the Jarvis brothers of Hope and named *Progress*.

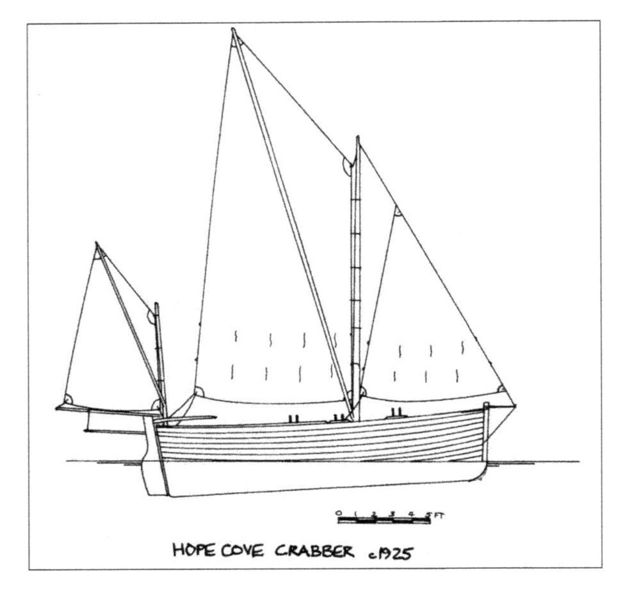

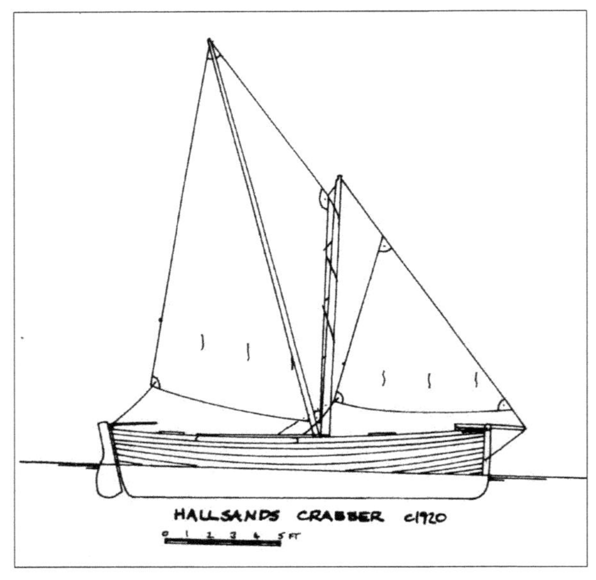

Challaborough or the Warren, Bigbury, or were taken by boat over to Bantham. The fishing was pursued alongside the smuggling, for which this coast remains renowned. French luggers dropped off many a keg of brandy for the fishermen to collect under cover of darkness, but as the coastguards stamped out 'free-trading' in the early nineteenth century, fishing became more of a full-time occupation.

Hope Cove, a few miles east, had beach boats around 1750 when the then-priest of the South Huish parish described the place thus: 'There is at Hope or Bolt Stay a sand that affords the neighbouring farmers a very good dressing and where great quantities of fish have often been taken, particularly lately upwards of twenty thousand mackerel were taken at one draught.' Nearly a century later, in 1820, Risdon wrote that 'it is a place noted for plenty of pilchards at times there taken'. As the harbour was not built until 1924, it is assumed that all the boats were hauled up the beach away from the waves. They were small rowing boats – some 15 feet long, yet, after the construction of the breakwater, the boats remained afloat and became bigger. These South Devon boats were sprit-rigged, the Hope boats having two sprits – main and mizzen, and a foresail set on a short bowsprit. They were up to around 18 feet overall and were nearly all built by Chant of Salcombe, whose reputation was unsurpassed in the area. They evolved through working off this exposed part of the coast. *Sarah*, built in Torcross in 1870, was typical of one of these beach boats; she was carvel-built, had a slightly raked sternpost, generous freeboard, and carried a single spritsail, topsail and foresail.

Just around Start Point was the village of Hallsands, perhaps the saddest of all fishing settlements in Britain. Like Bigbury Bay, Start Bay was renowned for its crabs, and the villages of Slapton, Torcross, Beesands and Hallsands all had fleets of beach-based craft. These were similar to the Bigbury Bay boats, albeit smaller for beaching. They had a single spritsail, and many were built by Chant, and some by Dornum, also of Salcombe, who built one of the last, the *Sylvia*, in 1921, which was based at Hallsands. Around 1900, there were twenty-two boats based at the village, and there was a healthy fishery, much of the catch being taken by smack to the Hamble. Then came the dredger to remove thousands of tons of gravel to make concrete for the new Plymouth Naval Dockyard. The villagers' protests that the dredging would undermine their houses were ignored. The dredger removed some 650,000 tons of gravel, reducing the beach level by seven feet by the time operations ceased in 1901. Although the fishermen were given meagre compensation for the loss of their fishing grounds, the authorities were not interested in the fact that their beach had disappeared. Then, over a period of bad weather during the winter of 1903/04, some of the houses collapsed after the seawall was breached. The damage was repaired, but now the beach level was some 12 feet lower, and the high-tide mark some 40 feet further up the beach. Over the years the seawall held, and fishing operated throughout the war until a gale in early 1917 coincided with a very high tide. This decided the fate of the village as, one by one, the houses fell as their foundations were undermined. Within twenty-four hours some twenty-nine houses, belongings and all, had gone. One house alone was left standing. Over the twenty or so years from the start of dredging some thirty-seven houses were destroyed. Compensation was finally paid by the Board of Trade, and new houses were built set back from the cliffs. Today many of the ruins remain, but the area is closed off as it is still regarded as being dangerous. The new village is quiet, and only two tiny fishing boats were to be seen recently.

Hallsands was also home to seine boats. One such boat was measured up by P. J. Oke in 1935. This 17-foot 7-inch boat was built by Chant in 1905, and like other seine boats, had its maximum beam well aft of amidships. Like all the crab boats of the area, the seine boats were clinker-built.

In 1863 the number of boats and fishermen working in Start Bay was reported. At Slapton there were five or six hook boats, three seine boats and twenty men and fifteen women engaged in the crab fishery. In Torcross there were three seine boats, twelve to sixteen hook boats, six mackerel seine boats, six pilchard drift boats and twenty men and fifteen women engaged in the crab fishery. Beesands had twenty-eight crab fishermen and thirteen seine boats, 22–24 feet long, and Hallsands had twenty-eight crabmen and three similarly-sized seine boats. During the winter the men either dredged for oysters or worked on the land.

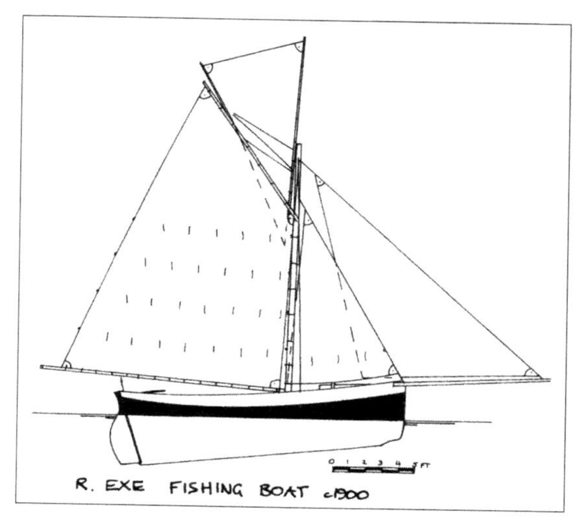

The fisheries of Torbay revolved mostly around Brixham as we've already seen. Torquay boats did land mackerel and herring when in season and sprats during the winter but the fishery was of no great consequence. Tourism figured higher in the boatman's life. Teignmouth and Dawlish had a few fishing luggers.

The River Exe had various quays from where several fishing boats worked. Lymphstone, Topsham, Starcross, Powderham and Cockwood supported boats, yet the quays were more likely to be crowded with trading vessels as the nearby city of Exeter grew up around the wool industry. Lymphstone had several cutters, 21 feet overall, that worked in the bay. These boats set a mainsail on a long boom, a topsail and two foresails on a bowsprit. Similar boats worked out of Exmouth and a 26 feet of keel boat worked out of Budleigh Salterton in 1869. Smaller boats worked the salmon fishery in the river Exe. The biggest of the cutters at that time seems to have been a 31 feet of keel boat out of Starcross. But these cutters were in a minority, as most of the boats working all along this coast were lug-rigged. The number of boats registered in Exeter in 1868 were as follows:

	Second class	Third class
Budleigh Salterton	30	10
Ladram	–	2
Sidmouth	37	13
Exmouth	14	33
Starcross	–	5
Cockwood	–	3
Powderham	–	2
Topsham	14	–

The beach at Beer towards the end of the nineteenth century with herring being barrelled on the beach before being taken away by horse and cart. (*Author*)

At this time the boats from further east were all registered at Lyme (LE), until all were moved onto the Exeter registry in 1902.

THE BEER LUGGERS

The saying goes that 'Beer made Brixham, Brixham made the North Sea'. The first men to trawl from Brixham were in fact Beer men in their little luggers. The Beer fishermen had a centuries old heritage (it was known as Shipcombe in 1005), and they were regarded as bold men; they probably dabbled in a bit of smuggling at times. Their craft were renowned for being the last of the three-masted luggers in Britain, and records tell us that there was a fleet of eight or ten boats at Beer around the mid-eighteenth century. They were used for drift-netting for herring and mackerel, lining for mackerel and potting for crabs and lobsters. Herring was particularly abundant with spawning grounds lying just off Budleigh Salterton, and the fishermen left the beach in the early afternoons during the season that spanned between November and March. Many winter nights were spent merely searching for the shoals, and often the men returned empty-handed in the early morning. At other times the herring was prolific, enabling them to secure a good wage. Herring fishing was always like that – sometimes good, sometimes bad, and occasionally a good haul of around 32,000 fish would be landed at one time. This was cured on the beach, taken off by horse and cart, and sent direct to Billingsgate market.

The boats were typically 28 feet overall, with plumb stems and narrow transoms. They were built at Beer up to 1900, there being two builders – Charlie Chappelle and Tom Restaric. After the turn of the century they were generally built by Lavis or Dixons of Exmouth. The Beer-built craft were clinker-built and cost £29. However, by this time the majority had done away with the mainmast, and the two-masted luggers kept the foremast stepped abaft the very short foredeck, giving plenty of space to work.

One of the last of the three-masted luggers was *Beatrice Annie*, E80, and she continued fishing until being broken up in 1918. A couple of years earlier, engines were fitted for the first time to Beer boats, *Little Jim* being the first in 1916. Brit engines from Bridport were one of the favoured units.

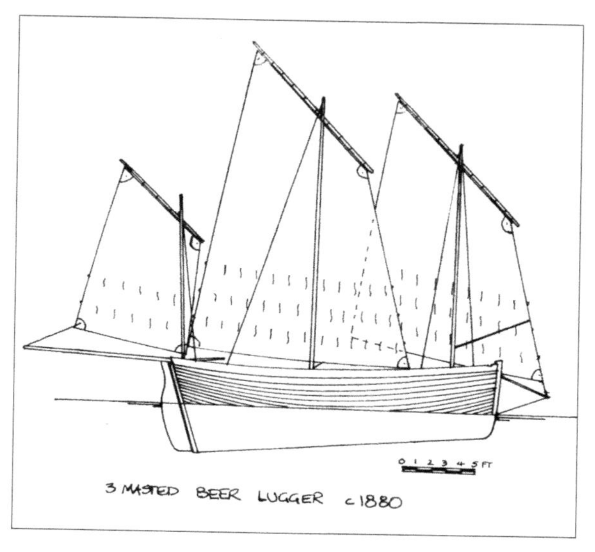

The advent of motorisation resulted in smaller boats being built by Lavis – between 21 and 25 feet – but these all retained the lugsails. Similarly rigged boats worked from Sidmouth, Branscombe, Seaton, Axmouth, Lyme Regis, Charmouth, Seatown

and Eypes Mouth, the latter being the start of Chesil Beach. The Sidmouth luggers were smaller at about 18 feet overall until mackerel and herring luggers were built in the 1880s, these being up to 24 feet long. Again they were undecked boats, clinker-built and set only one dipping lug. Later boats set two standing lugs, with the mizzen sheeted to a bumpkin and some even set a foresail on a short bowsprit.

All along this part of the coast the boats were hauled up the beaches. In 1997, I watched the boats being hauled by capstan up the steep Beer beach, listening to the creaking wire cable, and envisaged how the beach would have been a century earlier with upwards of a dozen boats working there. Greased timber is used today as it was then, but, instead of the winch, the hauling was done with block and tackle and many hands. The boats now are fibreglass, yet the fishing methods have hardly altered, and the process of launching and beaching is a time-consuming job. These fishermen are still set apart from those that merely speed into harbour, throw a couple of lines ashore and go home. The beach fishermen still retain a pride in their tasks.

A few relics from the past do remain, and these handful' of luggers race during the summer on a Monday night. Although motors are a necessary requirement to be eligible to enter, they are never used during the event. Other, more modern, yet wooden, craft remain solely for the purpose of being hired out to trippers. But the numbers of these by no means match those of fifty years ago, and the pastime of hiring 'self-drive' boats is dwindling so much that it is now hard to justify their continuance, spelling another major blow for the beach life.

Left: A three-masted lugger from Beer in 1877. This elm-on-oak-frames clinker-built lugger was built locally.

However, I spent a lovely evening several years ago sailing with these folk, and the mere sight of these luggers forging around the short course was enough to evoke visions of the heyday of these fine little craft. And, although it is easy for us now to be nostalgic about these hard times, it

Two-masted Beer luggers on the beach, *c.* 1910. (*National Maritime Museum*)

Beer luggers being raced under sail in 1997. Although the design has hardly changed, one rule of the race is that all boats must have an engine installed. This has tended to fill out the hull shape, although they still sail extremely handily. (*Author*)

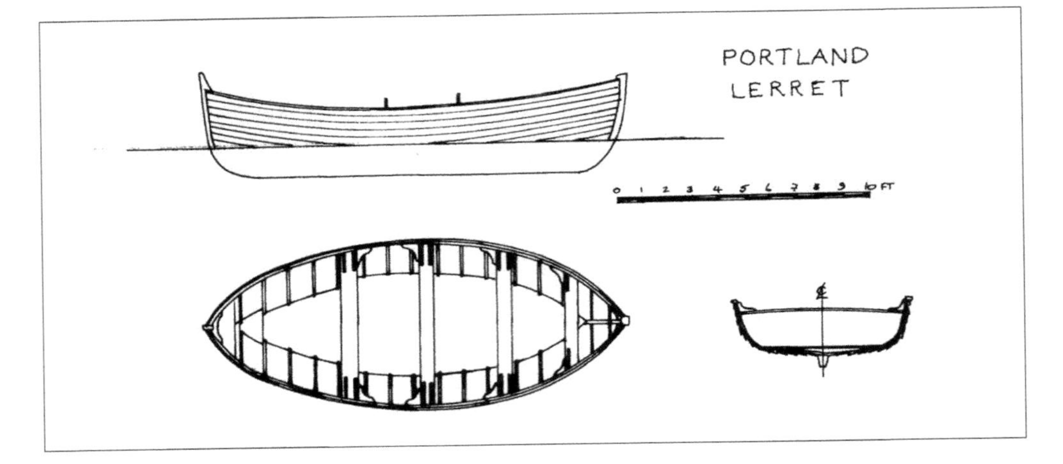

The lerret *Sunday-at-Home* at Church Ope Cove, *c.* 1900.

is an emotion that is shared by those few fishermen old enough to remember those days, and that, surely, shows that their shadows still retain a corner in the hearts of those that do care.

THE CHESIL BEACH LERRETS

The final type of craft along this coast is another beach boat, and one that has evolved through generations to fulfil the task of beaching on a coast that has dangerous cross-currents. Apart from West Bay, Bridport's harbour, where some transom-sterned luggers are still moored in the tidal harbour, the lerrets worked all along the coast from Eype to Portland.

Supposedly originating in the fifteenth century, the lerret evolved as a double-ender because of its ability to be readily launched and beached from the pebbly Chesil Beach. The Fleet is the stretch of water that lies inshore of the beach between Abbotsbury and Portland, and the boats were often rowed across this, then carried over the bank of the beach into the sea. The Fleet trow later evolved for this task. The lerret had to cope with the surge of the waves onto the beach and the cross-current, and the double-ender, pretty much the same shape either end, was found best to deal with this. They were also perfect for smuggling, many sailing over to the Channel Islands to 'free-trade'.

Although mostly a rowing boat, the lerrets that crossed the Channel had sails. Normally these consisted of two spritsails, although some had a lug mainsail. Sizes of boats differed from the smallest two-oared version, to four-oared, six-oared and the biggest eight-oared boats more often had the lugsail. The largest vessels were 24 feet overall, but these seem to have faded out in the mid-nineteenth century, whereas the six-oared boats, at around 20–22 feet, were the most widely used for fishing. The four-oared, at under 20 feet, was the last type to be widely used up to the 1930s. Details of the two-oared boats are non-existent.

One reason for the decline of the lerret was the need for up to fourteen men to launch it and sail it before the advent of winches. These men were collectively called a 'seine companey', and each village along the Beach had its own 'companey'.

Today two lerrets exist on display, one in the Weymouth Timewalk Museum and the other in the collection of ISCA at Eyemouth. Two or three are reported to be still in use upon the beach, one being the 1923-built *Vera* from Langton Herring. Gail McGarva has built a replica of this, the *Littlesea*, which was launched in 2010. Also surviving are a couple of the unique Fleet trows that were used to cross the Fleet – that piece of water between the hinterland and the beach itself – to get to and from the lerrets are too heavy to carry over the stony shore. While passing through some years ago, however, I saw none of these. Whatever number is remaining, the days of them seining for mackerel and whiting must surely be over. The bigger boats will have grabbed the catch long before it manages to get close inshore!

CHAPTER 13

East Dorset and Hampshire Coast: Portland Bill to Selsey Bill

Nice oysters!

—The cry of the oyster-sellers

At first glance the fishing boats of this area seem to be a mishmash of differing types. However, on closer inspection, they do all fit into a definite pattern of evolution through local influences.

The main influence is undoubtedly that of the Navy, with its strong presence around Portsmouth Harbour and Spithead. The transom-sterned wherries, one of the earlier types, appear to have evolved from naval vessels, which mostly seem to have

A Southampton Fishing Hoy off Calshot Castle – an unusual fishing smack of considerable beam and having heavily raked stems and sterns. (Edward Cooke, 1828)

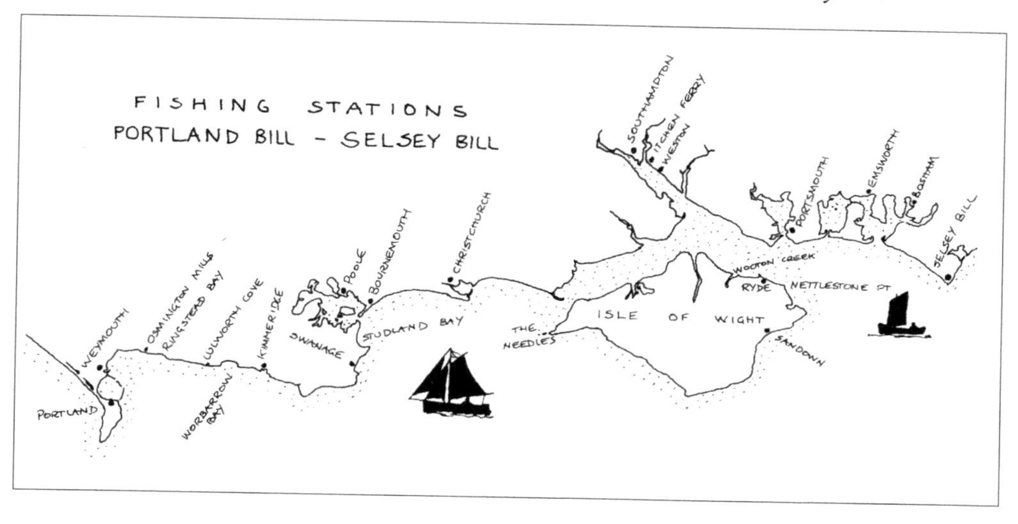

been transom-sterned. The second influence is from those racing yachtsmen of the Victorian era who built and sailed their powerful boats during the summer, usually with fishermen as crew. During the winter many boat-owners operated a small fishing boat of the Itchen Ferry type. It was through them that the cutter rig was adopted as it was generally regarded as being not only more efficient but also handier when sailing in the confined tidal waters of the Solent.

Different types, then, developed over 200 years or so, to produce those craft that were still fishing up to the 1950s. Poole, although not immediately in the vicinity, was within reach of these craft, and they soon became accepted around this large harbour, and consequently around the smuggling haunts of Purbeck. At Portland Bill, the influence ceases with the lerret and its possible Mediterranean roots, and at the other end, Selsey Bill, the beach boat is immediately more suited to the shingle beaches of Sussex.

Fishing within these boundaries has been practised for centuries. One of the first types to fish offshore was the Cowes Hoy, or Southampton Fishing Hoy, a variation of the Dutch Hoy. These boats fished in the late eighteenth century, and were used for trawling. They were clinker-built, had considerable beam with a semi-circular midship section and had a bluff, rounded stem and raking sternpost. It has been suggested that their name comes from the trading vessels that plied the area, in direct contrast to the fishing luggers of the time.

The three-masted luggers have already been described, and the later two-masted boats were fishing around the Solent, yet few seem to have been based in the area. Smaller boats, taking some influence from these luggers, were more suited to the confined waters.

THE WHERRIES

Sprit-rigged wherries fished these waters, and many sailed over the 80 miles or so to French ports in the practice of free-trading. Further west, the coast

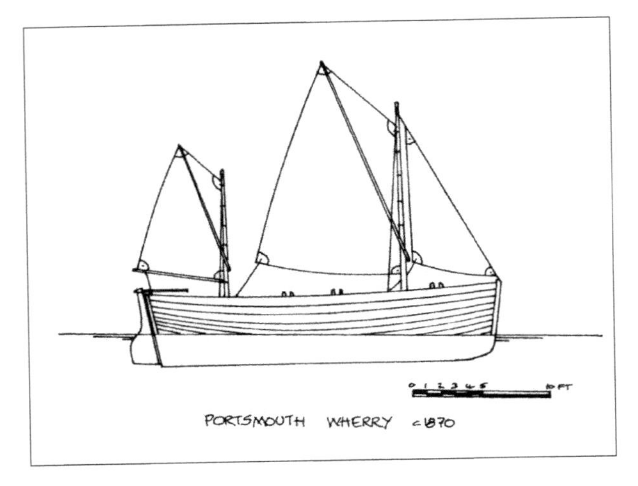

between Portland and Poole is especially renowned for its smuggling tales, and it still remains one of Britain's most unspoilt coastlines.

The Solent wherries were working in the days of Napoleon. Some of them copied the bigger luggers and set three masts, although most had two. Their development stemmed from the naval presence around the Solent, and they spent much of their time attending the warships and associated vessels that lay anchored off Spithead. Fishing, it seems, was only for the slack times when the navy were away, usually during the winter, when they netted or used hand lines.

The majority of the wherries were double-ended, in contradiction to earlier text, the stern only just having more buoyancy than the bows. They sailed extremely well in the short, steep seas of this part of the Solent where tidal streams run strongly and when against a countering wind can set up quite a chop; they also performed well to windward.

They were mostly built of oak, a timber that was not in short supply owing to the proximity of the New Forest and the amount of shipbuilding in the area. They were strongly built so that they had long working lives, and the largest of the first-class boats were up to 35 feet. These were half-decked and had a tiny cuddy with a stove that allowed a limited amount of sleeping and cooking in cramped conditions. The remaining part of the boat was open with benches along both sides and a wide after-thwart for the helmsman.

Most of these wherries were working out of Portsmouth, and second-class wherries were also to be found there. These were much smaller and were mainly used around the harbour itself. Transom-sterned wherries were also to be found, most of these originating from Cowes and Ryde, where the transom was favoured. These were often used for drifting for herring and sprats. During the summer they ran errands and passengers out to the boats anchored in the Cowes Roads, or even as far as Southampton. They were normally rigged with two loose-footed spritsails and a small foresail, yet were rowed around in the strong tides possibly as much as they were sailed.

Other than attending shipping, smuggling and fishing, and avoiding the press gangs, many local seamen found work in the summer aboard the many racing yachts in the nineteenth century. Yachting was a growing pastime for those with the money to play with these magnificent toys, and many a fisherman forewent his summer fishing to work aboard one. As well as providing more pay, it gave him the experience of high-level sailing that, on top of his fishing knowledge, helped to produce one of the handiest forms of fishing boat ever seen in British waters – the Itchen Ferry.

THE CUTTER-RIGGED BOATS

During the eighteenth century, Itchen Ferry was a small fishing village on the banks of the River Itchen, opposite Southampton. A ferry across the river had existed for centuries, and continued right up to the building of the Itchen Bridge in the 1970s.

Small, 12–14 feet overall, sprit-rigged craft worked off the beach, fishing the waters of the Solent and Southampton Water. These boats were clinker-built, and had a transom-stern. By the mid-1700s they had grown to 23 feet and had a loose-footed gaff mainsail with a vertical leech, and the largest boats set a mizzen. By this time they were commonplace all around the Solent, as the rig was deemed efficient for towing a trawl.

In 1851, a bowsprit with two headsails was introduced into the fleet, and the next year the main boom was extended over the stern to create a much larger sail area. Throughout their period of development between 1840 and 1910, though, the hull remained hardly altered, and by the turn of the century the boats were up to 30 feet, and had long, straight keels with a slight rake and were generally copper fastened. The smaller 23-foot boats did continue to be built, and, at regatta, were raced as a different class. Carvel-built boats were adopted during the second half of that century.

The boats were three-quarter decked with a small forecastle under the foredeck that stretched one-third of its length. In the cuddy were two berths, a cupboard and a coal stove. The cockpit, or stern sheets, had narrow side decks, and in front was the fish-hold. The boats spent most of their time fishing for shrimps and oysters before racing back to land their catches as quickly as the experienced skippers could. The boats were a common sight in all the small creeks and bays around the Solent, and along the coast to Christchurch and Poole. In 1872, according to the fishing registers, there were 570 second-class fishing boats working the Solent, and another sixty-one in Poole.

The best-known nineteenth-century builders of the Itchen Ferries were Alfred Payne of Northam, Southampton, and Dan Hatcher of nearby Belvedere. Both were noted for the racing yachts they built, although Hatcher is perhaps renowned as the overall master of these craft. Other builders were Luke of Itchen Ferry, who later moved down to the Hamble, and Fay of Southampton. The average cost of one of these craft in the closing stages of the nineteenth century was about £30, they often being built from offcuts from the bigger ships, and sometimes built for the captain of these ships when they brought these in for repairs. A sort of gift in recognition of their patronage!

One of the best-known boats is still sailing – Dan Hatcher's *Wonder*, SU120, built in 1860, yet mostly rebuilt – and many other Itchen Ferries have been saved from the

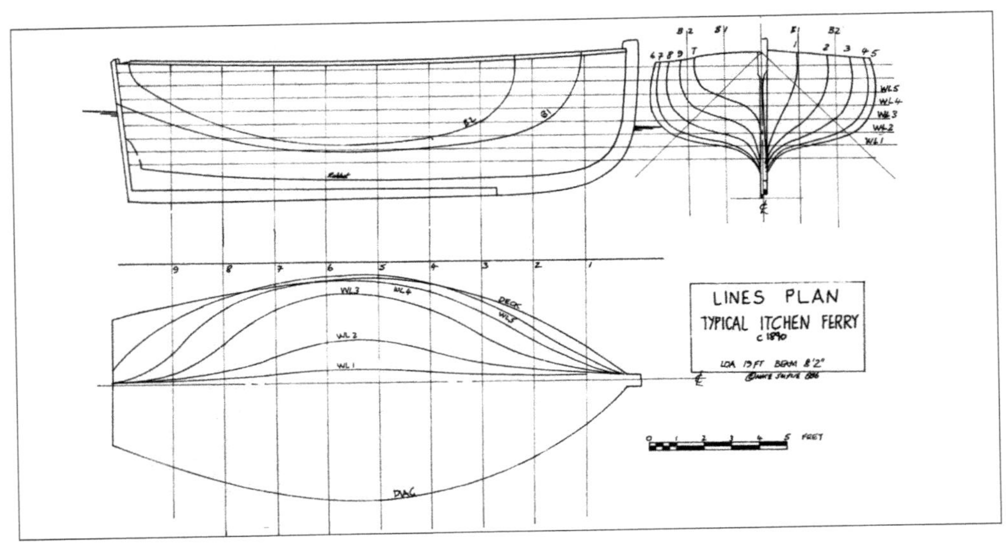
LINES PLAN
TYPICAL ITCHEN FERRY
c 1890
LOA 19 FT BEAM 8'2"
FEET
DECK
DIAG

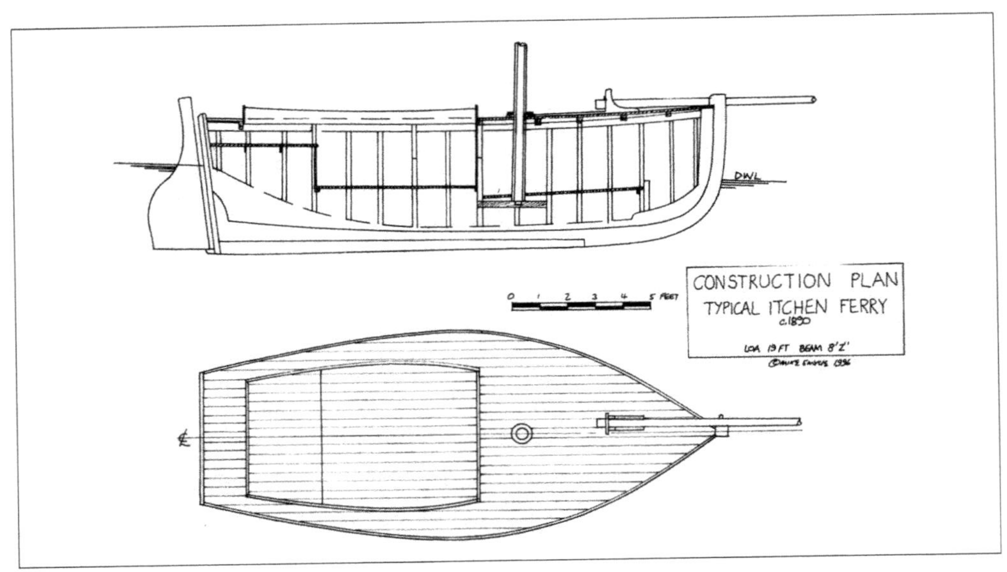
CONSTRUCTION PLAN
TYPICAL ITCHEN FERRY
c 1890
LOA 19 FT BEAM 8'2'
DWL
FEET

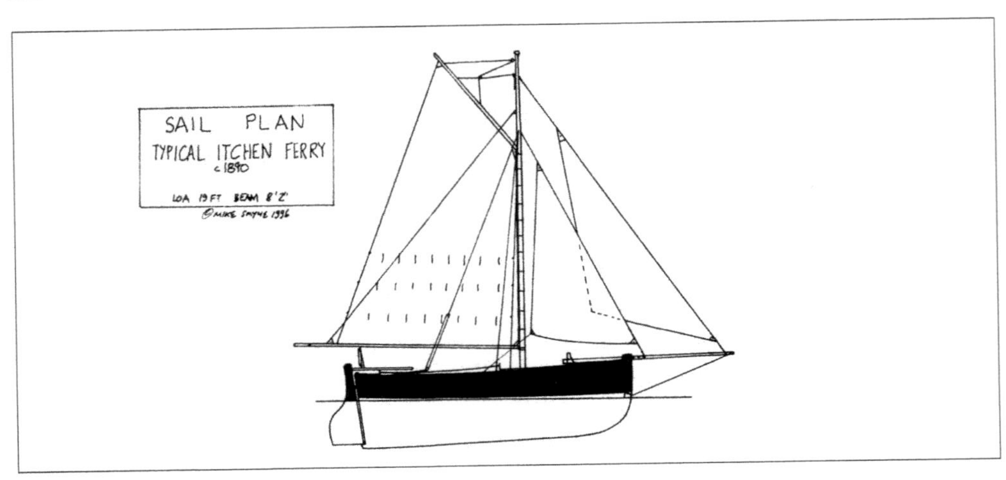
SAIL PLAN
TYPICAL ITCHEN FERRY
c 1890
LOA 19 FT BEAM 8'2'
©MIKE SMYLIE 1996

ultimate demise and have been lovingly restored, a sure sign that their reputation for long life is not misplaced. Perhaps the best example of a true original, unmodified Itchen Ferry is the Hatcher-designed, Fay-built *Freda*. Built as a mirror of *Wonder* in the best and lightest of materials for Captain Sam Randall, a renowned yacht-racing skipper of that era, she has worked in and around the Solent most of her life. Typically she had two rigs: her winter working rig of a boomless, high-peaked narrow working main on a short mast with a small jib and as such used for trawling, often single-handedly, and her racing summer rig of around 500 square feet of canvas that required several hands to ballast and operate. Today, according to her owner (to whom I am indebted for his help), she embodies the perfection that the Itchen Ferries had reached about 1890 at the pinnacle of their development, 'achieving the finest expression of performance available to its design limitations'. Luckily, *Freda* was saved from an untimely demise after being damaged in the 1987 storms that ravaged the south coast. At this time she was named *Viking*, CS110, had a Thornycroft Handy Billy installed and was owned by two old fellows who felt they were unable to repair her and so were about to burn her there and then on the beach. A local yacht broker appeared in the nick of time, and so began a new chapter in her long life that involved repairs, restoration to her originality and constant race winning ever since. It seems she is most definitely the fastest Itchen Ferry in commission.

Another fine example of a genuine Itchen Ferry is the Harry Feltham, 1870-built *Black Bess*, CS32, now owned by the Friends of the Classic Boat Museum of Newport, Isle of Wight. She worked for three generations of the same family and is still used occasionally for 'trawling, trammelling and dredging'. She is pitch-pine planked on sawn oak frames and has a lead ballast keel. The Itchen Ferry still remains one of the most widely appreciated types of small sailing fishing boat.

Smaller sprit-rigged vessels – usually just referred to as 'fishing boats' – worked off the Southampton Water shore, from beaches such as Weston. Many of these were built at Southampton, and were usually about 14 feet overall. The mainsail had a long boom, and a foresail was set on a bowsprit. They were open boats more suited to rowing when at work, and they fished further out into the Solent, around Spithead and into the various river estuaries using seine-, trammel- and peter-nets in shallow water, and fishing for whelks and netting mullet at different seasons. They also drifted for herring and sprats in autumn and longlined at other times. Later boats became lug-rigged.

Above: *Freda* ashore at the Baltic Wharf Boatyard, Totnes, in 1996, being prepared for a new rig. (*Earle Bloomfield*)

Left: The Itchen Ferry *Freda* sailing into Brixham harbour after being the first gaffer home in the classic race at the Brixham Heritage Festival of 1997. (*Earle Bloomfield*)

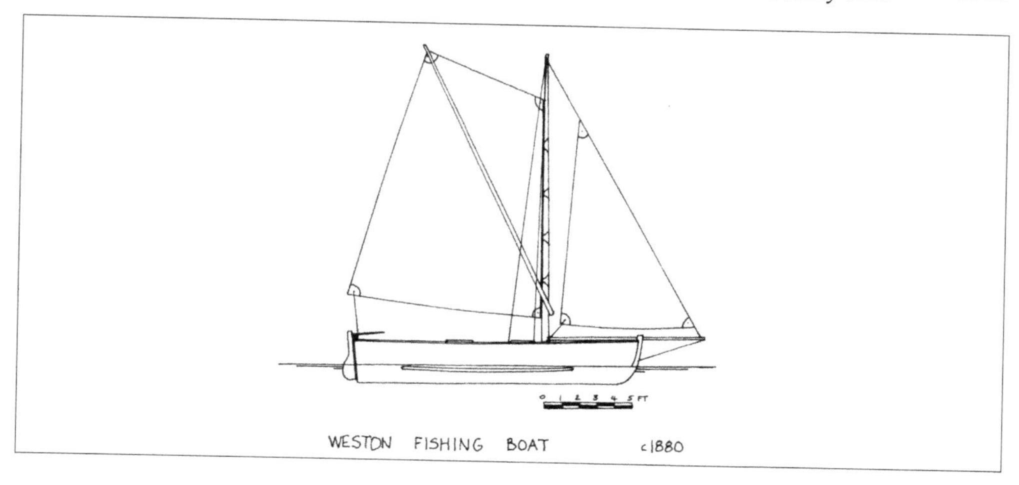

The herring season was from October into November, when the shoals moved into Southampton Water as far as Netley. Here, of a night, the water literally teemed with small boats netting, often bumping into each other with a curse as the tide played with them. Nevertheless, large amounts of herring were landed in the morning, and were taken directly to market. With the railway stations of Southampton and Portsmouth close by, the fishermen had access to huge markets for their catches.

Today's Solent is still renowned for its yacht racing, and the seeds of this must have been sown by the fishermen of old. The spirit of regatta was strong in all the working waterside communities, and each had its annual racing. The many varied fishing craft, often joined by the watermen's boats, and perhaps even the smugglers, all contended with a serious mind. It must be assumed that this fierce competition ignited a spark that led to the great Victorian yachting era that, in turn, led to the building of some of the finest craft ever to have been built. Today, as we all know, Cowes keeps this tradition alive, although seemingly somewhat aloof of the memory of those working craft of old.

THE ISLE OF WIGHT

As well as the small wherries used by the Ryde and Cowes watermen for winter fishing, there were fishing boats working out of each river and many a beach around the island. Fishbourne and Wooton Creek were home to several large smacks, some built locally. These generally dredged for oysters, as we shall see. Also, as already mentioned, many Itchen Ferries worked from Wight harbours.

Sprit-rigged boats worked from Bembridge and Nettlestone Point (now called Seaview), but these were mostly for pilot duties. Along the 'back of the Wight' there are various beach landings all along from Nettlestone to the Needles. Mostly the boats were transom-sterned and clinker-built. In 1790 Robert Wheeler of Chale reported that two boats took 3,000 mackerel to Portsmouth, and the beach there later became a mackerel centre. Sandown was renowned for its herring fishery of old.

THE DORSET COAST

Weymouth has long been remembered for George Ill's visit in 1789, during which he went bathing. This immediately popularised sea-bathing, and the wealthy flocked to Weymouth in the same way as they did to other watering places around the country to give themselves the medical benefit by taking themselves down the steps of the bathing machines and immersing themselves in the Channel water. Many of these bathing machines were operated by fishermen.

Prior to this fashionable trend, fishing had been prosecuted since the Roman times. The harbour at Portland had been a military base since the days of the Armada, and the area remains renowned for its smuggling past. Consequently fishing remained a minor industry with no substantial commercial fishery ever being established, yet still small fishing punts remain in the harbour today, and the few fishermen there have a good local knowledge. The Portland fishermen are best known for their cranes, or 'whims', for lowering their boats into the water close to Portland Bill. The currents here are treacherous, yet the bass are prolific, so these small punts are still lowered off into the sea some twenty feet below. Only one or two cranes remain in use, one of the last wooden cranes having been destroyed during an argument between two fishermen.

When the railway came to Weymouth in 1857, instead of opening up markets for the fishermen, it served only to bring in tourists. This created employment for fishermen taking trippers out, and hence fishing never did take a place of importance within the local economy, although the fishermen did have their own fleet of small, sprit-rigged lobster boats.

Further east, however, fishing played a more important role. Here the coast remains today much as it was one hundred years ago. Smuggling was rife among the secluded coves between Portland and Swanage. The inaccessibility was perfect for escaping the Revenue men and nowadays still only a few coves are reachable by road. Osmington Mills still has the remains of a small stone pier, yet eight fishing boats worked from here at one time. Further east again, Ringstead had five fishermen in 1841. Lulworth Cove was home to a thriving lobster fishery, several boats being moored in the Cove at the turn of the century. Worbarrow Bay, where the ghost village of Tyneham still stands silently, was the base of a mackerel fishery at the same time, where 5,000–6,000 fish were netted in one catch. The boats were all small, clinker-built, transom-sterned craft, often rowed but with one or two spritsails. They adopted the cutter rig by the 1930s, somewhat later than their Solent counterparts. A small fleet of similar vessels was based at Kimmeridge Bay. However, all along this part of the coast it is said that the herring was a prolific winter fishery until the 1920s.

Swanage was an important herring station in the late eighteenth century and it had its own fleet of small sprit-rigged, two-masted, clinker-built boats. Leland in 1788 reports 'a fishartown called Sandwiche, and there is a peere and little freshwater'. In October of that year it is reported that the herring fishery was promoted by Mr William Morton Pitt MP, and that smoking and curing houses were erected for the

The beach at Swanage with a small two-masted lugger similar to those from Devon.

duration of the fishing. In 1886, some 2,000 herrings were taken at one catch, with some 30,000 herrings coming in a week. The port declined in 1887 when the railway arrived, although this did briefly help the fishermen. Tourism became the backbone of the local economy, probably due to Dorset having some of the finest beaches in Britain. Studland Bay, to the north of Swanage, had a small fleet of boats that also adopted the cutter rig in the 1930s.

Poole, by contrast, followed hard on the heels of the Solent boats in changing over to cutter rig during the 1890s. Poole had grown up on the fishing; in 1583 the town sent ten or twelve ships to the Newfoundland fishery, leaving in the early spring and returning after the end of the fishing. Traditionally the season finished on 20 September. These ships had a crew of ten men and two masters. After about 1700, many local families emigrated to Newfoundland not only to save the journeying over each year but to escape the constant threat of the press gangs.

In Henry VIII's time, the local boats were double-ended and very beamy, and they tended to work the oyster trade that had existed since the eleventh century. By 1500 Poole was renowned for its clinker-built boats and it became an important boatbuilding centre.

In the nineteenth century the town expanded through the export of local clay, and the import of coal and timber.

The fishing boats of the eighteenth century were sprit-rigged, and were some 14 feet long, and clinker-built. They remained thus until the late nineteenth century when they grew in length up to about 19 feet. These later boats had a tiny cuddy under the short foredeck that extended to the mast, and they fished mainly for shrimps and oysters and trawled in winter. Sometimes they joined the spratting boats off Christchurch ledge where a thriving sprat fishery grew up. The season lasted for up to five months, beginning in early November. In 1872, as we have seen, there were sixty-one of these sprit-rigged vessels, and within twenty years they were all cutter-rigged. Many

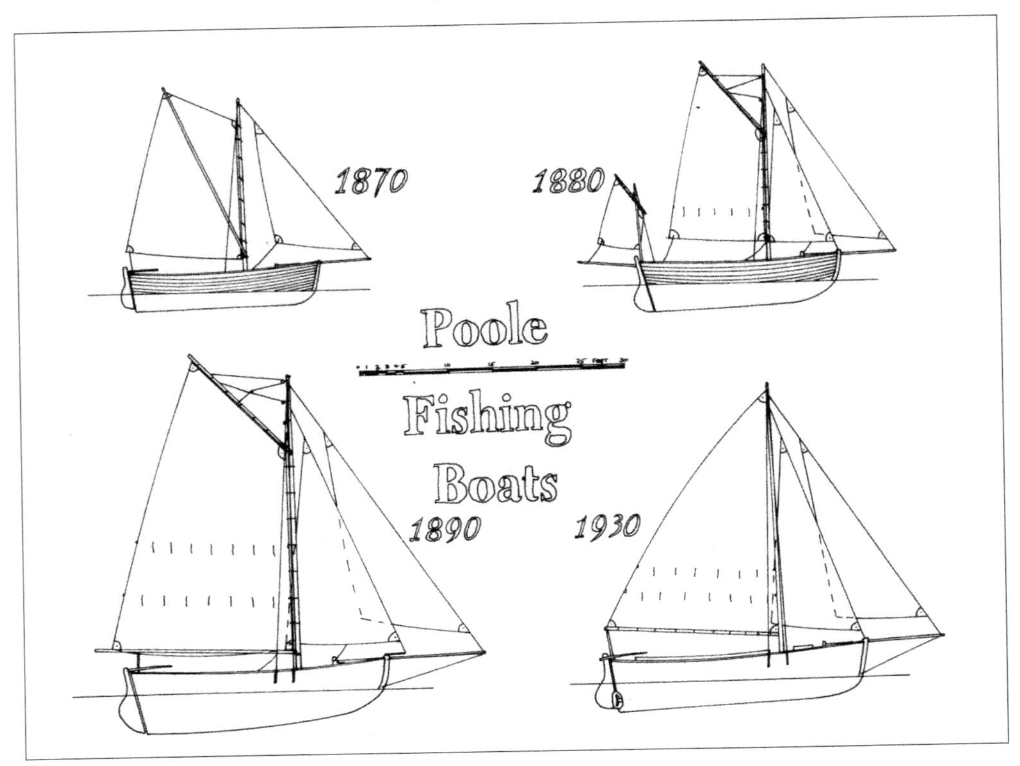

added a sprit mizzen on a short mast stepped against the transom and sheeted to a bumpkin so resembling their counterparts on the Solent, and they, too, adopted carvel construction.

By the early part of the twentieth century there were some forty boats moored up off the town, when they had been moved by the local council who wanted their previous home at the quayside at Hamworthy to be redeveloped. Their size increased again up to 25 feet, their floors became flatter and the forefoot was rounded to produce a handy form of boat. As they sailed even further afield, sizes reached up to 30 feet, and by the 1930s there were still some thirty boats fishing full time, mostly having adopted Bermudan rig by this time. They were 4 feet 10 inches in the beam and drew 2 feet 6 inches to 4 feet 6 inches of water and generally had no centre keels, as these were deemed unsuitable for hauling up the beach, and were instead fully bilged to suit sitting upright. A few, though, had centre plates. When, in 1936, the Fishermen's Dock was installed, they were able to land directly ashore.

Because of the double high-water that is unique to this part of the south coast, and also to the enormous area enclosed by Poole harbour (it has a shoreline of 16 miles), the port enjoys the advantage of having a 'stand' of high water for about seven hours and thus for about fourteen hours out of the twenty-four. But because the harbour entrance is narrow, the flood, and particularly the ebb, streams in the entrance channel run very strongly. The fishing boats trawling in Poole Bay could thus fish for longer, but had to pay the price of greater difficulty in getting in and out.

A typical Poole fishing boat, the *Polly*, PE69, the last of her type, on a trailer and about to be launched in 1937. (*Ernest Bristowe's Photographic History of Poole*)

Seine-net boats also worked from the beach at Sandbanks, especially when unfavourable weather prevented trawling. Using four oarsmen, they used the ebb to get there before they set the net. The men wore large leather corsets on their backs, which were laced up the front, and a belt carried a large hook to which the hauling rope was used. The men trudged backwards up the beach hauling the net rope – a particularly laborious task. This method of fishing also produced large amounts of immature, dead fish which were left on the beach both for the screaming gulls and the enterprising school kids. However, motorisation and its subsequent overfishing soon destroyed the breeding grounds in the bay.

These later boats sported a bigger cuddy with an engine fitted here. Many boats were built by Newmans of Hamworthy, and were fitted with a Brit single-cylinder 10 hp engine, or sometimes a Parsons, Barker or Kelvin engine. By 1936 there were only six boats left, and, by the following year, only one, the *Polly*, PE69, remained.

Another type of boat found in Poole is unique to the harbour: the flat-bottomed 'canoe' is a transom-sterned and hard-chined punt. At about 16 feet long, and with the modern ones having a box-like outboard motor well, they were rowed or motored around the harbour for a myriad of uses. They primarily collected cockles and winkles, and sometimes oysters, yet were just as likely to be out bait-collecting or shooting ducks. With their flat bottom they could easily be run onto the mud when collecting shellfish or to await a passing duck. A few early boats even had a small spritsail for sailing outside the harbour. Today, many canoes sit around the harbour,

The Poole fleet at anchor off the harbour, *c.* 1910. (*Ernest Bristowe's Photographic History of Poole*)

some of 20 feet in length. Viewing them critically, it is hard to imagine them sailing out into a breezy Channel chop, yet it certainly seems that many did.

Further east, Bournemouth was a tiny fishing village in the eighteenth century where fishermen and smugglers worked side by side. Nearby Branksome Chase was the same, although today it has been smothered by the huge mass of Bournemouth as we know it. The resort grew out of the building of one single grand house at the end of the eighteenth century, and within fifty years it was more profitable for the fisherman to offer trippers a row around the bay than it was to catch fish. The town's growth into a major resort contributed to Poole's development of the fishing to supply the needs of these tourists.

THE SOLENT SMACKS

Various fishing smacks worked out of Solent harbours, such as those mentioned from Fishbourne. In Chapter 13, mention is made of the Hamble smacks that travelled to Devon to collect crabs from Bigbury for Norway. Other smacks were based in Portsmouth and Cowes, there being seventeen registered in 1872, including many bought in from the east coast of England. They were of a typical design that could be found almost anywhere around the British Isles.

CHICHESTER HARBOUR

The natural harbour of Chichester had an abundance of fish, and fishing there dates back beyond the *Domesday Book*. Offshore of the harbour entrance huge shoals of mackerel were caught in the late summer, and the herrings then appeared from Michaelmas to Christmas. Shellfish – oysters mostly – were trawled in winter. In 1671 William Spriggs of Emsworth died in ownership of 'a hoy and 2 small boats with other fishing craft such as nets and draggs', and these were valued at £30. 'Draggs' is in fact one of the earliest mentions of trawling gear. These hoys, as at Cowes, were used for the offshore fishery.

The oyster trade flourished when massive oyster beds were discovered off Selsey in the 1820s. Immediately the local boats were joined by others from Colchester, France and Holland, who dredged and wholly denuded the beds.

Bosham and Emsworth each had its own fleet of smacks, and others were based at other harbourside locations such as Fishbourne (not to be confused with that of the Isle of Wight), Dell Quay, Birdham, Itchenor and West Wittering. These were 25–50 tons, and they worked away from home for the most part.

James D. Foster was a local timber merchant, boatbuilder and fisherman who built up a fleet of smacks between 1880 and 1902. His first boats were typical smacks of about 55 feet; *Evolution* was the first overall that he built himself in 1888. By 1901 he had ten smacks, the biggest of which was the *Echo*, P76, a steam auxiliary ketch of 52 tons and measuring 110 feet overall by 21 feet 6 inches beam and 9 feet draught. This innovative vessel was regarded as the ultimate in British fishing smacks, and was the result of years of experience at the oyster dredging. She had raking clipper bows and a long overhanging counter. She was constructed of 3-inch pitch pine on 6-inch oak frames. Another similar smack followed before the outbreak of war in 1914, but she was never finished and remained berthed at Emsworth until being destroyed in the 1970s.

Echo, like the other smacks, had a 'wet well' to keep the oysters alive until they were landed. This well held 90 tons of seawater, which was continually changed. The boat had a crew of eleven men.

In 1902 oysters supplied to a banquet of civic dignitaries caused the death of the Dean of Winchester. The pollution was traced to raw sewage at Emsworth and a total ban was imposed over the whole of Chichester Harbour. Foster, though, continued working out of Newhaven until the war in 1914, after which fishing recommenced and continued, up to 1939, although in a much-reduced state. He died in

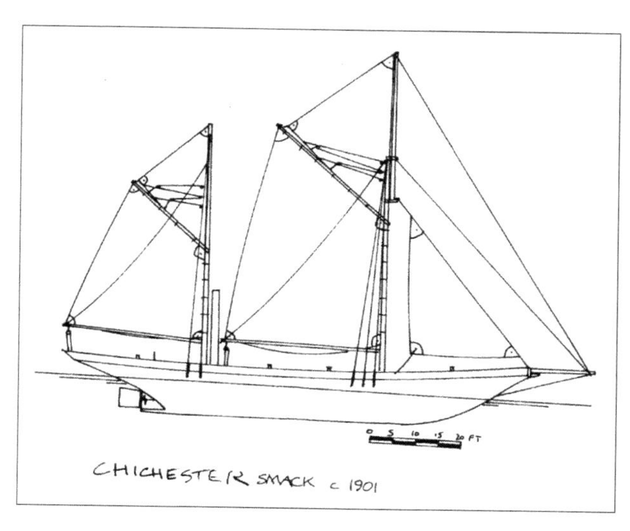

Fishing boats at Emsworth, *c.* 1905.

1940, and what was left of his fleet remained laid up at Emsworth. The *Echo* was finally towed across the harbour in the 1960s and unceremoniously, and disgustingly, set ablaze. Within years the council had cleared the remaining smacks on the foreshore so that by the late 1970s there was nothing at all left of a fishery so important to the local economy one hundred years before. As burning fishing boats is now accepted Government policy, the fact that the *Echo* – one of the most progressive and historically important of British fishing vessels – was torched in a similar manner is hardly surprising. It merely follows a pattern that, with hardly any wooden fishing boats left, will soon extinguish itself through the lack of boats to burn.

THE INSHORE FISHERY

Working alongside the offshore oyster dredging, the fishery inside the harbour was worked by a fleet of luggers. 'Jerkies', as Emsworth luggers were called, were 15-foot clinker-built boats with a square-headed dipping lug with four reef points to enable reefing while towing a trawl. They had an open foredeck, two rowing thwarts – the mast being stepped against the forward one – and a fish tray aft with space

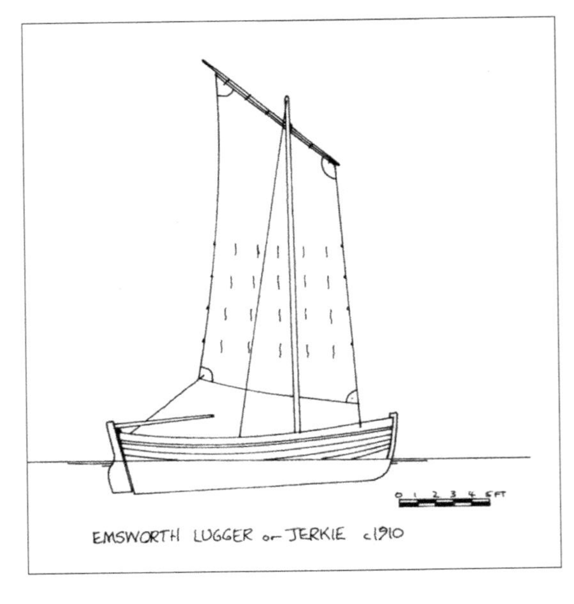

A typical Emsworth lugger, the *Mathilda*, *c.* 1920.

The beach at Selsey with small open craft for inshore fishing, particularly the herring.

underneath. These jerkies trawled for fish and dredged the oysters inside the harbour, and sometimes set drift-nets outside the harbour for mackerel and herring. Before 1898, the fishermen never set a sail, but after that date, sailing, especially when trawling and dredging, was normal. The last boat built was the *Matilda* by Feltham of Portsmouth in 1945, and by the early 1950s their use seems to have been phased out. A couple of similar seine boats still survive in Portsmouth.

Similar luggers were based at Bosham, these being called Bosham punts – pronounced 'Bozzum'. They, too, had a dipping lugsail, but some also had a centreboard to allow them to beat to windward.

Some of the Chichester luggers worked into Langstone harbour, usually to dredge for oysters. It seems that Langstone had few fishing boats actually based there, possibly because the harbour is approachable both from within Chichester Harbour as well as from the sea itself.

Although Selsey technically lies beyond Selsey Bill, the cut-off point for this chapter, the small inshore boats are worth a mention as they resemble the Bosham punts. They will be described in the next chapter. Today most are gone, although a few still sit upon the beach, and sometimes, in adverse weather, they can be found holed up in either West Wittering or West Itchenor, just inside Chichester Harbour itself.

CHAPTER 14

Sussex and South Kent Coast: Selsey Bill to North Foreland

Here, for the nonce, we take our stand,
Where Deal confronts the Goodwin Sands

—Unknown

The general coastline changes somewhat in its build up at about Selsey Bill. After the coves and cliffs, hidden bays and stark appearance of the western portion of the south coast, and the rivers, natural harbours and creeks of the Solent and Spithead, the coast becomes more uniform by way of a near-continuous shingle beach, in most parts backed by white cliffs. The main rivers are the Arun, Ouse and Rother, and each was navigable to its ancient port – Arundel, Lewes and Rye, the last still remaining so. The other eighteenth-century fishing stations were developed into resorts after the onset of popular sea-bathing in the latter half of the eighteenth century. Once the wealthy began to travel to the seaside, especially from London along this coast, bathing machines appeared so that the men could plunge naked into the waters; the women, more decorously, in their flowing robes. Bathing machines and tripper boats alike were operated by fishermen, as, prior to tourism, these folk were the only local inhabitants.

For example, in the mid-eighteenth century Eastbourne consisted of the four hamlets of Bourne, Southbourne, Meads and Sea Houses, the last having been a fishing station since the fourteenth century. Along the coast, Worthing was the fishing hamlet of Broadwater in 1750, and by 1801 had a full-time population of 100 and was becoming a popular watering hole.

At Dungeness the coast becomes low, but the beach continues to support small fleets of boats. The beach cliffs return around Hythe, and from here to the Foreland, communities have mostly stemmed from fishing. At Folkestone, fishing gave way to the Channel ferry, and the same occurred at Dover, while Deal survived with its famous boatmen serving the shipping of the infamous Goodwin Sands. At the tip of Kent, tourism flourished alongside small beach-based fleets.

It is hardly surprising, then, that similar boats are to be found along the coast, and they have the one common factor – that they all are able to work off the shingly beach.

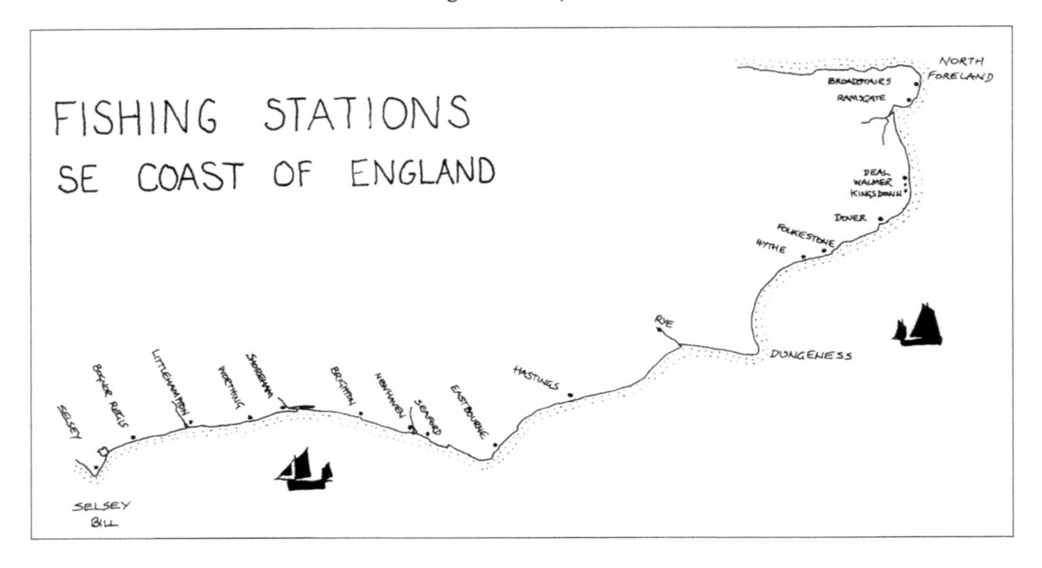

THE EARLY LUGGERS

Hastings has been associated with the Yarmouth herring fishery since the eleventh century, it being the most influential of the Cinque Ports. In 1619 Manship wrote that the men of the Cinque Ports were 'the principal fishermen of England' in Saxon times. Records tell us that during the twelfth century the herring fishermen were using boats similar to the Saxon and Northmen's (Viking) boats. This stemmed from the connection between the town and the Normandy coast where direct influence flowed into Britain a long time prior to the invasion of 1066.

These earliest of boats, then, were rigged with one square sail and were rowed by teams of oarsmen. They were open boats, yet they sailed into the North Sea to fish out of Yarmouth. These fishers were the force that created Yarmouth into the herring capital it became.

During the next few centuries the size of boat increased, so that they became single-masted seagoing vessels that were easily converted into warships. By the sixteenth century they had adopted two further masts, and these eventually developed into the herring busses. By the end of the eighteenth century, the transition to lug rig had occurred, and the first of the lug-rigged vessels as we know had arrived. That the Hastings men were some of the first to adopt this rig stems from the direct link between the French fishermen with their *Chasse-Marée* boats and the Sussex smugglers. These French craft were evolved from naval vessels, and it has been said that the French actually copied the design of their powerful luggers from English smuggling vessels. I tend to agree with the notion that the evolution came about simultaneously and was developed in both countries by the people concerned, i.e. the fishermen and the smugglers.

Washington produced a lines plan for one of these Hastings luggers in his 1849 report. He gave the principal dimensions of this three-masted lugger as: length 48 feet, keel 38 feet, beam 14 feet 8 inches and the displacement was 17.9 tons. However, the actual measurements taken off the plan indicate higher figures, with a length of 54 feet. As the majority of the luggers were nearer to the 40-foot mark around the 1830s, we must assume the former figures to be correct.

The rig consisted of three big lugs, one topsail and a jib. These were handled easily by the crew of eight men and one boy. The boats were powerful sailers, and this is attested by the fact that they fished during the winter in the Channel. After leaving Hastings in the New Year, they progressively moved south-west, being off Land's End in the early spring, catching mackerel. They continued mackerel fishing even after returning home, until they beached the boats for an annual overhaul. Nets were changed as well, before they 'fleeted off' to join the other huge fleets of luggers fishing the east coast herring. Sometimes they sailed right up to Scotland before finding the shoals; at other times the shoals were further south. They were usually lucky enough to get back to their families by Christmas time.

All these luggers were beach-built, many in Hastings being built by George Tutt, who had a yard near the beach. Another builder was Robert Kent, who began building in 1835 and who built one of the largest Hastings boats, the *Jane*. She measured 50.4 feet and was 25 tons. Cost of the luggers was said to be £250, but presumably the bigger ones were more expensive. There seem to have been no more than twenty-eight of these luggers working from the town's beach.

Although these Hastings luggers were the only ones identified by Washington, there is no doubt that more worked from other Sussex bases. Brighton had a few luggers that sailed to Plymouth for the mackerel fishery, and several worked out of Rye. Further east, Ramsgate and Folkestone had their own fleets, as we shall see.

Smaller lug-rigged boats, though, also worked from the beaches. They were flat-bottomed to enable them to sit upright on the beaches, and were bluff-bowed, like the earlier herring busses. This seems to have had the added effect of preventing their bows from digging into the shingle as they were hauled up the beach. They were entirely open boats, and had flat transom sterns. Many were fitted with leeboards that suggest some Dutch influence, and they either had two or three masts. These lugsails were extremely square-headed, and remained so well into the nineteenth century.

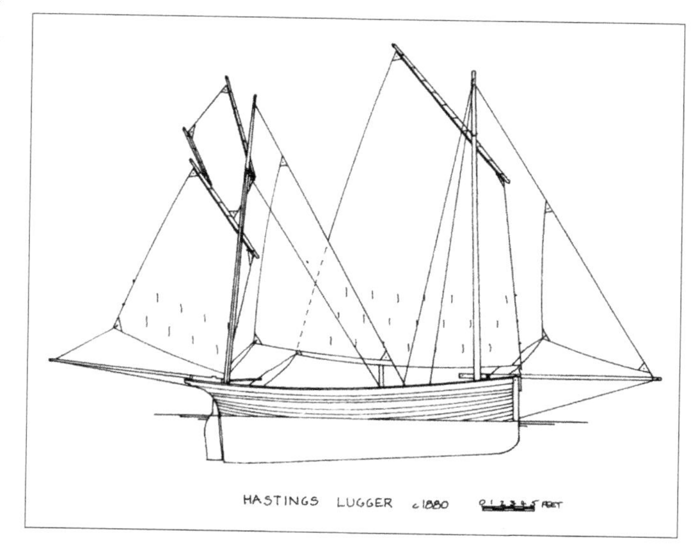

Sizes ranged between 20 and 30 feet, and not all the largest boats had three masts. The size depended on the fishing they were working at. The bigger boats drifted for mackerel and herring, the former from May to August, and

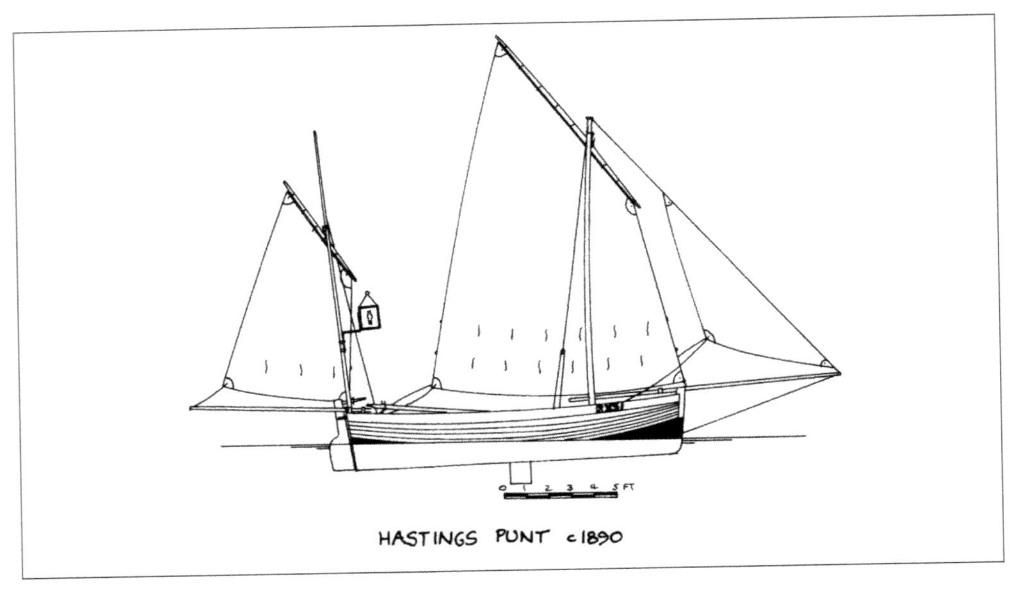

HASTINGS PUNT c1890

the latter from Michaelmas to Christmas (October to December). In between times they would trawl for plaice, sole, turbot and dabs. The smaller boats would drift for sprats or trawl for shrimps. Cod was also caught by longline.

These smaller boats adopted a sprit-rigged mizzen sail, and some single-masted boats just had a single spritsail. The sprit was a popular sail for small working boats, yet was seldom set on larger boats.

BRIGHTON

Further along the coast, however, a distinctive larger type of spritsailed craft did evolve in the eighteenth century. Brighthelmstone, as it was called then, had, around the beginning of the seventeenth century, dispatched up to thirty boats to the North Sea to fish for 'linge, codd or herrings'. The fishery then declined, for a while, before improving by the end of the century when there were upwards of fifty boats working there. During the eighteenth century the wars interrupted fishing, many boats being captured by the French and Dutch, and only the advent of tourism invigorated an ailing trade during the latter half of the century.

The Brighton 'Hoggie' or 'Hog-boat' emerged as the common boat, there being some seventy such craft around the close of the century. Rigged with spritsails, the hoggies were unique to the town, although some were said to have been based further along the coast at Shoreham. This is contested, and whether they were in fact Brighton boats based there or Shoreham-built boats is not yet clear.

These hoggies were fully decked, unlike many other fishing boats of that era and were up to about 14 tons around 1789, and up to 35 feet overall. They had a forecastle for living quarters, although none seem to have had a stove. They were extremely beamy, having an extraordinary length/beam ratio of 7:4 – reminiscent of

Brighton beach, with the King's Road Arches, and with typical beach boats, *c.* 1890. (*Brighton Fishing Museum*)

Dutch beach boats at Katwijk. They were designed to be hauled up the beach, and had a full bilge and bilge keels to enable them to remain upright when out of the water. They had bluff bows and a small transom-stern, giving them that venerable look of old, and were built in good clinker fashion. One builder of the time was James May & Co. of Brighton. In the nineteenth century he moved to Shoreham where he joined forces with another local builder to become May & Thwaites. Almost equidistant but to the east, Charles Geer had a shipyard at Newhaven at the same time.

The bigger boats set two spritsails, while the smaller, inshore boats (8 tons and 23 feet) had only one. They both set a large foresail, and occasionally a jib was carried, but rarely a topsail. Because of the full shape of the hull, excessive leeway necessitated the use of

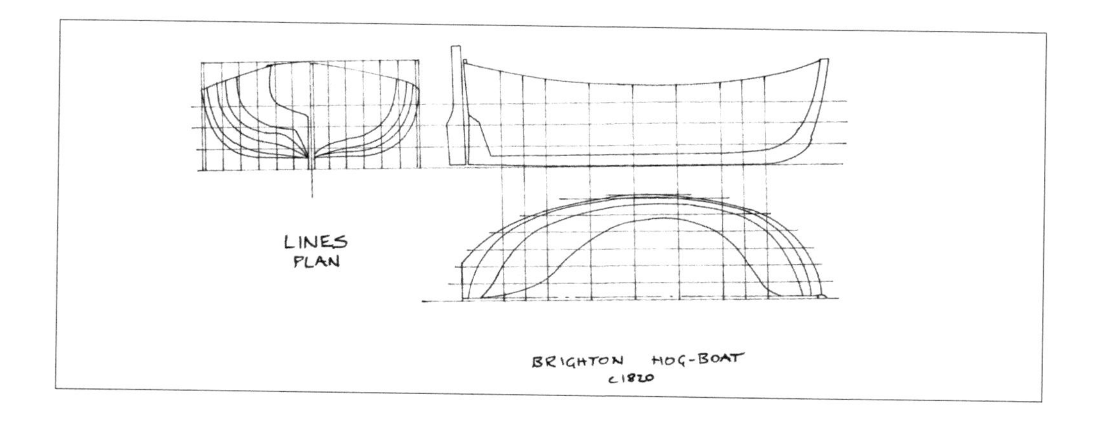

A lugger near the Blockade Station, Brighton, 1830. (*Drawing by Edward Cooke*)

leeboards. Some hoggies were fitted with sweeps so that they could be rowed, especially when coming into the beach, thus giving the crew powerful control over the boat.

The hoggies were regarded as good seagoing vessels and many worked out in the Channel, drift-netting for mackerel. Others trawled close inshore. Before the mid-century, another type of boat was gaining popularity among the mackerel fishermen – that of the lugger. To compete, the hoggies were adapted, some being suited with lugsails, others having their arrangement below decks altered. The cuddy was enlarged to include a stove, and the holds were increased in size. Unfortunately, the resultant hoggies were still not deemed efficient enough in comparison to the luggers, so that by the 1880s only two remained. As we have seen so many times before, the fate of the

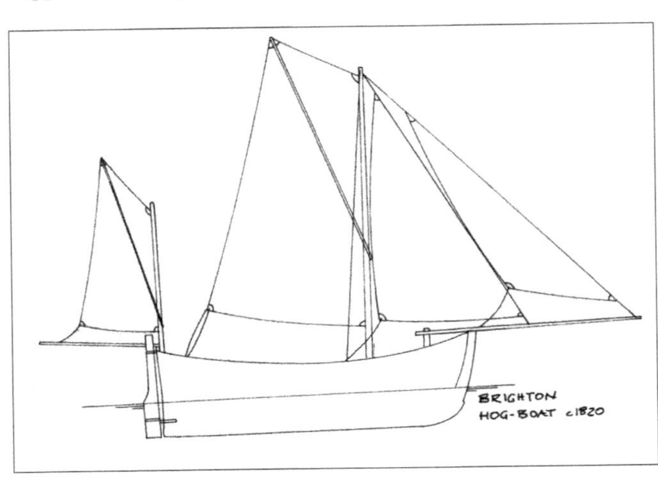

last remaining hoggy was to be torched – as part of the Guy Fawkes Night celebrations on Brighton beach.

The lugger had taken over from the hoggy by mid-century, these having about 26 feet of keel, 12 feet beam and being clinker-built. Having little overhangs at either end, the length overall was only a couple of feet or so more than the keel length. As well as having upright bows, these transom-sterned luggers adopted the concavely overhanging counter stern, or lute stern, as was typical at Hastings. This had the effect of increasing the overall length to around the 30-foot mark. They were all two-masted, the three-masted luggers having done away with the foremast all around the south and east coast by this time.

These luggers were as full-bodied as the hoggies, and still retained the leeboards because of their undue leeway, yet these were not used traditionally while drifting for mackerel. Beneath the waterline, they had huge bilge keels that enabled them to be dragged up the beach, much in the same way as the hoggies were, the keels reducing the scraping of the pull.

The earliest of the Brighton luggers did possibly fit three masts, but if so, this was short-lived. The two-masted luggers were much easier to handle, and the absence of a foremast created much more deck space for working. They all set a foresail on a bowsprit. As the luggers gained in popularity, they increased in size, so that 20-ton boats were joining the fleet and these were crewed by up to seven men. Smaller 6-ton luggers fished the herring and these were only crewed by two men.

Centreboards were fitted to some luggers at the end of the nineteenth century, at first slotted through the keel, and later through the garboard strake on one side to

Lugger on the beach at Brighton, 1830. (*Drawing by Edward Cooke*)

prevent weakening of the keel. But, as motor power developed soon after, these were short-lived, the space being taken up by the engine. At the same time, the elliptical stern was introduced from Hastings, and later vessels adopted this. By the time of the Second World War, Brighton's fishing had mostly faded into obscurity, the town's place among the foremost of British resorts dwarfing the fishery. Yet some boats did survive, and these can increasingly be seen on the beach, by the King's Arches, where the Brighton Fishing Museum is to be found. As the collection grows, so does popular awareness of Brighton's fishing heritage, something that, until now, has remained much in the dark. Which is a little odd, considering that Brighton's fishing past, after all, ranks high in importance in comparison to many other ports of Britain.

WORTHING

As previously mentioned, Worthing was a fishing hamlet until its rise to popularity occurred in the early eighteenth century. Here was the home of another fleet of luggers, similar to the Brighton luggers, and, like them registered at Shoreham – using the letters 'SM'. These luggers sailed down Channel chasing the mackerel and herring shoals, and up as far as the North Sea for the herring there. These fully decked luggers were up to 50 feet overall, and were built locally.

Towards the start of the second half of the nineteenth century John Belton set up business as a boatbuilder, and he built many of these luggers as well as smaller craft. His business continued up to the First World War.

Brighton boats, *c.* 1930.

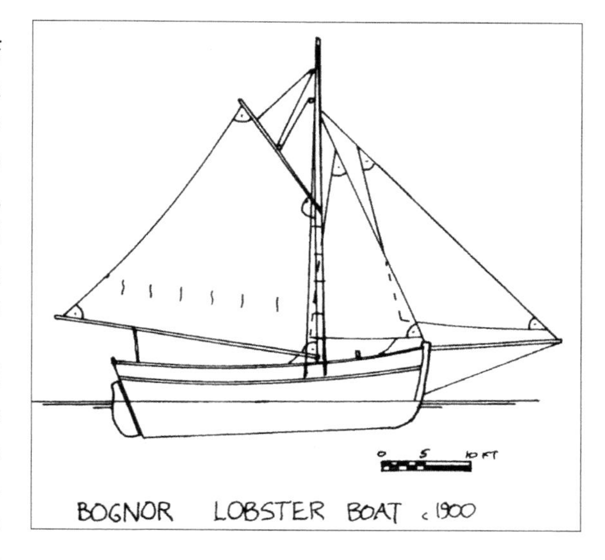

Fishing continued in the latter half of the nineteenth century, although somewhat subdued compared to earlier years. The town's growth as a resort gave employment ashore, and those still going to sea found work during the summer pleasing the trippers by taking them on a trip around the bay.

Further west, Littlehampton had one of the few natural harbours along this coast. Few fishing boats worked from the harbour, but a nineteenth-century photograph shows small cutter-rigged transom-sterned vessels of about 20 feet. This was possibly built by the local boatbuilder 'Old Stowe', as he was called. The craft resemble the Solent types. Another photo shows a typical Sussex beach punt with lute stern. The onset of tourism around 1760 and the arrival of the railway and the Cross-Channel packet boats some sixty years later brought different work for the Littlehampton men. Fishing took second place.

Offshore of Littlehampton lay a good lobster fishery, and both Selsey and Bognor men joined the Littlehampton fleet catching these lobsters. One Littlehampton lobster boat was the *Gwendolene*, LI49, built by old Granby Hopkins in 1893. This boat was 24.5 feet overall with a 23-foot keel and 8-foot 4-inch beam. She was cutter rigged with a 37-foot mast. She set a gaff main, topsail, foresail and jib. Oysters were also landed from the prolific beds offshore, and mackerel in season. Smuggling, however, seems to have been the mainstay of the economy, many Littlehampton luggers bringing barrels of brandy from across the Channel.

A typical Selsey fishing boat. These were all registered at Littlehampton, and spent most of their time lobstering and oystering.

Similar long-boomers were found in Bognor for the lobster fishery. In Selsey, other gaff-rigged smacks, similar to the Chichester harbour smacks, worked off the beach. Smaller beach punts were built by either Lowers of Newhaven or the Feltham brothers, Harry and George, of Portsmouth. Like elsewhere, these boats had engines, usually Thornycroft 'Handy Billies', and were gaff rigged with dark-tan sails. One or two still remain on the beach. Lowers continued building fishing boats up to their closure in 1974, the last boat being *Nguyen van Troi*, now in the Brighton Fishing Museum.

EASTBOURNE

Returning east by way of Brighton, the coast beyond has several beach landings with small fleets. Edward Cooke, etching in the early nineteenth century, shows various beach punts, all lug-rigged and clinker-built, lying on the shores. A few luggers worked out of Newhaven, but the ferry terminal overshadowed any fishing. Eastbourne was renowned for its fleet of luggers, which travelled far and wide in search of the mackerel and herring. These two-masted luggers were 35–40 feet long, were sharper than other Sussex boats, and they sailed as far as the Irish and East Scottish coasts, perhaps one reason that they adopted similar hull shapes. Other luggers at Eastbourne were two-masted craft called 'Shinamen', some 28 feet long on 22 feet of keel. They were smaller versions of the bigger luggers, and set a dipping main, a standing lug mizzen sheeted to a bumpkin and a large foresail on a long bowsprit, and they had a centreboard. All the boats had to be beached as Eastbourne had no harbour.

Many were built locally by George Gausden, who joined up with another builder, the company becoming Gausden & Sisko. Boats were elm-planked on oak frames, and they had rounded counter-sterns. Later boats had elliptical sterns.

Smaller inshore punts were built for the lobster and oyster fishery, but they sometimes spratted and longlined. They were rigged with single lugsails as were the other Sussex beach punts.

HASTINGS

By the mid-nineteenth century, all the Hastings luggers had discarded the foremast. Fishing had declined somewhat in the first part of the century, but the coming of the railway in 1851 brought with it a sense of revival, and this led to the rise of trawling.

To participate in this fishery, a new type of Hastings boat developed, specifically suited to trawling. They were called '28 boats' – or bogs – because of their dimensions: 28–30 feet overall and 26–28 feet on the keel. Rather than evolving from the original luggers, it appears that they were influenced by the smaller punts. These punts were approximately 15 feet overall, and had tiny foredecks with storage below. They had single lugsails, and were used inshore for spratting, long-lining – or hooking as it was called – and lobster fishing.

Although called trawlers, these '28 boats' did drift for herring at certain times of the year, and, too, sailed as far as Yarmouth for the autumn herring. They were again clinker-built, fairly beamy and full so that they could be easily brought ashore.

A word about the lute stern. This innovation made these Sussex beach boats unique among British fishing boat types. The lute was an extension of the transom stern, but it reacted smoothly in the face of oncoming waves when being launched or beached. The lute itself was designed to deflect water away from the hull, ensuring that the boat was not pounded in the surf. In 1892, the elliptical stern appeared on the *Clupidae*, RX126, following several successful designs on yachts. While the lute was a concave extension, the elliptical stern was a convex continuation of the planking and was thought to give the boat more lift when on the water's edge. Today, as one walks along the beach, almost all the boats still have these elliptical sterns; the lute became unpopular because the sea tended to flood through the gap around the rudder.

The last Hastings lugger to be built was the *Enterprise* in 1912. This boat was 27 feet 9 inches on the keel, 11 feet 8 inches beam and 4 feet 6 inches deep. Her other claim to fame is for landing one of the largest catches of herring: twelve lasts of fish in one catch (28 tons or 160,000 herring). She survived until being laid up in 1954 and is now in the Hastings Fishermen's Museum.

These boats, like many others around the coasts of Britain, had basic accommodation in the forepeak ('foreroom'). Two bunks and two sleeping shelves on top, a stove against the net room ('chay') bulkhead, and little space for stowing food and clothing, made life aboard just as difficult as it was for any other fishermen. Although mostly clinker-built, a carvel boat, the *Swan*, was built in 1916. Those built after the *Enterprise* were not really of the traditional type as they had engines and were fuller in shape. Even though they retained the rig which was often used, they were regarded as halfway boats. They filled the gap between the last of the luggers and the modern motor boats. One such vessel, the *Edward and Mary*, built in 1919, now stands outside the Fishermen's Museum.

The Eastbourne lugger *Our Lassie*, NN120, *c.* 1890.

Although the sailing luggers were superseded by motor boats, a visit to the beach today still evokes the memory of this era. If one discounts all the aerials and modern fishing gear and propellers, the boats appear as they might have done

A typical Hastings lugger on the beach *c.* 1890.

when George Woods was photographing the beach. With the renowned fishermen's net stores as the backdrop, the sight and smell of these boats must surely awaken emotions in all of us. Today few beaches remain where one can still find such a scene, and that it has survived must be partly due to the foresight of the fishermen of old. Yet, I cannot help recalling the Hastings lugger *Industry*, RX94, built in 1870, redundant by the end of the Second World War. She joined the growing list of fishing boats that ended their days on the beaches in flames, and she, too, as part of the 5 November celebrations. As today's Government policy takes its toll on our fleets, I can't help but think that Guy Fawkes himself, that loather of Westminster who was prepared to see it all disappear in a puff of smoke, would be shocked to see these boats burned in his memory in this way. Some call it the 'Westminster Revenge'!

Seated by the *Dove*, 232RX, on Hastings beach. (*George Woods Collection*)

Launching the *Mary*, 52RX, at Hastings by pushing her into the sea while the rudder is raised. (*George Woods Collection*)

Two luggers – the elliptical-sterned *Our Pam and Peter* and the lute-sterned *Bloodaxe* on the beach at Hastings in 1995.

RYE

Although not a founder member of the Cinque Ports confederacy, Rye had become a member by the thirteenth century, and accordingly gained certain privileges, among them being the right to land, dry nets and sell fish at Yarmouth. This obviously encouraged maritime activities, especially those of a fishing nature, over in East Anglia. Herring was the dominant fishery, with other fish gaining popularity. These included plaice, whiting, sole and conger, Rye Bay being especially rich in these species. Yet Rye seems never to have excelled at fishing, much of the fish there being caught by fishers

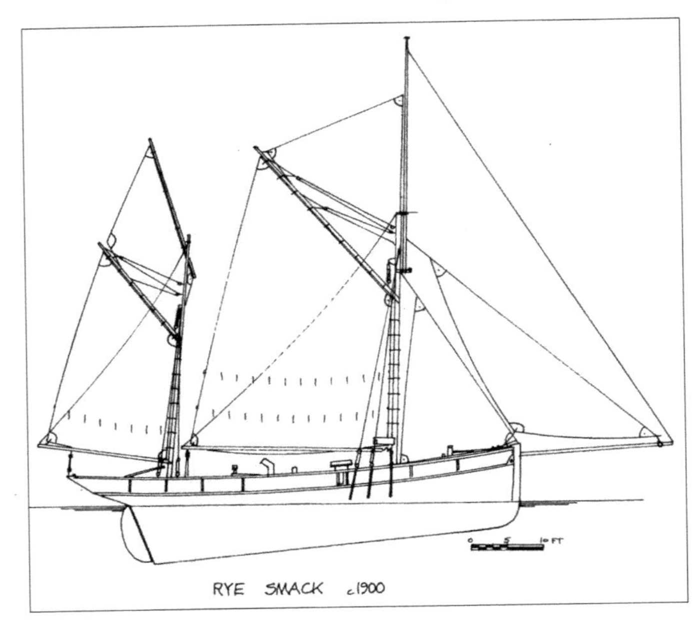

from away. Perhaps due to nearby Hastings, or its geographical position a few miles upriver, the fishing industry never developed to anything like the extent that it did with its neighbours (centuries earlier Rye lay on the coastline).

Having said that, Rye became renowned for its smacks, which it supplied to both Lowestoft and Ramsgate. The boats that were kept at Rye were trawlers, twenty-seven being reported there in 1878. One of these is the *Keewaydin*, built by Geo & Thos Smith of Rye in 1913, which fished out of Lowestoft as LT1192 until 1937 before being sold to Sweden. Today she is back in British waters under the ownership of Paul Welch after he found her languishing in Malta in the mid-1990s and is working as a charter boat out of Cornwall.

Rye was the port of registry of Hastings, a bone of contention among the Hastings fishermen, so much so that even today the boats have painted on their sterns as homeport 'Hastings, port of Rye'. Rye used to use the letters 'RE', but this was changed to 'RX' to avoid confusion with Ramsgate – 'RX' representing the first and last letters of Rye, Sussex.

The Rye smacks were about 50 feet overall, and were full bodied. They were mostly built either by the Hoad Brothers or G&T Smith. The last to be built there, in 1896, was the *Three Brothers*, RX153, which was typical of the smack of that period. Her lines were taken off by P. J. Oke in 1936, but her whereabouts after about 1946 are unclear.

THE KENT COAST

Beach boats still work from the shore at Dungeness. People first settled at this barren coast because of smuggling, it being one of the last smuggling haunts in Britain. Fishing then became the main occupation, mainly from small beach boats. As markets were so far away, fish caught was only for local consumption and hence there was never any 'fishery' as such. Boats were small, open, much like the Hastings punts. Today, the decked boats resemble the Sussex beach boats that are omnipresent along this coast, and especially numerous at Hastings.

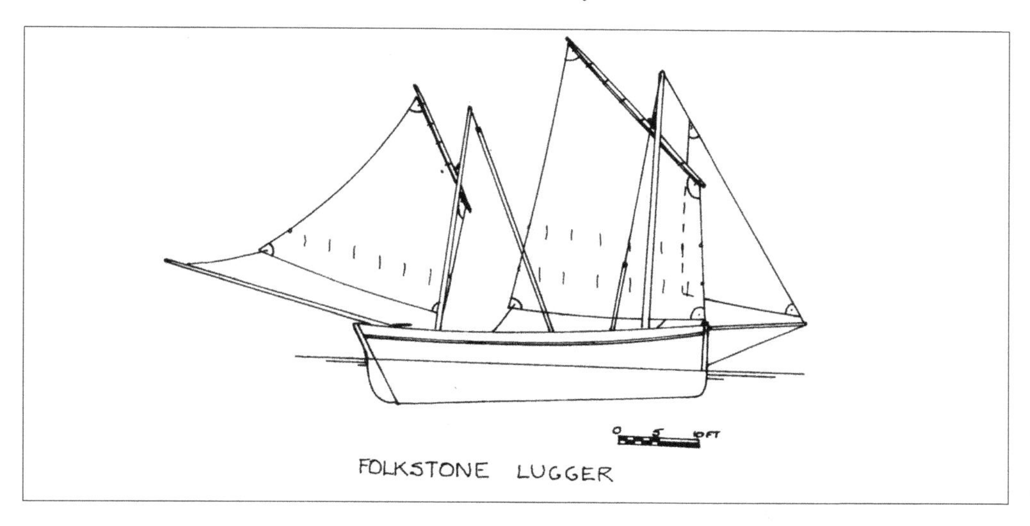

Hythe is an original Cinque Port yet little is known about its medieval fisheries. Typical foresail and mizzen sprit punts were worked during the nineteenth century here, being typical of the beach boats of this coast. The mainsail was dropped to create a bit more space aboard the small 12- to 16-foot vessels. Several decked boats now work off Hythe beach.

Folkestone had its own fleet of luggers, many of which were built in East Cornwall. They were rigged with a large dipping main, a small foresail and a standing lug mizzen sheeted to a bumpkin. They were moored in the harbour on legs to keep them upright at low water. Because of the tide, and the growing cross-Channel trade, and perhaps also the increase in tourism, fishing never developed comparably with Hastings. Today only a few boats remain working from there.

Two Folkestone boats that do remain are *Three Brothers*, FE93, and the *Happy Return*, FE5. The latter was built by Kitto of Porthleven in 1904. She was saved from the chainsaw by the Mounts Bay Lugger Association and has since been restored back to how she was when she was a working lugger (with modern navigation and safety gear).

In the early nineteenth century up to fifty fishing boats worked out of Dover, but the boats were owned by Torbay men. The harbour, though, was not favoured, mainly because of the enormous flow of cross-Channel ferry traffic that disrupted the harbour from a fisherman's point of view. Also the strong tides outside and the poor facilities inside the harbour kept the fleets away.

In the latter years of the last century some smacks still remained based there, but most of these, too, seem to have come from away.

Dover did, however, have its own fleet of small luggers. These were small open punts, mostly built at Deal. They were used to fish the rich grounds of St Margaret's Bay in the closing decades of the last century. Here they drifted for mackerel, herring and sprats, and at other times went lobster and crab potting. An example of one of these Dover luggers is the *Argonaut*, DR56, built by Nicholas of North Deal in 1880 and at 14 feet 7 inches she was a very small vessel. These punts were all clinker-built

The old harbour at Folkestone with a typical lugger entering.

Folkestone Harbour at low water, 1831. (*Drawing by Edward Cooke*)

Luggers leaving Folkestone. The majority of these boats were built in Cornwall although some resembled the Hastings boats. (*National Maritime Museum*)

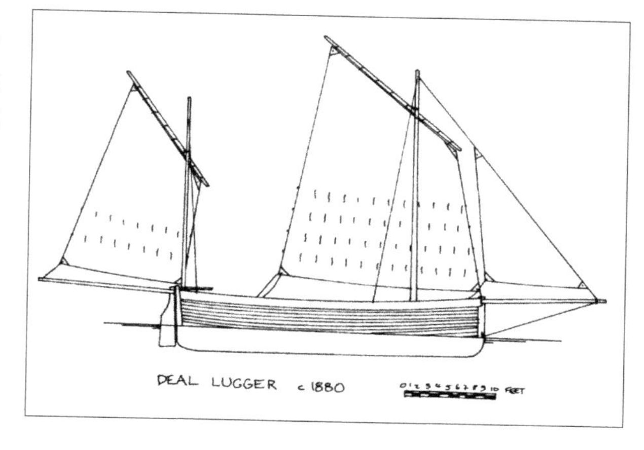

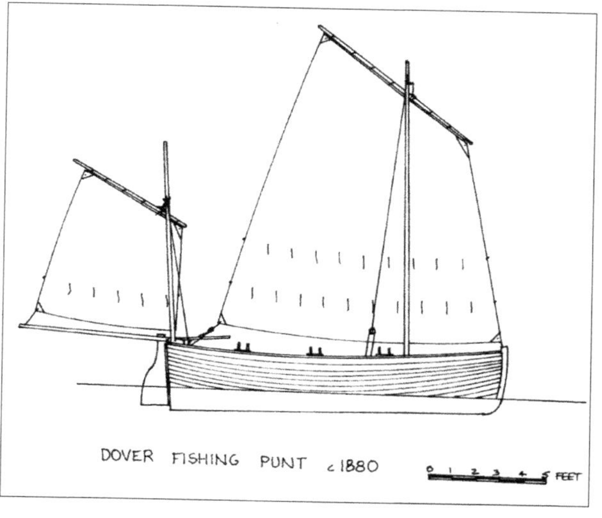

and had transom sterns. They were rigged with a dipping lug main and a small standing lug mizzen, and were often rowed.

Washington also shows a Deal lugger in his report. Deal was renowned for its boatmen, who made a living from the busy shipping lanes that lay off the Goodwin Sands, a treacherous bit of sea that was always catching sailing ships out on its ever-shifting sands. To the south lay routes up and down the Channel, while to the north lay the Thames and its ships, which were constantly sailing to and from the capital. The Deal men acted as salvage boats, supply vessels and pilotage boats all in one. Again their craft were clinker-built, up to 40 feet overall and very fast. The smaller ones were called 'cats'. As elsewhere, the Deal men dispensed with the foremast to produce a two-masted lugger, setting a dipping lug main and standing lug mizzen with a jib set on a bowsprit. They had a small forecastle with berths for six and a stove so that they could sail up and down Channel for periods of a week or more. They were often encountered off the Lizard looking for a vessel in need of pilotage up-Channel. When all else failed, they went fishing, it was said. Yet many joined the herring fishing every year, a mention of this being made as early as 1810.

Walmer and Kingsdown lie to the south within a couple of miles of Deal and both were home to fishing luggers. These boats were again often built at Deal, mostly by generations of the Hayward family, who had a small boatbuilding shop in South Street, and the sails were, too, made in the town. These particular luggers fished for mackerel off Hastings and Dungeness, and drifted for herring during the season. They were anchored off the beach during the summer, and were hauled up the shore when the weather was threatening.

Although these luggers were primarily concerned with salvaging, it seems that after about 1890 they concerned themselves more seriously with fishing due to a decline in hovelling. 1892 was said to have been an exceptionally good year for the fishing – mostly for sprats and herring – and the local curing stations flourished accordingly.

However, the fishermen themselves then became victims of the Goodwin Sands, as many of their small craft were run down by screw steamers. One such vessel was the *Fawn*, a second-class lugger belonging to North Deal, and renowned as one of the fastest. On 4 August 1864, with four hands aboard and riding her nets, she was run down by the steamer *Biddjed* on passage from Bordeaux. It seems that the steamer's crew mistook the light and rode straight over her. All the crew were drowned, leaving three widows and eleven children unprovided for.

The Ramsgate fishing fleet was one of the finest in Britain at one time. In Elizabethan times it is said that there were fourteen boats working from the town, employing seventy men, although there were only twenty-five inhabited houses there! The harbour was built in the eighteenth century, and later fishermen worked in three-masted luggers similar to the Deal luggers, these generally drifting for mackerel and herring.

A typical Deal lugger lying on the beach, *c.* 1890. (*National Maritime Museum*)

The harbour at Ramsgate. At one time this harbour was home to one of the best trawler fleets in Britain.

When trawling commenced in earnest at the beginning of the nineteenth century, the Ramsgate fishermen soon bought in Brixham smacks and proceeded to trawl. Many Devon fishermen based themselves in the small harbour as they found a ready market for their catches and the harbour facilities were good and cost them nothing. When harbour dues were levied in 1862, some of these boats returned to Devon, and others went north. However, the lull was short-lived as Ramsgate reached its zenith around 1890, when there were some 185 smacks working from the harbour.

Up to about 1870, all the smacks were imported, but after that date, local builders began building them. The region had huge supplies of oak, and the town had various shipyards. One of those to concentrate on smacks during those latter years of the nineteenth century was the Moses family, whose last smack was the *New Clipper* in 1913.

A smack in 1880 cost about £700, and these were mostly dandy-rigged. Another £100 would supply a steam capstan. Later ketch-rigged smacks cost another £100 or so and were upwards of 50 tons. Looking through the register of smacks up to 1890, it seems nearly all were built at places such as Galmpton, Rye, Porthleven and Lowestoft.

Other than the smacks, Ramsgate had its own fleet of toshers, smacks under 25 tons. These were built after the 1894 Merchant Shipping Act that made it illegal for any trawler over 25 tons to go to sea without a certificated skipper and mate.

Another fleet in the harbour consisted of lug-rigged punts, much like the Deal punts. Although they were all registered at Ramsgate, many were based at Broadstairs, a few miles to the north. These small punts were used mainly to drift for herring and sprats in the autumn and early winter, lobsters and crabs infilling the rest of the season. The punts were rigged in the normal manner with a dipping lug main and standing lug mizzen sheeted to a long bumpkin. During the summer, when they often as not were taking trippers out, they adopted a loose-footed gaff main with a single brail rope, and retained the mizzen lug. A jib and foresail were set upon a bowsprit. These boats were up to about 20 feet overall, and were to be found as far away as Margate, around the Foreland.

Broadstairs also had its own fleet of larger luggers in the early part of the nineteenth century, but these had mostly disappeared by about the 1870s. The only other type of craft found in the tiny harbour was the wherry, and we shall discuss that a little in the next chapter.

Today few boats work from this coast, but I did manage to photograph a small foresail and mizzen punt in 1996. The *Girl Julia*, R1, looked smart with her rig of red sails, although the mizzen appeared to be a gunter rig sail. Unfortunately, I could not get close to the boat as she was sailing offshore on a beautiful summery afternoon, but surely a fleeting glance is enough to evoke visions of the fleets that once worked off these shores.

CHAPTER 15

The Thames Estuary: North Foreland to the River Stour

Large Silver Eeles, a Groat a Pound, Live Eeles. Who's for an Eele Pye?
—Costermongers' and hawkers' London cry

And so around the 'Bloody Foreland', where the mighty and regal Thames flows, bringing with it the charm of the city and the muddiness of England. Cast into its coast are numerous rivers – the Swale, Medway, Crouch, Blackwater and Colne to name the biggest ones. Each river has its own creeks and these hide in the low coastline, where the tide flows over muddy flats that stretch for miles at the ebb, and offshore lie continually shifting sandbanks that are just waiting to catch any unwary boat.

Each creek was home to a few smacks, for this is smack territory. Here the cutter rig found eminence among the fisher communities. It is easy to imagine that these men invented the cutter rig, although they didn't, and it is equally hard to envisage the days when square-rigged vessels sailed up and down these channels. The sprit-rig, whose appearance in the fifteenth century among Dutch vessels seems certain, led in time to the development of the gaff. The sprit was handier in comparison to the square sail, which had the additional disadvantage of no windward ability. In confined waters, such as those of the Low Countries, manoeuvrability was all important, so the seaman sought ways of improving the sail. Bowlines (vargoods) were attached to the luff of the square sail to keep the leading edge tight. Eventually the sprit emerged as the final development, with the luff becoming laced to the mast, and the yard becoming elongated as the sprit. The sprit had its heel against the mast almost at deck level and ran diagonally across the sail with its top end supporting the peak of the sail.

Then came the original, or standing, gaff which was in fact the same sail, but one where the sprit was shortened and moved progressively up the mast until it was attached along the head of the sail, and supported by halyards at the peak and throat. At first the gaff itself was short, but soon grew in length to heighten the peak of the sail.

In the confined waters of the Thames and the Essex backwaters, the fishermen and bargemen needed powerful rigs that were easily handled. The gaff rig, with increased boom length, was seen for its advantages, and quickly adapted to suit local

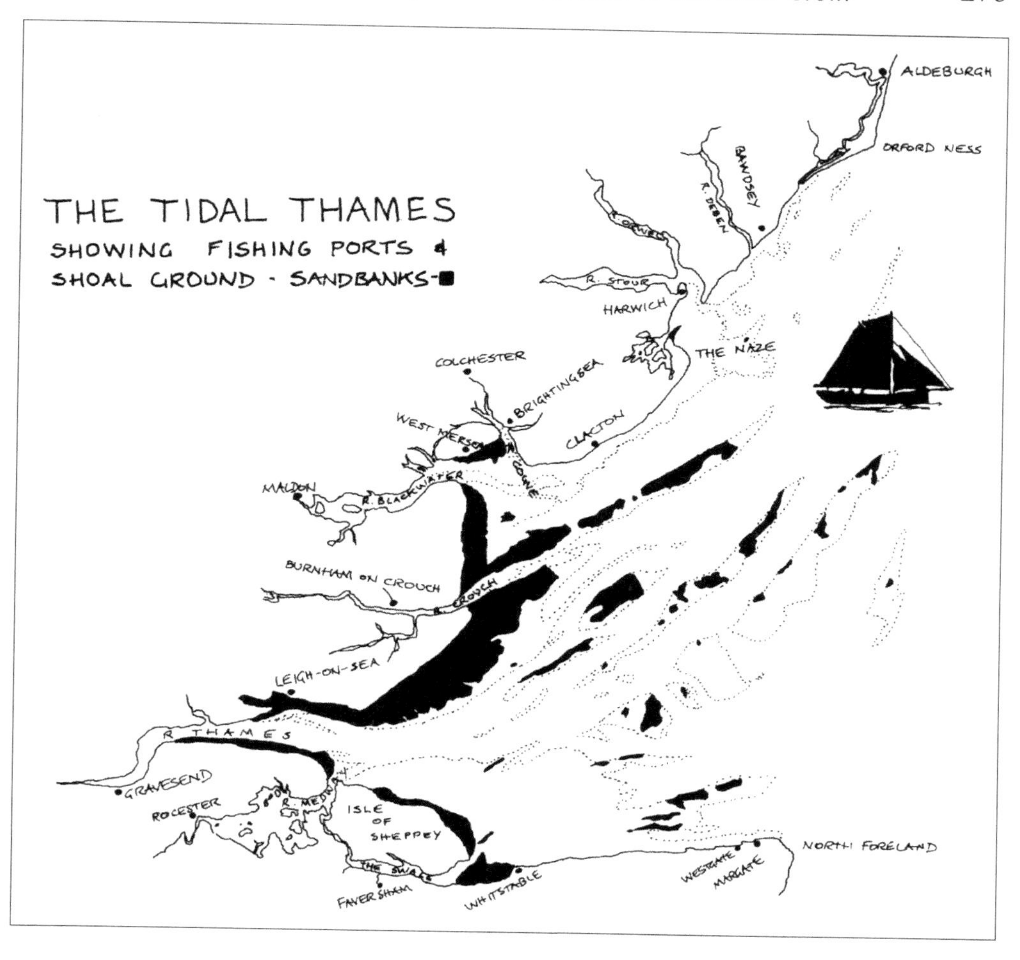

Peter boats at Greenwich, 1828. (*Drawing by Edward Cooke*)

conditions. Towing a trawl needed even more power, and the ever-changing channels and swatchways needed the handiness. Gaff rig, as we know it, had arrived.

This may be an over-simplification of the evolution of the fore- and aft-rig, but space is short and many good books have been written on the subject. Likewise, the same good books tell of the Essex fishing boats, and, not wishing to be repetitive, it is the intention here only to show these vessels in the context of the general picture to ensure continuity in this purview.

THE PETER BOATS

The Peter boats *were* the original fishing boats of the Thames – some call them Cockney boats. The name appears to have come from St Peter himself, that patron saint of fishermen. The story goes something like this: two cathedrals were erected by the East Saxons, one at St Paul's and the other, St Peter's, on the island of Thomey which at that time was just off what is now Westminster. Small boats were needed to convey the congregation over, and these took the obvious name, and in time they were used for fishing. Sounds feasible!

The boundary between Danelaw, the territory of the Danes after 878, and that under Saxon control lay in a rough line between Chester and London, and passed through the capital. Viking incursions, then, must have influenced the fishermen of the Thames, the latter not necessarily adhering to a definite frontier, so that the Peter boats evolved in true Viking fashion, although their construction does show certain similarities to that of the Saxon Graveney boat. Of course, both have many similar characteristics.

Actual documented evidence of the Peter boats does not come until the sixteenth century. Prior to this, it must be assumed that the boats were primarily small rowing boats that fished as far up river as Windsor and as far down to Gravesend and around. These craft were 12 feet, clinker-built and were extremely handy when working the tidal waters where double-enders were easily turned under oars. As the boats grew in size and ventured further downstream, a spritsail was fitted – perhaps earlier a square sail. They were built with a wooden well amidships that was flooded by way of small holes drilled in the planking. They were, in fact, probably the first type of fishing boats in Europe to have a flooded well to keep the fish fresh until being landed. This well divided the boat into two compartments so that the oarsman sat in the forward section and the fisherman in the stern sheets on a platform that enabled him to stand higher up when the net was being shot.

As their name implies, they worked mainly with peter-nets – small seine-nets – and at other times with trawl- and whitebait-nets. Some even set small stow-nets for catching sprats.

Similar double-ended boats were found along the coasts of Kent and Essex, where, in time, larger Peter boats sailed for the fishing there. Boats were then up to 25 feet, and some even 30 feet in the early nineteenth century at Leigh, that ancient of fishing stations on the south Essex coast where shrimps began to be landed around 1830.

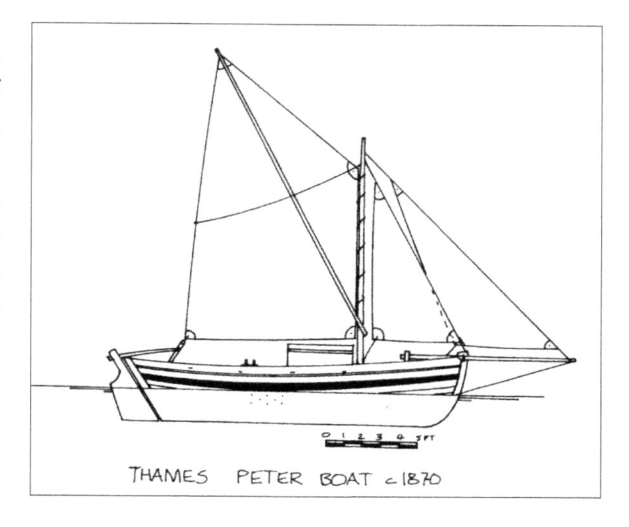

THAMES PETER BOAT c 1870

These boats were locally called 'pinks' or 'pinkers' because of the pointed stern. This was to distinguish them from some transom-sterned Peter boats that were already being built at the village. A transom stern gave increased stability and more deck space, the pointed stern not being necessary once the rig had overtaken the oars.

Some of these larger boats worked away from home for a few days at a time, and so the fishermen had to sleep aboard. For this they hoisted a canvas cover – or 'tilt' – on three semi-circular hoops over the boom with the mast lowered to create a shelter. Some of these shelters even had a stove whose chimney was poked out from under the canvas.

In 1820 a Peter boat was built at Strood on the Medway. This 27-foot 6-inch-long boat had a boomless cutter rig with a topsail, and represented the first of a new type of boat to surface – the bawley.

By the 1840s, the Peter boats had a small forecastle, but this development came too late, as within a few years the new bawley was gaining favour within the shrimping fishery. Use of the Peter boats petered out, especially away from the upper Thames!

By 1890, when pollution on this part of the Thames had severely damaged the river fishery, the remaining smaller Peter boats had on the whole disappeared. The last, it seems, were smelt fishing, at Chiswick at this time, and by the turn of the century only one or two remained.

Consequently none survived far into the twentieth century, and the Peter boat remains one of the lesser-known secrets of British fishing boat types, and perhaps one of the oldest.

THE THAMES HATCH BOATS

Before we continue with the new breed of bawleys, mention of the other older types of Thames boats is worthwhile. The Thames Hatch boats were double-ended, clinker-built and sprit-rigged. They were also fully decked and had a small cabin together with a long, narrow cockpit. Similarly, they also had a wet well. Later boats had a topsail and jib set on a bowsprit. It is supposed that they developed from the Thames wherries, and not the Peter boats, as they had flared bows like those of the wherries. The Thames was teeming with these Hatch boats in the early nineteenth century, there being 635 in 1854; yet nine years later numbers had fallen to only ninety-seven boats, due to the new type that was quick to catch on. Similarly, open wherries were extensively

Thames Hatch boats off Gravesend. (*Drawing by Edward Cooke*)

used on the Thames for all manner of work, including fishing, and especially so by the watermen of Gravesend, who were always in search of an incoming ship to serve.

BARKING

At the same time the men of Barking Creek were continuing to develop the trawl boats they had been using since the days of Queen Elizabeth I, although, as we saw at the beginning, the men of Barking have been catching herring since AD 670. Barking was at one time the biggest trawling station in Britain, possibly the world, it being well situated to supply the London market. Well-smacks were introduced into the town in 1798 after they had proved successful at Harwich where they were first tried in 1712, and in 1833 Barking had 120 of its own cutters, many being clinker-built. Twenty years later it had 134 smacks of around 50 tons trawling and another forty-six long-lining for cod off the Norfolk coast. But, as filth polluted the Thames, the fishermen began to land their fish at Yarmouth, so that by 1854 most had based their operations at that port and Barking ceased to be a fishing station of any importance.

THE THAMES BAWLEY

The name bawley is probably a corruption of 'boiler', as the installation of a boiler to cook the shrimps is the only difference between these boats and the large, cutter-rigged

Peter boats. The bawleys were shallow-draughted because of the shoal water, and they became well adapted to sailing around and over the sandbanks of the estuary. Consequently they were commonplace at Gravesend, Leigh, Southend, Strood, Chatham, Faversham and Margate.

They became unique in their shape: a long, straight keel, transom stern with a little rake, high freeboard, initially little sheer like the Peter boats, although this later became much more pronounced at the bow, and a short mast with an extremely long topmast, and a very long gaff so that the leech of the boomless main was almost vertical. They also used a long bowsprit.

They were well built and cheap, using oak for the centre-line, frames and beams, pitch pine for the underwater planking, and pine for the planking above the water, the deck and mast and spars. They had a forecastle that extended behind the mast with two berths and two sleeping shelves above, and a coal stove. This accommodation was entered by way of a small hatch on deck. Abaft the cuddy was the fish hold with its coal-fired shrimp boiler.

In 1872 there were some 100 bawleys working out of Leigh, and most were built by Aldous of Brightlingsea. Others were based across the Thames at Ladbury's Quay, Rochester, where the earliest actual bawley was the *Pearl*, built by George W. Gill in 1844. By 1895 this business had become Gill & Son. Other builders of the bawleys were Heywood of Southend, a town fast turning into a resort, E. Lemon at Strood, across the river from Rochester where the prototype had been built – who was responsible for the majority of the Kent bawleys, and other smaller builders such as Fiddle of Gravesend and Stone of Erith. Numbers had decreased to eighty-six by 1890.

The bawley *Mosquito*, 144RR, in 1915.

Thames bawleys at Gravesend, *c.* 1890

Similar bawleys were later built at Harwich, the fishermen being so impressed by their performance. George Cann had begun this building in the 1870s, and his two sons took over twenty years later, and continued the practice, their boats later being renowned for their beautiful hulls. Some sixty bawleys worked out of Harwich, shrimping in the summer and whelking during the remainder of the year to satisfy the constant demand from the cod smacks for whelks as bait for the Icelandic fishery.

Some Kent bawleys were said to have set a mizzen lug on a second mast, sheeted to a bumpkin, and spent the summer and autumn drifting for the herring.

Today several bawleys remain in a seaworthy state, and these can be seen at the many regattas during the season. Others remain ashore under covers while restoration fettles new life into them. The type still remains as the best example of the boomless fore and aft rig, yet it is hard to imagine that their evolution started maybe a thousand years ago at a cathedral on the Thames.

THE LEIGH COCKLE BOATS

Today the beach at Leigh is a mass of broken cockle shells, a sure sign of a fishery of importance. Yet this fishery has only been in existence for 200 years, and is today as vigorous as it was at its start.

The first cockle boats were mostly naval punts that were surplus to requirement, and these were rigged with a boomless main and jib, similar to the pinks. The pinks were no use for cockling because cockle boats had to take the ground to enable the fishermen to

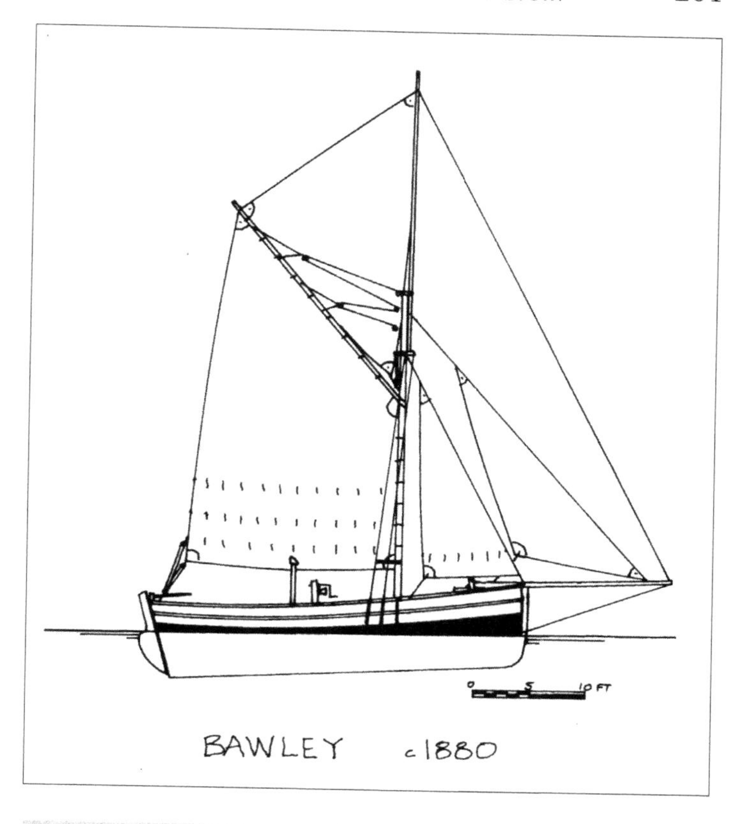

The Thames bawley *Good Intent*, owned by Don Windley, which he charters around the Essex coast from his base at West Mersea. Clinker-built in Faversham in 1860 and fished until 1927, she was rebuilt between 1988 and 1994 and is now carvel-planked. (*Don Windley*)

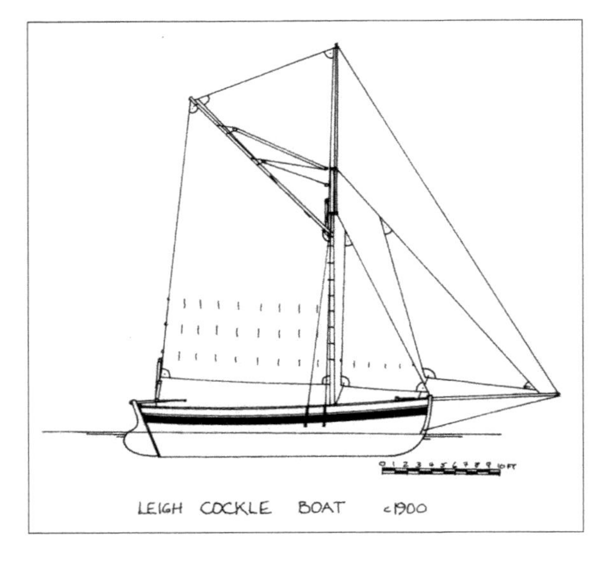

climb onto the sands to rake up the shellfish, before loading aboard the boat. These punts, or galleys as they were called, were suited, being of a shallow draught yet fast to reach the sands as soon as possible after high tide to be able to anchor in the shallow waters. Once at anchor, the three crew would wait for the tide to drop and the sands to appear before jumping out and collecting cockles until the tide returned.

Other odd assortments of craft were used, until purpose-built boats appeared in 1901. These 28-foot carvel-built boats were again bawley-rigged and generally open except for the narrow waterways and the cuddy that could accommodate up to four crew. There was a centre-plate to improve windward ability, but that did not hamper the necessary shallow draught. The boats were capable of carrying 2 tons of cockles in the open hold amidships, with a steering well aft. By 1912 some 34-foot boats had joined the fleet, and subsequent boats had motors fitted. One of these was the 1914-built *Mary Amelia* although she didn't have an engine fitted until 1920. After a substantial refit she was launched once again in 2009 and sails from the River Deben. Later boats were up to 40 feet and today most of them are steel monsters.

THE ESSEX SMACKS

In the same way that the Peter boats were essentially the boats of the Thames, the Essex smacks became the smacks of Britain, the forerunners of a tradition that spread to many corners of the country.

Where their beginnings arose is unclear, but that they evolved through centuries is evident. Wivenhoe built fishing boats in 1690, but whether these were herring busses or smaller vessels is unknown. Like other types of fishing craft, the Essex smack evolved through a process of shaping to its purpose, of meeting the demands of its working environment, where the tidal rivers and estuaries call for handiness, and offshore, the channels and sandbanks mean that a boat must be capable of good sailing performance, although the waters are relatively sheltered by these offshore hazards. Whereas the lug rig is perfect for drift-netting in the open sea, the fore-and-aft rig suited these tight channels and their fast running waters. The fishermen developed a sense by which they could navigate in fog or at night and know exactly where they were. Hence the smacks, with their long booms, underwent certain refinement, yet no substantial alteration in the nineteenth century.

The early types were clinker-built, and, like other craft, were bluff-bowed and square-sterned. The first carvel-hulled boat appeared in 1836 – the *Tribune*, built at Ipswich – and at the same time they became finer in the bow, low in freeboard and lute-sterned. Awkwardly shaped sails, too, changed when the diagonal cut was invented.

Boadicea, CK213, built at Maldon in 1808, is an exceptional boat, and remains one of the oldest craft afloat in Europe. She is a fine example of one of the early smacks in her shape and size. Today she is carvel-planked, although she was originally clinker-built, a change that occurred during a refit in 1890. She has remained in the ownership of the Frost family since 1938.

What is different about *Boadicea* is her transom stern. These were prevalent at the time of her build, later craft having a lute stern as an extension of the square stern. A further development of the idea brought in the counter stern, which was adopted generally by about 1860.

Nineteenth-century smacks came in three sizes, and were all registered at either Colchester (CK) or Maldon (MN). The smallest were under 35 feet and 12 tons, and these mostly worked in the oyster dredging and trawling. Some of the very small smacks were open boats that were used around the shores, the 'bumkins' being West Mersea-based small oyster boats, and the 'winkle-brigs', the small winkling boats that were also based there after about 1906.

The midsize smack was up to 50 feet and 20 tons, and these fished inshore, along the coast, generally spratting with a stow-net or oyster dredging, trawling or musselling. The biggest boats were found mainly on the rivers Colne and Crouch and these fished away from home for extended periods. Many of these became dandy- or ketch-rigged. They dredged for oysters in such far off places as Luce Bay, the Menai Straits, and at Swansea and the River Fal, in the Firth of Forth and off the Fife coast, and nearer to home off Shoreham and along the Norfolk coast, and the locals were so impressed that many adopted similar rigs. They even worked off the French and Dutch coast, the boats working off the latter at Terschelling being nicknamed 'skillings'. The catches here were superb, but the area was dangerous and three out of fifteen boats failed to return in the worst disaster of 1883.

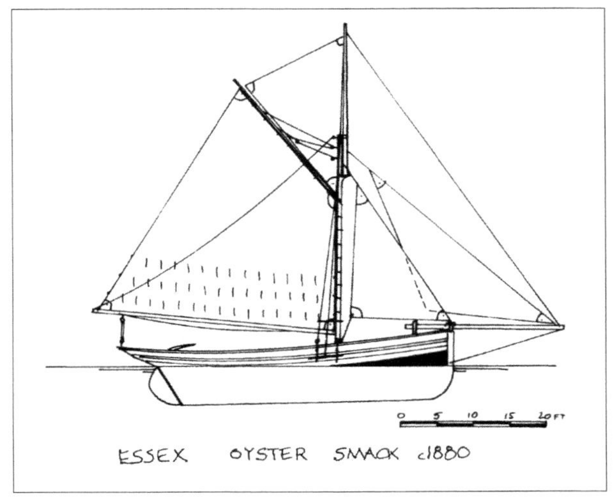

ESSEX OYSTER SMACK c1880

Some smacks fished within the River Blackwater, where a particular herring fishery grew up. This particular species of herring was another sub-species like the Welsh Cleddau herring, the fish having one fewer vertebrae than its bigger brother. It was fished between November and February, and is still caught today, albeit in much lesser numbers. Many fishermen locally still regard it

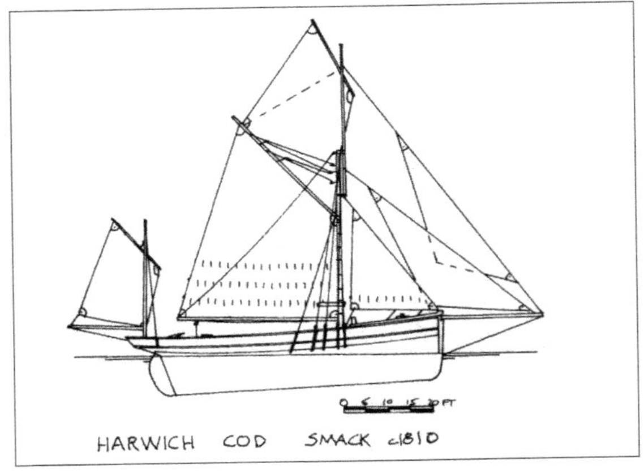

as a delicacy. There were huge numbers of smacks around the Essex backwaters, each creek being home to a few. The majority, though, were based at Brightlingsea, with West Mersea, Tollesbury and Burnham having smaller fleets. Aldous again was the main builder at Brightlingsea, with Root & Turner and John James producing a fair amount. Cost was about £100 complete with mast and spars – and sometimes even a Bible!

Harwich, at the northern extremity of the Thames estuary (it is actually deemed to end at the Naze) had its own small fleet of large smacks that spratted and trawled. Here, too, were the Harwich cod smacks, or 'cod-bangers' as they became known. Documented evidence of long-lining for cod off Iceland goes back to 1482, but it wasn't until 1712 that the first of these well-smacks was adapted in 1712 from an earlier smack and fished successfully in the North Sea cod fishery. By 1730 there were twenty-four such smacks, and numbers increased to 100 at the end of the eighteenth century. The boats grew too, up to 80 feet, and many adopted the dandy rig with a small mizzen lugsail, and later the ketch rig. Vaux of Harwich built many, and others were built at Aldeburgh. They fished off the Iceland coast with a crew of ten until the cod fishery declined during the late 1800s, although one or two remained up to 1900.

The advent of yachting gave employment to many boat skippers during the summer months, encouraging them to lay up their boats. Like the Solent and Clyde fishermen, they lived aboard the racing yachts in similar cramped conditions to those on their own boats, but they gained experience in the art of tuning their vessels, so that once they returned to their own boats they perfected their ability and speed. Each return to port became a regatta in itself; not only did they race home to land first, but also to show off their skills.

The railways opened this area up in the mid-nineteenth century, serving to increase the numbers of trippers to the blossoming resorts, yet at the same time serving to help the transport of fish. London was eating 144 million oysters a year around this time, and, until the decline that followed pollution in the late nineteenth century, oysters were deemed part of the national diet.

The Essex smack, then, is an all round workhorse of the rivers of this low-lying coast. They were as suited to trawling in deep water as they were to navigating around the sand banks with an oyster dredge behind. Whether these Essex boats were the forerunners of all the other smack-rigged vessels in Britain is not certain. But judging by the enthusiasm for sailing these smacks today, and the fact that there are possibly more ex-sailing fishing boats off this coast than there are in the whole of the rest of Britain, it is certainly feasible that these Men of Essex did have the skill and knowledge to have influenced many of Britain's other fishing fleets.

THE KENT COAST

The Kent equivalent to the Peter boat was the Medway doble, the latter most likely being a localised version of the former instead of an actual variant. A name change occurred during the nineteenth century, with the name coming from 'double boat', which correctly describes the boat, as the wet fish well did indeed divide them into two. The dobles were almost identical to the Thames craft – similarly decked with a central wet well, single spritsail, and double-ended in Norse style with rounded stems and sterns. They were strongly built in oak, mostly by Gill of Chatham, Lemon of Strood and Abbot at Chatham Ness. They were nearly all of similar measurement, being 18 feet in length, perhaps plus or minus a few inches, and 6–6.5 feet in beam. Many had a centre-plate.

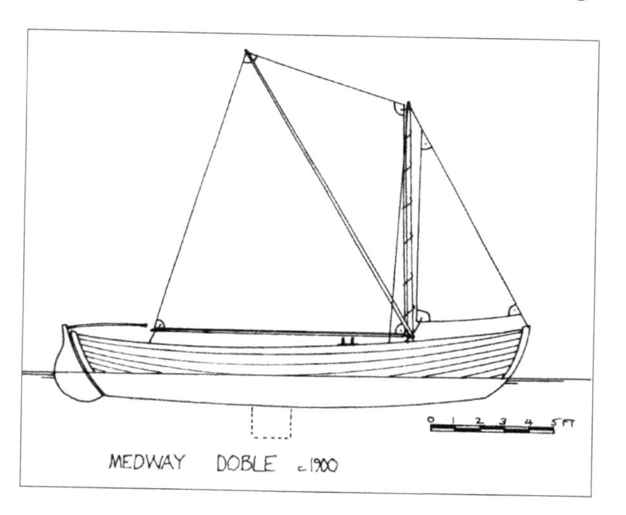

They were used for harvesting the smelts in the winter, and one or two dobles survived at this into the twentieth century. They also trawled for brown shrimps, and sometimes dredged for oysters, but the larger bawleys had always outnumbered the dobles. A few worked well into this century fitted with motors.

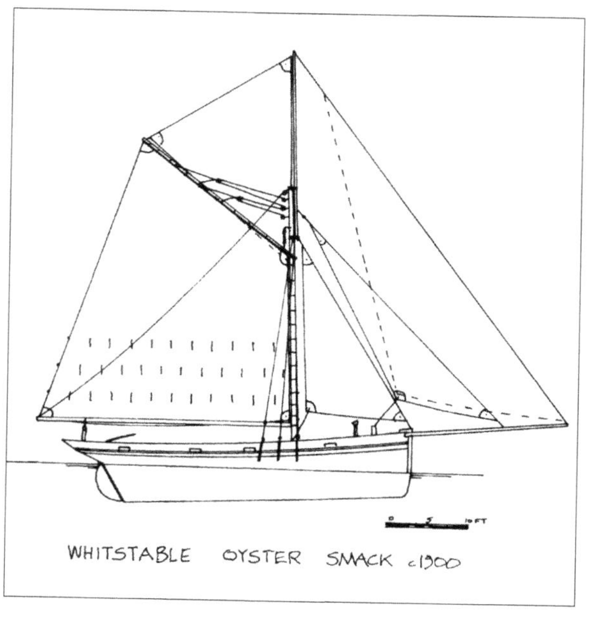

THE WHITSTABLE OYSTER YAWLS

Further east, the oyster fishery at Faversham is believed to be the oldest in the world. Nearby Whitstable was home to a large oyster fleet. Whitstable, originally Whitenestaple in 1086, consisted of a group of fishermen's huts and inns around the creek at Horseferry in the sixteenth century where Canterbury's trading vessels unloaded. By 1700, hoys traded there, and it had become Canterbury's main port, importing coal and exporting grain and hops. A century later there were some eighty houses and seventy fishing smacks that anchored in the sheltered bay. The harbour was built in 1832, yet twenty years later there were 100 smacks, few of which used this harbour. The Whitstable Company of Oyster Fishers & Dredgers had been founded in 1793, and by 1850 there were 300 members.

The smacks were similar to the Essex boats, clinker-built, fore-and-aft-rigged with a small cuddy. Carvel planking was introduced to the fleet about 1850, reputedly by the foreman of the Oyster Company, Johnathan Morday. No doubt he had seen the carvel Essex boats, and he, too, extended the counter stern and deepened the hull so that the boats could take the ground. Locally they were called yawls, a trend previously seen much further north.

These yawls worked on the oyster beds that lay just off the coast hereabout. Rarely did they ever leave the Thames Estuary except for a few that carried oysters over the Channel to Calais. Many were actually laid up throughout the summer season, and the fishermen used their smaller boats for taking the trippers out.

The railway arrived in 1860 from London, enabling the oysters to be taken there quicker. Again, it also encouraged tourism, so bathing machines, tea booths, swings

Small whelk boats at Whitstable *c.* 1910. This fishery was dominated by Norfolk families, and the design of these double-enders will be seen to show similarities to the Sheringham crab boats (see Chapter 16). (*Author*)

and boating all became summer occupations for the fishermen. The day-trippers came mainly in their thousands on bank holidays and regatta days, so this did not interfere too much with the fishing.

Most of the Whitstable boats were built in one or other of the five yards along the shore, the Collar Brothers producing the most. Sizes varied enormously from 10 to 25 tons. Bigger smacks sailed further afield, often following the Colchester smacks. A typical smack was the *Rosa & Ada*, F105 – they are all registered at Faversham – and she was 46 feet 8 inches overall, 13 feet 3 inches on the beam, and she had a large cabin extending aft to the mast, with the usual layout and access from a companionway hatch on deck to port. Her hold had two hatches.

By 1912 Whitstable was landing some 19 million oysters a year, a large proportion of the 33 million being landed each year throughout England and Wales. The smacks continued to dredge, and they were gradually all fitted with motors. Within ten years, however, disease from America decimated the beds and the fishing went quickly into a spiralling collapse. Smacks were sold off, broken up or merely left to fall apart. Fortunately, however, several remain, including the *Rosa & Ada*, whose lines were lifted in 1936 and which still remains sailing. Others include *Shamrock* (now *Elele*), built by Collar (1901); *Gamecock*, F76, the last one built at Whitstable by Collar (1906); *Ibis* (1894); *Stormy Petrel*, F71 (1890), and others which are under restoration. *Favourite*, F69, built by the Whitstable Shipping Co. in 1890 remains awaiting restoration after she was holed and beached in the 1950s. It appears she has become some sort of static exhibition. *Thistle*, F86, built in 1908, was relaunched in the late 1990s after a lengthy restoration project, as was *Emeline*, another Collar-built in 1904 and restored in the 1990s.

Margate was a popular resort for Londoners in the nineteenth century. To supply the ready market for cockles, whelks, shrimps and jellied eels, its small harbour was often full of small, open fishing boats. Bawleys, as we've seen, worked from Margate, distinguishable by their standing lug mizzens.

Herring punts were similar to the Broadstairs herring boats described in the previous chapter. These were rigged with a dipping lug main and standing lug mizzen. They often copied the bawleys by setting a boomless gaff main in the summer when taking trippers out. All the Margate boats were registered at Ramsgate.

Many of the boats were built at Margate by William Huggett, in the 1880s. They were 17 feet or more long, with a beam of nearly 6 feet and a depth of 3.5–4 feet. They were all clinker-built. After the summer season they went drifting for herring, and then for sprats in December.

Similarly built wherries drifted for herring from the resort, and these, too, took trippers out around the bay. They also fished for sea bass and potted for lobsters. They were generally rowing boats, although some had a lugsail fitted for fishing. They were long and narrow and clinker-built. Again built locally by Huggett or L. C. Brockman, they cost £1 per foot of keel. They were constructed of pitch-pine on oak frames, had a transom stern and were popular up to the 1930s. The last was built in 1927, and by the 1960s few remained. One of the last, the *Haughty Belle*, was given to the National Maritime Museum as late as 1978.

CHAPTER 16

East Anglia: River Orwell to River Humber

Then up jumped the herring, the king of the sea,
Says 'Now, old skipper, you won't catch me.

—East Anglian song

The Suffolk, Norfolk and Lincolnshire coasts are shores that are constantly on the move from the continual action of the North Sea tides and winds that make this a coastline under siege. The erosion has continued for centuries, and man has reacted by building sea defences to try and prevent this ingress onto his lands. Over the years most of these attempts have failed – one thousand years ago the coast was indented with streams and small rivers, yet today these remain silted up by the lateral movement of sea and sand, possibly even aggravated by these past failures. Floods have reopened parts of the coast, and the waters lie shallow inland. A few rivers still drain off the low-lying land and at their mouths today's large ports have grown out of tiny fishing stations. Yarmouth and Lowestoft are two prime examples and both have been synonymous with the herring fishery for which East Anglia is famous. For the most part, however, this coastline is one of beach landings that fishermen have worked off since the days of the Viking incursions, and before. The beaches are gentle, edged with low cliffs that, too, are under the constant threat of the sea. These have eroded away at many points, and in Dunwich Bay they collapsed so much that they undermined the town of Dunwich itself. This old medieval town was once an important fishing station that sent ships to the Icelandic fishery. Now the town sits under the sea, and the shingle beach that remains is home to several beach boats.

Together, Suffolk and Norfolk probably have more beach landings than any other part of Britain – landings, that is, that have no protection from the sea, are completely exposed so that the boats must be hauled high up away from the attacking waves. The only exception is the Wash, whose fishing is shared by both Norfolk and Lincolnshire, and where King's Lynn and Boston have grown into important fishing ports, geographically placed as they are so conveniently only a few miles upstream. The remaining part of the Lincolnshire coast is mostly bleak and quiet, although Skegness and Mablethorpe are two resorts of significance. Skegness did have its own

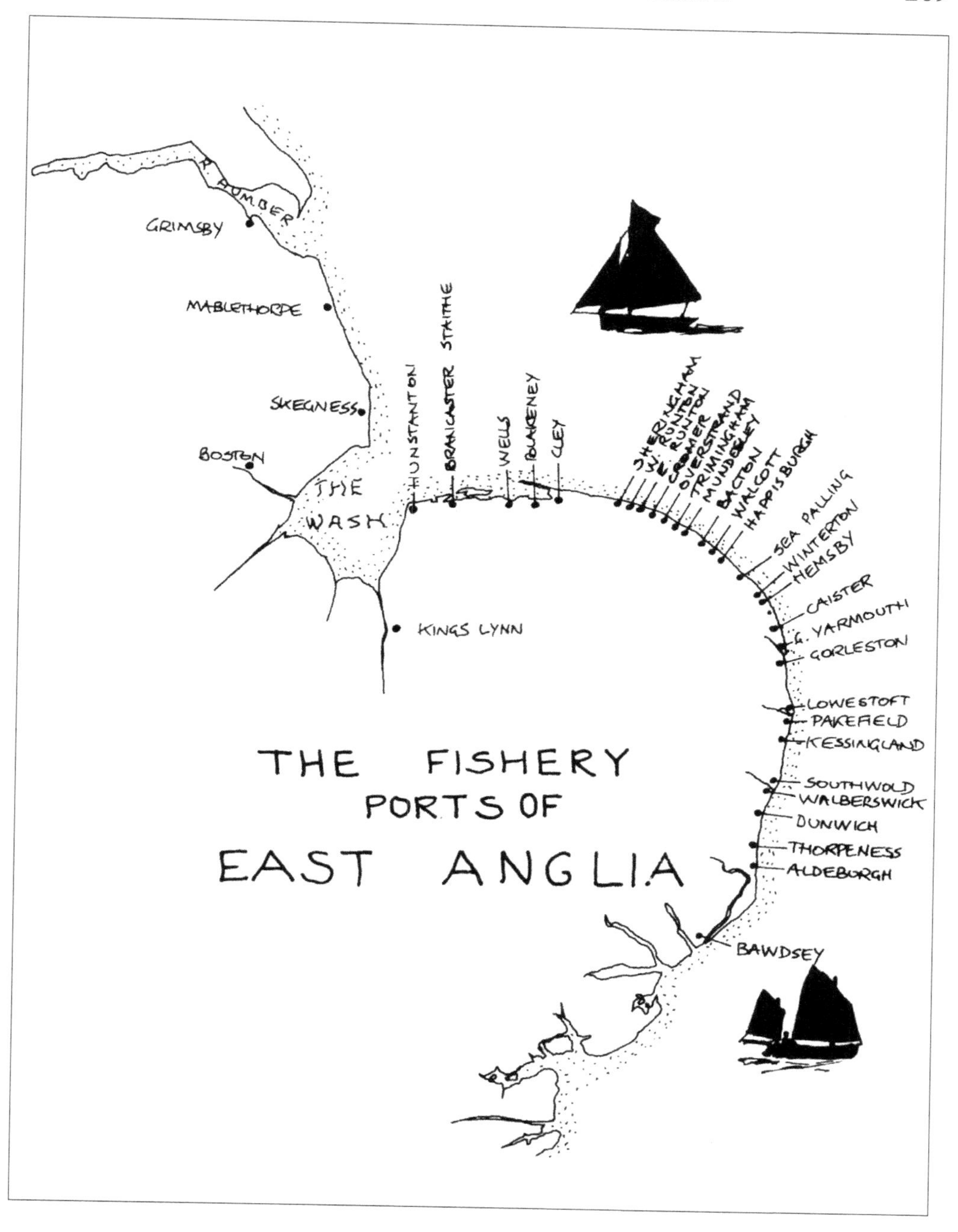

fleet of Yorkshire cobles in the last century that normally took refuge in Wainfleet harbour in the lee of Gibraltar Point, and Norfolk crabbers are known to have worked off this coast, especially around Mablethorpe. Otherwise between the Wash and the Humber, fishing was not generally exercised, the land being so fertile, although small indentations in this otherwise sandy shore – such as Saltfleet or Anderby Creek – were perhaps home to tiny fleets of beach boats.

THE HERRING FISHERY

Yarmouth was at one time the busiest herring port in the world. Its roots go back to the Saxon settlements that were later overrun by Danes more than a thousand years ago, and its association with the herring must be as old, because it is known that these Danes, and the Norsemen, caught herrings as their chief catch. Since then the association with the fish has remained active right up to this century, with its peak being reached in 1913, when an estimated 835 million herrings were landed by 1,163 vessels. The previous year some 750,000 barrels of pickled herring were exported. Although the First World War saw a massive decline in the herring, and one that was never to recover, the town of Yarmouth remained a landing port for whatever herring was caught. Today, though, the quays are full of timber and imported cars, and hardly any fish.

We have already seen the growth of the buss fishery that was centred on Yarmouth, and the evolution of the three-masted luggers after the Napoleonic Wars. Washington shows both a Yarmouth and Lowestoft lugger and a Yarmouth punt in his report of 1849, although the Yarmouth boats were one of the first to dispense with the third mast in the early 1830s. It was also in that decade that the port of Lowestoft was first built to harness the overflow from the now hectic Yarmouth. Both ports had large fleets of the two-masted luggers during the nineteenth century, there being over a thousand at Yarmouth in 1896. These drifters adopted the dandy rig around the 1870s, and after the introduction of steam capstans in the early 1880s, they then built bigger boats and the dipping lug main was swapped for a loose-footed gaff sail that was deemed much handier. Many converted to trawling when the herring was out of season. But, with the advent of steam, the old boats soon found themselves redundant. Today, only one example of a Lowestoft drifter remains in Britain – *Integrity*, which was built by Chambers & Colby in 1895 – although one or two others exist.

YARMOUTH 2-MASTED LUGGER LYING TO NETS WITH MAINMAST LOWERED INTO CRUTCH.

A Yarmouth lugger sailing out of the harbour, *c.* 1915. (*National Maritime Museum*)

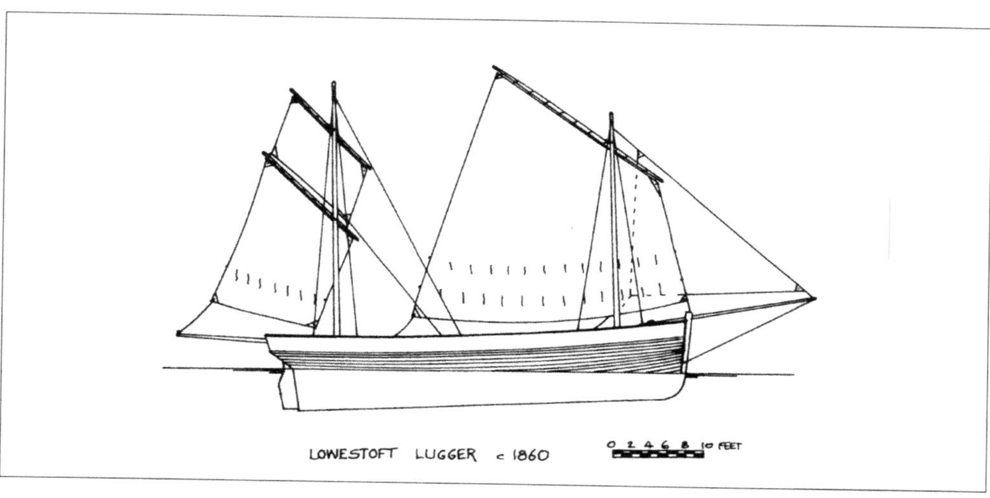

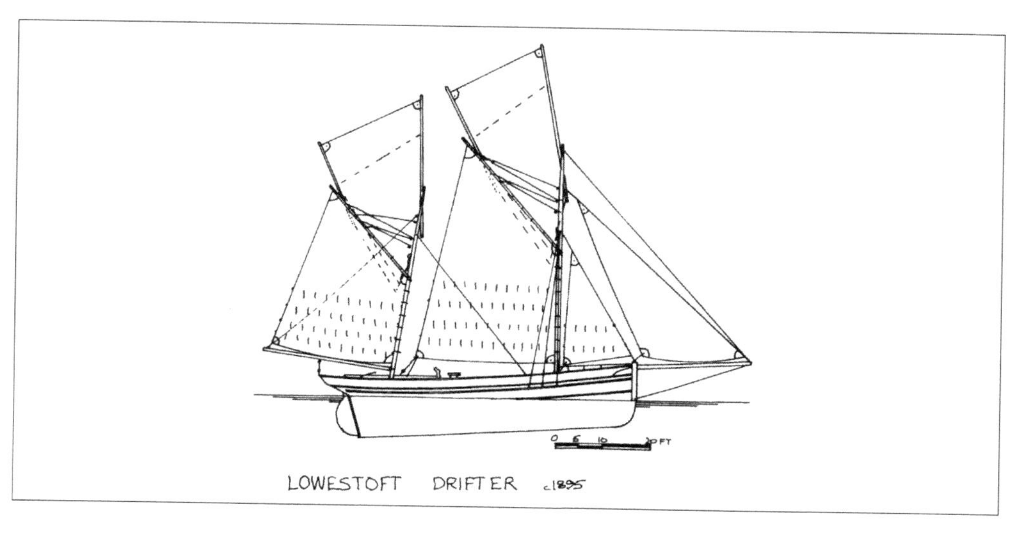

Yarmouth herring boat unloading at the quay, 1829. (*Drawing by Edward Cook*)

Southwold, a bit to the south, had its own fleet of the two-masted luggers, drawing crews from both there and from across the river at Walberswick. On the North Norfolk coast, fleets of luggers called 'pinks' or 'pinkers' worked off the beach at Cromer and Sheringham. Cromer is reported as having sent thirty ships to Iceland to fish and trade in 1527, although some came from Cley and Blakeney, with another thirty coming from Yarmouth, ten from Lynn and six from Wells. These were called 'crayers' and were heavy, three-masted square-rigged vessels. The pinkers were half-deckers with pointed sterns and they fished for cod, while bigger luggers called 'great boats' fished for herring. The latter were similar to the three-masted luggers of Yarmouth with many having been built there. In 1860 there were forty of these luggers working off the beaches, the majority coming from Sheringham. Thirty years later their numbers had fallen to about a dozen and of these, seven had been converted to dandy rig. Until they disappeared with the onset of the steam era, many were to be seen fishing for herring off Southwold or off the Yorkshire coast, landing herring at Scarborough. At other times of the year they longlined for cod or fished for crabs and even mackerel in the late spring.

TRAWLING

Trawling flourished at Yarmouth alongside the herring drifting, especially after the railway arrived in 1844. Although herring was already booming, the Devon and Barking fishermen were soon landing huge quantities of fresh white fish. The smacks that they used were mostly built at Ipswich, Wivenhoe, Rye, and Ramsgate or in the

West Country. These were powerful boats that fleeted away from harbour for several weeks at a time, and most were ketch-rigged by the 1880s.

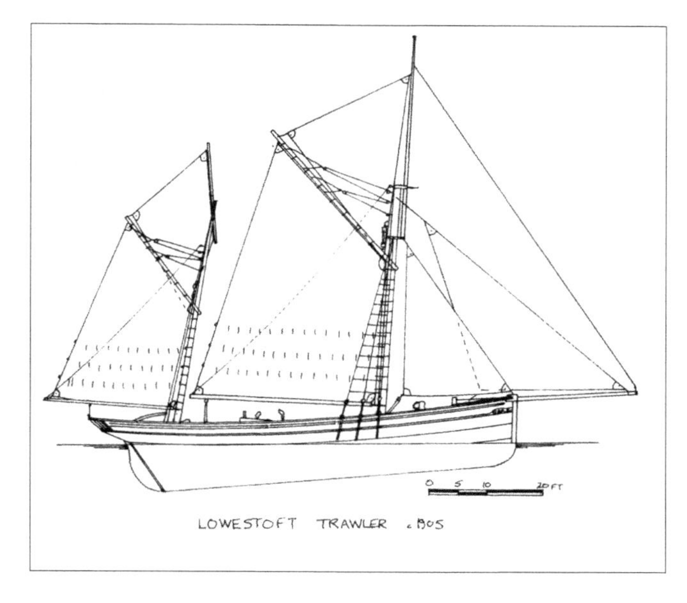

8 miles south, Lowestoft was also flourishing as a trawl port. The well-known companies of Richards and Chambers were both building trawlers. These tended to be smaller than their Yarmouth counterparts, as they were generally not away from harbour for as long, perhaps only a couple of days at the most. They were of 45–50 tons, whereas the Yarmouth boats were up to 65 tons, yet they were powerful. In 1876 there were 348 trawlers working out of Lowestoft. Smaller trawlers called toshers worked closer inshore and were up to 25 tons, akin to those from Ramsgate. The Yarmouth fishermen called these boats 'wolders' because they tended to fish in the area known as the Wold – inshore of the Haisborough Sands, offshore between Cromer and Caister.

Today a handful of Lowestoft trawlers remain, perhaps the best example being *Excelsior*, LT472, built by John Chambers in 1921. She is constructed in oak on oak, and has been owned by a sailing trust since her last refit in 1989.

THE BEACH BOATS

Small beach punts worked all along the Norfolk and Suffolk coasts from the Wash to the River Deben. The Suffolk fishing punts in the nineteenth century were generally transom-sterned unlike the beach yawls that were double-ended, although the punts, too, had been originally double-ended. These were clinker-built and evolved from the same Viking influence that is apparent all along the coast northwards. Prints of 1815 show these punts as double-ended at that time, but soon afterwards they were built with transom sterns because of the need for more space and the fact that they were seldom rowed. The beach yawls retained the pointed stern because they were often rowed and because it made launching easier in rough weather. The beach yawls were found all along from Caister to Aldeburgh, and they attended the constant shipping at anchor in Yarmouth Roads. They acted as salvage boats, pilot boats, supply boats and lifeboats, and were owned by companies of longshoremen. Although their use was not primarily fishing, some are presumed to have fished at slack periods, and many

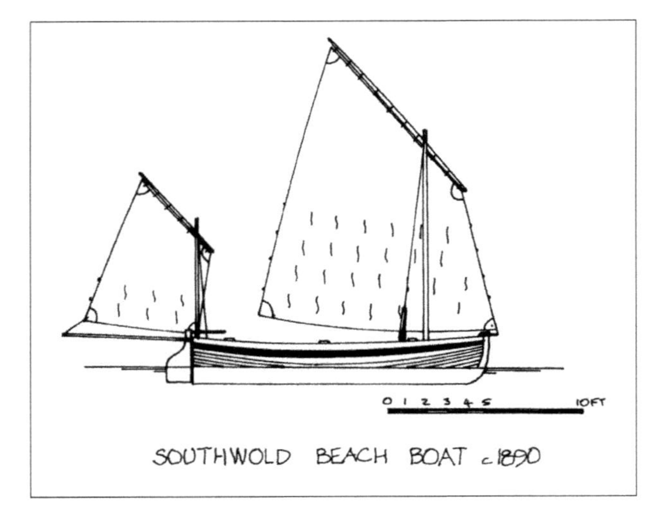

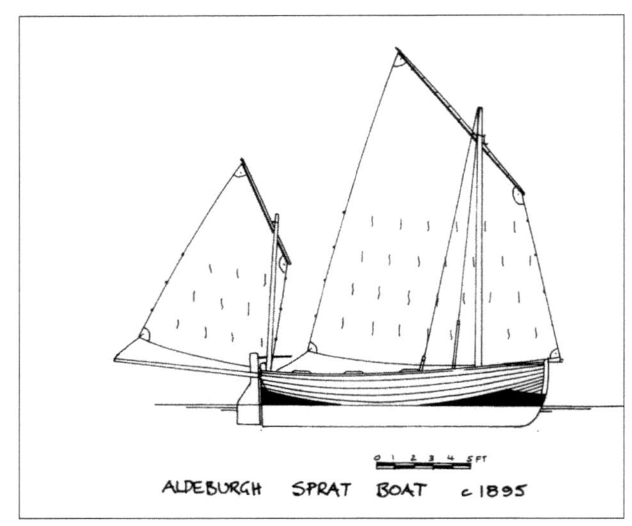

Above left: An Aldeburgh sprat boat fishing with a trawl down, *c.* 1901. (*National Maritime Museum*)

others were used to carry herring ashore from luggers at anchor. The best example of a Southwold beach yawl is the model of the *Bittern* that is in the Fishermen's Reading Room there. The largest example ever built was the *Reindeer* at over 70 feet long, which is an incredible length for an open boat.

The Aldeburgh sprat boats were typical of the Suffolk beach punts. In Aldeburgh they were simply called 'boats' as against punts at Southwold. Aldeburgh was home to a fleet of cod smacks, as we have already seen, but higher numbers of these small punts worked from the beach there. Similar punts also worked from Bawdsey, Hollesley and Shingle Beach to the south and from Thorpeness to the north. At about 20 feet long these punts have been described as being chubby. Smaller versions of around 16 feet were also built. Generally the larger punts were crewed by three men when drifting for herring as they did in the autumn, and when spratting in winter. In the spring and summer they trawled and occasionally shrimped.

Critten of Southwold built the *Ossie*, IH77, in 1893 for Alfred Smith of Aldeburgh. At 15 feet 2 inches she was one of the smallest and cost £30. Her clinker planking was oak,

Southwold beach about the turn of the century, with typical beach punts pulled up above the tide line.

as were her frames and centre-line. Although she was one of the smallest examples of these punts, she sported two masts with a dipping lug main and standing lug mizzen sheeted to a bumpkin.

Pet, a 1902 lobster pot boat, was rebuilt in 1981 by Frank Knights of Woodbridge using the plans of *Ossie* that are in the Science Museum. She had been discovered in a semi-derelict state by Robert Simper, and he then set himself the task of having her restored. He has since rebuilt *Three Sisters*, IH81, built in Thorpeness for a Mr Ralph in 1896. Along with the 1904 *Bessie* and the 1893 *4 Daughters*, out of 350 or so of these punts fishing in the last quarter of the nineteenth century, it appears that these are the only four survivors.

Southwold punts were slightly larger by a couple of feet or so. Like most of the Aldeburgh, Thorpeness and Dunwich punts, they were built by Critten, although Ladd was another builder of them at Southwold. The fishing had flourished at Southwold for centuries, with the nearby Buss Creek so named from its association with the early herring fishery and the vessels used within it (see Chapter 2). The trade continued to grow in the nineteenth century, especially after the opening of the railway from Halesworth in 1879. This was a unique 3-foot gauge line, and a further extension to Blackshore Quay was built in 1914.

YARMOUTH

As well as the larger drifters and trawlers already mentioned, Yarmouth was home to a large fleet of shrimpers, and a smaller fleet of beach punts that worked off the beach outside of the harbour.

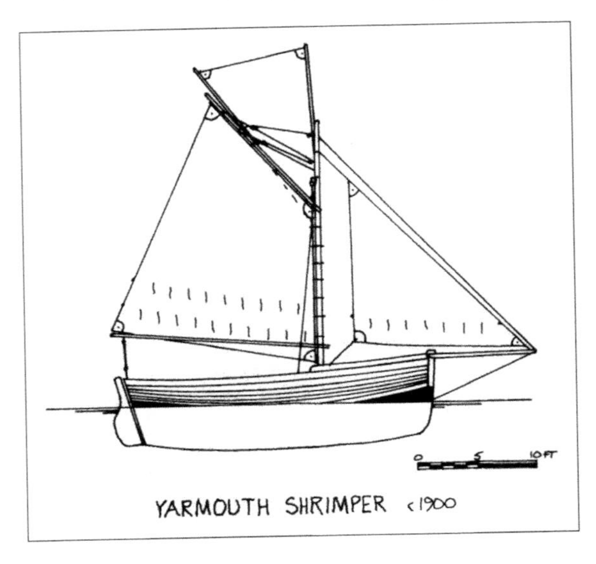

YARMOUTH SHRIMPER c1900

The shrimpers were similar to the beach punts in hull form, clinker-built, and ranged generally between 19 and 22 feet in length, although the longest boat built was *Mazeppa* at 26 feet.

Around the 1880s there were some eighty of these small craft working from the port, but their presence was normally dwarfed by the hundreds of larger craft. Their appearance came from the growth of tourism around the golden sands of Great Yarmouth and its adjoining coast. These Victorian trippers enjoyed the ritual of 'shrimp teas' as we've noted in other parts of the country, most particularly around Lancashire and North Wales where the Morecambe Bay nobbies landed the succulent shellfish.

The Yarmouth shrimpers were half-deckers, with a tiny cuddy under the foredeck that extended nearly to the mast. This cuddy provided storage lockers, a small stove and a couple of berths so that the crew could sleep aboard if necessary. The boats were handy at sea, a quality necessary given the strong tides in the river in the harbour and just at its mouth. To manoeuvre in these waters, they adopted the fore and aft rig, unlike the other beach punts, obviously encouraged by the modes of the larger trawlers.

The shrimping season began in the spring when quantities of brown shrimp were landed. Later on during the summer the pink shrimp was caught until October, after which the boats were generally laid up for the winter as the majority of the fishermen then crewed aboard the herring boats for the duration of the season. Unlike the bawleys, the Yarmouth shrimpers did not have a boiler aboard, so the shrimps were landed fresh and taken home by the fishermen to be cooked prior to being sold outside their houses to the promenading holiday-makers.

In 1900 there were still sixty-five shrimpers working out of Yarmouth, and a few from Lowestoft, but by 1931 the Yarmouth fleet had been drastically reduced to thirty-one, most of which had by this time been fitted with engines. By the outbreak of war in 1939 the number had further dwindled, and the remaining boats were requisitioned by the government and moored on Oulton Broad to prevent enemy seaplanes from landing there. By the end of hostilities most of these, it seems, had rotted away from neglect. Thankfully, however, one has survived – the *Horace & Hannah*, YH32 (now YH321), built by Beechings of Yarmouth in 1906, and today she is one of few remaining. Another more recent addition to the fleet is the 1957-built *Crangon*, while *Franasha*, YH147, built in 1912, awaits restoration. The first shrimper, incidentally, to receive an engine was *Dido*, YH484, which was given a Smart & Brown paraffin engine in 1909.

The Yarmouth shrimper *Horace & Hannah*, YH321, built in 1906 and still sailing strong. (*Robert Simper*)

Yarmouth punts, on the other hand, were double-ended craft, in true Viking fashion, and were similar to the famous Norfolk crab boats that we shall shortly discuss in full. The main difference was in the flatness of floors similar to the Suffolk punts, hence the term 'punt'. These punts were mostly found between Happisburgh and Pakefield, and perhaps could be best described as a balance between the northerly crab boats and the Suffolk punts, and they are collectively referred to as Norfolk punts. Unlike the shrimp boats, they were undecked, and they ranged between 15 and 22 feet, setting one large dipping lugsail. They were used for all manner of fishing by their beachmen owners, who often were members of the local beach companies that owned beach yawls on the coast between Caister and Winterton. Like their counterparts in Southwold, these attended the shipping in and out of Yarmouth and those anchored just off.

The beach punts longlined in summer as well as tending to their crab and lobster pots. This they combined with a certain amount of tripping on good days. In autumn they joined the local spratting fleets, sometimes fished for mackerel and then, later on in the year, went 'to the herring'. Some fishermen did not take their punts to the herring, but, like the shrimpers, crewed aboard the bigger boats.

Like the Suffolk punts, these were built mostly of oak. Spence of Great Yarmouth built a 20-foot punt in 1913 at a complete cost of £27. Reynolds of Lowestoft was another builder of these craft.

CROMER

Although we have already seen that Cromer and the surrounding neighbourhood had possessed its own lugger fleet since the Middle Ages, it was for its crab boats that it became best known. The earliest mention of these small fishing craft was in the will of one 'Clement Fysheman of Crowmer', who left his 'VI oore bote' to his wife 'Alys' and another to his son John, although John was to pay his brother, Robert *6s 8d*'. Whether this represents half the boat's value is unclear.

Daniel Defoe writes in 1724 that Cromer was famous for its lobsters, and by 1800 Edmund Bartell is telling us that 'lobsters, crabs, whitings, cod-fish and herring are all caught in the finest perfection' here. In 1875, Frank Buckland, Inspector of Salmon Fisheries, prepared a report for Parliament entitled 'The Fisheries in Norfolk – especially Crabs, Lobsters, Herrings and the Broads'. He identified fifty crab boats in Cromer, this being the main preoccupation among the fishermen, of whom there were 120 with the population being 1,145. He states that their crab pots had only come into use twelve or fourteen years before. Each boat had some twenty pots, and there were another 100 boats working from Sheringham and fifty from other North Norfolk beaches.

These crab boats developed from the early rowing boats to suit the local conditions. It is assumed that influence came from the Viking boats and in 1820 the Cromer boats had one square sail set on a mast positioned about one-third length from the bow. At that time the boats were some 12 feet in keel length, 16 feet overall. By mid-century they had adopted the dipping lug that was tacked directly onto the stem head. Ballast was carried in the form of shingle bags, which were shifted onto the windward side when sailing, and emptied when beaching. Once the boat was beached in the surf, the oars were passed through the 'orrucks' – oar ports in the top strake and across the boat so that it could be carried up the beach by four men. The steeper floors and pointed sterns made beaching on this exposed coast possible, and the boats of today have retained the same shape.

Builders of crab boats were Lewis Emery and Lown's, both of Sheringham. The former continued the tradition, by descendants of the family, until the yard closed in 1980, and the latter was taken over by Johnsons until that closed in 1950. Between the two of them they accounted for the majority of crab boats built over the hundred years up to 1950, although Gaze of Mundesley, Tom Dack and John Rogers, both

Crangon alongside *Horace & Hannah* at the Sail Ipswich Festival of 1997. *Crangon* is several feet longer and has an engine fitted. (*Author*)

A typical North Norfolk crab boat.

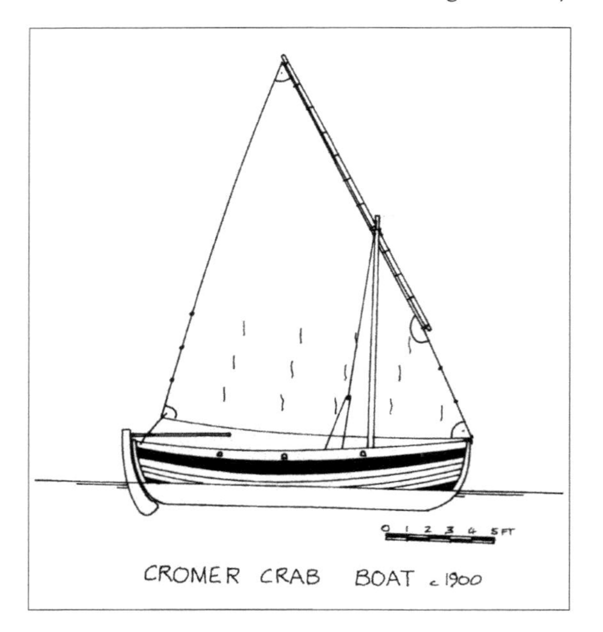

On the beach at Cromer. (*Drawing by Edward Cooke*)

Crab boats on the beach at East Runton Gap in the 1880s.

of Wells, William Starling of Blakeney and Howard Brett of Cley all built a few examples. More recently, they have been built at King's Lynn, and fibreglass versions at Blakeney.

The boats were built in traditional fashion; the oak keel was first laid, with the oak stem and sternpost then attached, the larch planking thereafter being steamed into place to suit the builder's eye, and the oak ribs then notched over these strakes. Knees and thwarts were then fitted to strengthen the hull. The wide strakes were always painted colourfully on the topsides but planking was tarred below the waterline. The rudder projected below the keel line to give added grip on the water, needed because of the beam of these boats.

Today's crab boats show a remarkable likeness to the cobles of Yorkshire, especially around the bow. Comparing photos of the two types as they are today on their trailers shows the Viking similarities between them. Considering the proximity to each other's coasts, this is hardly surprising, for Norfolk men went north to fish and the Yorkshire men came south. Today's crab boats are mostly bigger than was usual before motors were fitted, and, like the cobles, they are pulled up and down the beach by tractors, not carried by the men themselves.

Crab boats worked off the beaches of West and East Runton, Overstrand, Trimingham, Mundesley, Bacton and Happisburgh (a small fleet was also maintained at Whitstable for many years). Today these beaches are mostly deserted except for the holidaymakers enjoying the sun and sand there. Hotels, bed & breakfast houses, cafés and shops adorn the streets and beachside, where one hundred years ago fishermen's huts, decorated by drying nets, and their boats would sit. But, after our having travelled around most of Britain by now, this no longer comes as a surprise: it's the same story everywhere. Long gone is the smell of tar mixed with the stench of rotting fish. The fishing has generally gone from beach communities and what fishing has survived the onslaught of European politics is centralised in a few ports. This is the price of progression. And further to emphasise the point, recent legislation has designated ports for the landing of fish so that any pretence in supporting small coastal communities will finally have been ejected from the political arena.

Larger boats of a 15-foot keel length were known as 'hubblers' or 'hovellers'. Hovellers, as we know them, generally attended shipping, but these were different and were used to drift for herring and mackerel, long-lining for cod and fishing for whelks. Prior to 1804 they acted as lifeboats until the Greathead lifeboat station was installed that year. These hubblers were big enough to have a small cuddy so that they were found as far away as Southwold during the herring season.

Along the coast in a westerly direction, crabbers worked from Cley, Blakeney and Wells. Although similar in design these boats had deeper floors because they were not beached in the surf, but could be sailed up to the quays there to dry out at low water.

Brancaster Staithe fishermen were locally renowned for their oysters, cockles, mussels and whelks. Thirty-foot oyster smacks were brought in from away to dredge the Burnham Flats, until the oyster fishery declined about 1870. For cockling and musselling they used 14–16 foot canoes, flat-bottomed vessels, when working in shallow water and larger vessels, 21–24 feet, with a lugsail, when in deeper water. These were also used for white fishing. Early whelk boats were crab boats specially modified that were brought in from Sheringham. Later whelk boats were ordered from Lynn's renowned builder of fine craft, William Worfolk, who in 1912 built these 32-foot clinker smacks for about £100. These craft were similar to the Wash boats described below. The same smacks were sometimes engaged in collecting cockle and mussel spat from the Wash to be laid locally.

THE WASH

Although Hunstanton was the first fishing hamlet in Norfolk to be turned into a holiday resort, little is known about its early fishing. Today the beaches are full of tourists and devoid of any work boats of significance.

On the other hand, King's Lynn had three classes of fishing boat working out of the port in the closing years of the nineteenth century. The largest of these were the smacks that were common to Boston (St Botolph's town) as well as Lynn. These big carvel cutters were up to 60 feet long. They were used to dredge oysters and, later, for fishing for whelks that were sent to Hull to be used as bait by the longliners.

At Lynn, Thomas Worfolk had moved to the town in 1900, and his first boat was the Lynn yoll *Baden Powell*, which cost £60. She had only just sold her

A motorised Cromer crab boat being brought ashore in 1996. (*Author*)

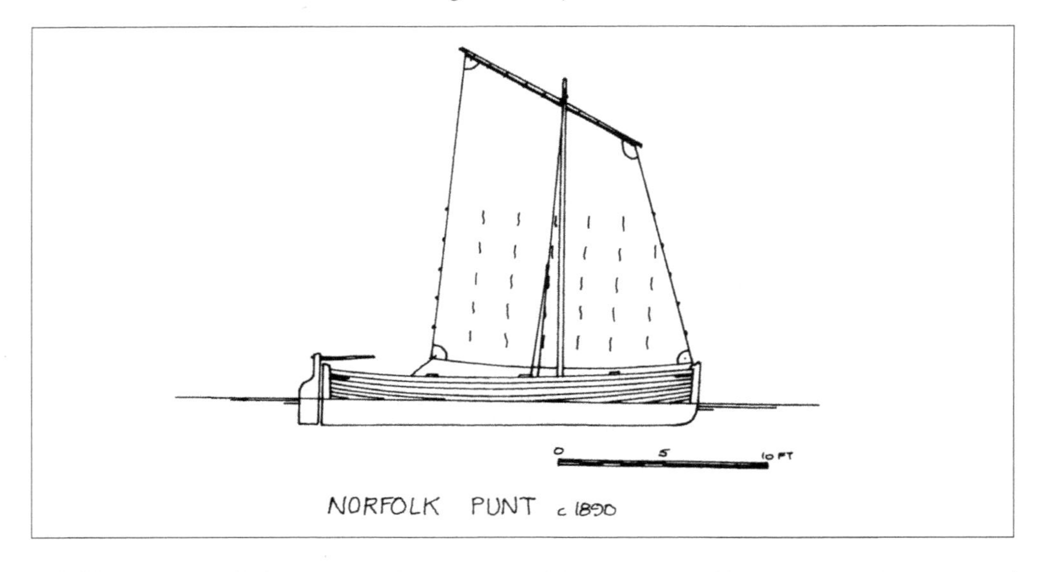

A typical double-ended Lynn yoll. Note the upright sternpost.

fishing licence, in 1996, and was being threatened with destruction under the British government's outrageous decommissioning policy. Thankfully *Baden Powell* has survived to proclaim a more honourable truth.

Thomas Worfolk was succeeded by his sons, who traded as the Worfolk Bros. In all they built about twenty fishing boats for the Lynn fleet, although in total they built about 650 boats, the last being a yawl named *Lady of Lynn*, LN107, which in 1997 was based at Brancaster. They also built the last whelker – *Britannia* – in 1915 at a cost of £290.

Lynn yolls were double-enders (yoll from the Norse *yol*). They were similar in shape to the Norfolk punts due in the main to a Cromer crab boat having first been converted to a cockler. They were sharp-sterned, clinker-built and flat-floored, yet were bigger at around 30 feet overall. They adopted the cutter rig for handiness around the confined channels of the Wash. They were used for the collection of cockles and mussels and for shrimping, with the catch being boiled on board. Their flat floors enabled them to be grounded on the sandbanks when out collecting cockles and mussels. If the wind was in the wrong direction, or failed them completely, the fishermen rowed the yolls home using a pair of long sweeps.

The third type of Wash fishing boat was the shrimper, a 40-foot

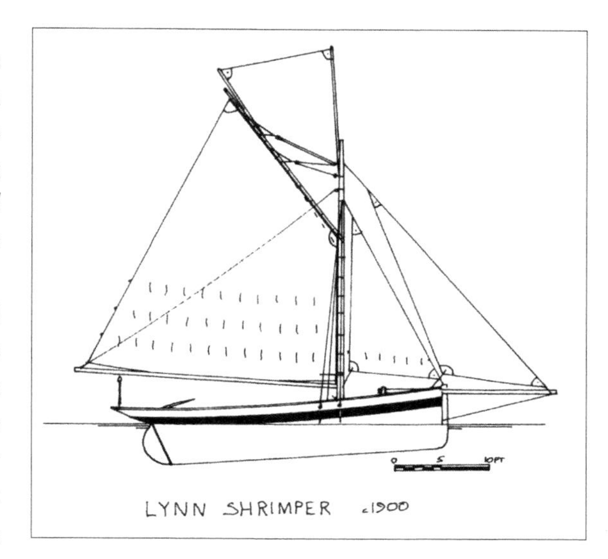

LYNN SHRIMPER c.1900

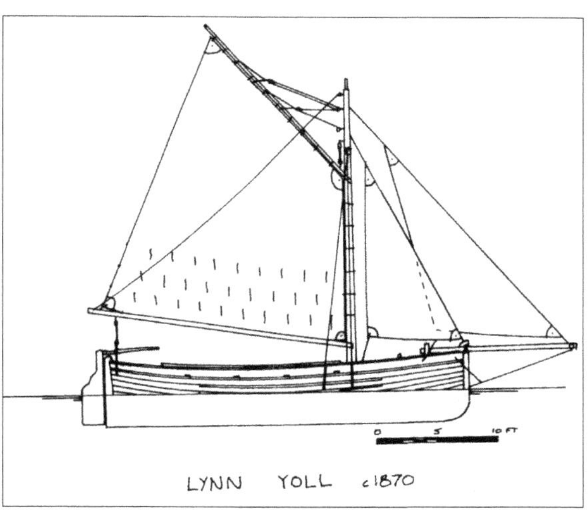

LYNN YOLL c.1870

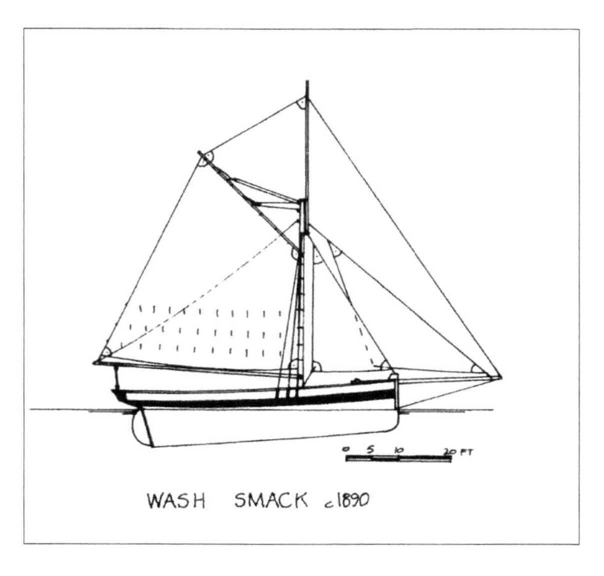

WASH SMACK c.1890

smack with a long counter and narrow beam. They worked from all the rivers of the Wash – at King's Lynn (Ouse), at Sutton Bridge and Wisbech on the Nene, Fosdyke on the Welland and at Boston (Haven). The Wisbech shrimpers were some ten feet shorter than the average boat because of the greater distance upstream. They were all half-deckers, with a little cuddy, narrow waterways and an afterdeck with stowage below. They operated two trawls, an 8-foot one forward and a 12-foot one aft, and they worked mostly in the Boston Deeps. They always boiled the catch in the coal-fired boiler prior to landing. Like the yoll, the shrimpers collected cockles and mussels at various times of the year, and they were, too, beached upon the beds for this purpose. These smacks hardly ever left the confines of the Wash, and were handy little craft adept at working in its shallow, tide-swept waters.

Thompson, Gostelow and Keightley were building these smacks from their riverside yards near the town centre at Boston. Thompson survived until about 1920 and Gostelow continued up to the 1950s, although the latter only built on a small scale.

In 1874 there were 111 boats registered at King's Lynn, eleven of these being first-class vessels, ninety-four at Boston – five first-class – and sixteen second-class only at Wisbech.

Some examples still remain within the Lynn fleet, the fastest of them all being *Queen Alexandra*, which was built by Worfolks at a cost of £110. The Boston smack *Albert* is undergoing a long refit and is hoped to be finally completed soon. Two others – *Mermaid*, LN32, and *Telegraph*, BN122 – still sail regularly and were both built by Gostelow in 1904 and 1906 respectively.

GRIMSBY

Grimsby grew up as a major fishing port after about 1860. From having one fishing vessel in 1858, there were more than 840 boats working from the port by the 1880s. The growth was attributed to its geographical position 25 miles downstream of Hull which saved the boats some 50 miles in the return trip – a valuable saving in time and money. With the arrival of the railway, that encouraged the building of good dock facilities, owners soon realised the advantages, and Devon fishermen were among the first to base themselves there. The Essex men followed very soon afterwards.

The Grimsby trawlers were among the biggest built, 80 feet long, 20 feet beam and up to 100 tons. They were built in various places around the country, and some came across from Germany.

Yet, as steam replaced sail, the sailing trawlers disappeared as quickly as they had arrived. By 1910, the port was empty of sailing craft, and those that had remained afloat had either been sold abroad or had been turned over to coasting. However, one example does remain – *Esther*, built there in 1888, is under restoration at the National Fishing Heritage Centre at Grimsby, having been recently brought back to Britain from the Faeroe Islands, where she had been for some years.

CHAPTER 17

North East England: The Coble Coast

Nowt's over bad for them owd dogs!
—Coble fishermen when catching a dogfish

In Chapter 2 we saw how the development of the Yorkshire yawls came about, and that they were adapted from the earlier three-masted Yorkshire luggers. It is from this point that the journey through the north east coast of England begins.

At that time, the early part of the nineteenth century, Staithes and Filey were the two most important fishing stations on the Yorkshire coast. Staithes was little more than a small river or beck flowing out into the sea, with tiny cottages huddled around the bend in this beck, as if hiding from the vicious North Sea winds. Filey was only a beach, somewhat protected from the same winds, but only totally from the north.

Scarborough boatbuilder Robert Skelton was the first to adopt the Yorkshire yawl in 1835, by dropping the middle mast, but retaining the lug rig. Influence for this change is said to have come from the Norfolk beach yawls, many of which did venture north. Nearly forty years later, in 1872 the first dandy-rigged boat appeared, such as was the trend all around the British coasts. This rig was deemed to be lighter and easier to handle, although at the expense of speed.

These 50-foot yawls were flat-bottomed for beaching. In the year 1860, there were 100 such yawls working the Yorkshire coast – thirty from Filey, twelve from Bridlington, twelve from Whitby and fourteen from Staithes, the rest from further afield. The cost of one was £600, with another £100 for gear.

The boats were primarily used for the herring fishing, and out of season they were laid up – like the five-man boats. Smaller boats were then used for the winter fishing. But in their search for the herring, to prolong their season, they sailed up and down the east coast with the shoals. In 1858, for example, Filey landed over 35 per cent of all the herring caught off the north-east coast and the last yawl there survived until 1916.

The Staithes yackers and Marshall luggers were variations of these yawls, the latter coming from the yard of Whitby boatbuilder Robert Marshall around 1850.

Before we consider the coble in detail, though, we must first view the River Humber.

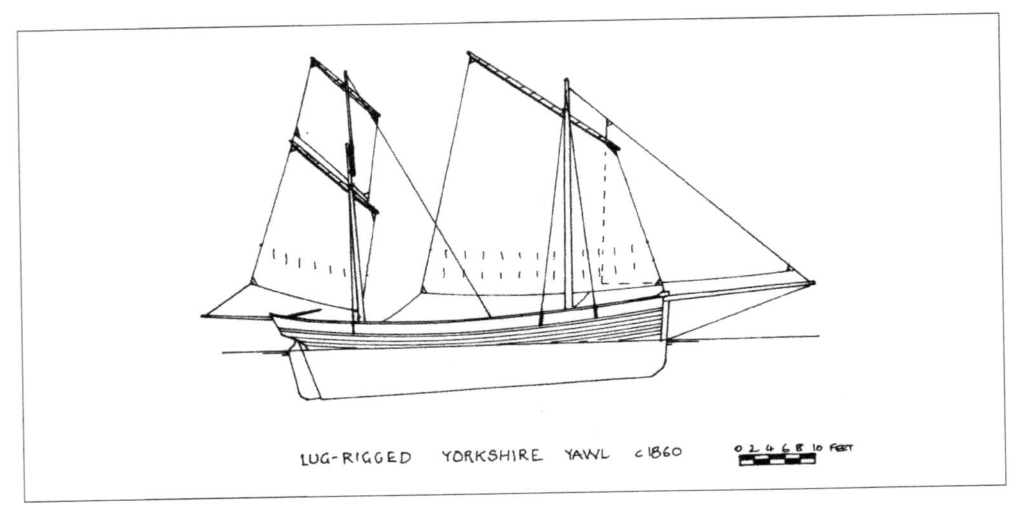

LUG-RIGGED YORKSHIRE YAWL c1860

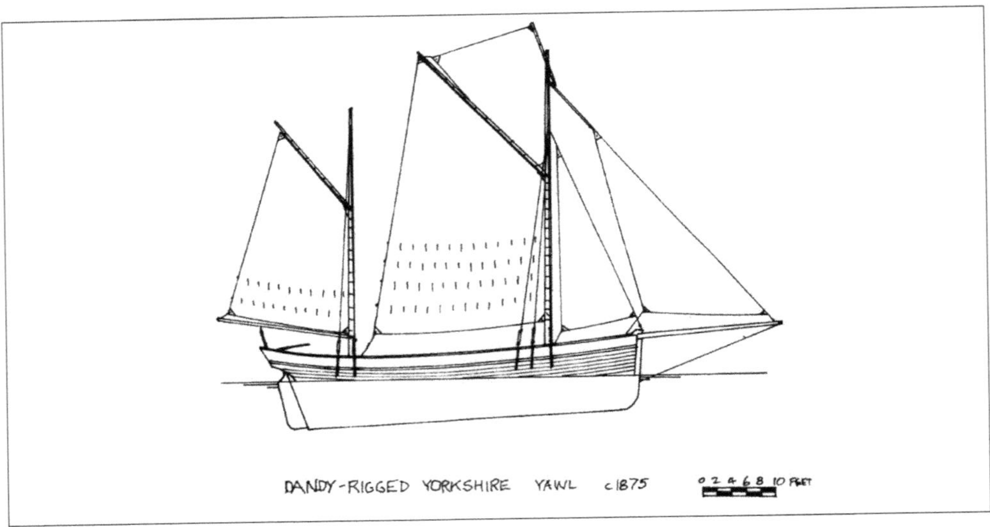

DANDY-RIGGED YORKSHIRE YAWL c1875

YORKSHIRE YAWL c1860

THE HUMBER

Fishing in Hull developed, as in Grimsby, from the introduction of trawling after the southern fleets of trawlers from places such as Plymouth and Brixham were blown north from the eastern waters of the Channel in 1843, and unwittingly discovered the rich fishing grounds of the Dogger Bank, otherwise known as the 'silver pits'. This, though, is not to say that Hull wasn't already an important port; it had been so since the Middle Ages. It just didn't have anything but a localised fishing fleet. By the turn of this century, however, it was considered as ranking among the best deep-sea harbours in the world. The vessels themselves were basically the same as those from down the river and were mostly built away until local yards began building in the latter decade of the eighteenth century. By this time steam had mostly overtaken sail, and the looming era of the internal combustion engine was only just over the horizon.

The small village of Paull was home to a very different fleet. The Paull shrimpers worked two trawls in the shallow water on the Lincolnshire side of the Humber during the period of March to September. Built at Hull and Winteringham, they were 28 feet overall, cutter-rigged and were open boats except for a tiny shelter under the foredeck. They worked 10-foot-wide beam trawls – two in echelon fashion – from ancient methods, possibly those in use since the fourteenth century. Often fishing in shallow water, their favoured grounds were on the Lincolnshire side of the river, off Killingholme and the shrimps they caught were sieved and boiled aboard on the journey home, and sold from handcarts when Paull was a popular resort for day trippers from Hull in the late nineteenth century. During the winter they worked longlines for cod and ling. By 1910, there were only twenty boats left, and then by the outbreak of the Second World War there were practically none left – the last one, *Venture*, ceasing fishing around 1950.

Between the mouth of the Humber and Bridlington, the coast is sparse, with one or two small beach-based fishing communities such as those at Hornsea and Withernsea. At Spurn Head, tucked quietly into the shelter of the point there, several cobles can still be found at anchor. A century ago there would have been upwards of thirty cobles on the beach. Some keelboats were there as well, these being sometimes referred to as 'Sheringhams' to distinguish them from the cobles (see previous chapter). Many fishermen from East Anglia settled in this part of Yorkshire, some even bringing their crab boats. Crabbers were also to be found at Hornsea and Withernsea.

PAULL SHRIMPER c.1900

The last Paull shrimper – *Venture* – at Hull.

Unloading a catch of herring from a Yorkshire yawl at Whitby.

The harbour at Scarborough with a double-ended coble alongside a yawl. Note the boatbuilder's yard behind.

THE COBLE COAST

The Northumberland coble – pronounced 'cowble', as against the Yorkshire 'cobble' – was first mentioned in the Lindisfarne Gospels of AD 860, and has been in constant use ever since. It is hardly surprising, therefore, that much has been written about this descendant of the Viking boats, both historically and of its use today. I do not wish to reiterate what meticulous writers have said before me, except that in continuing along this coast, the different local influences must be studied. For reasons of chronological order, we must jump ahead to begin our study in the northern part of the coble coast – Northumberland – before making our way backwards in a southerly direction to Yorkshire.

NORTHUMBERLAND

The most striking feature of this coast is its long beaches of golden sand with the natural coastal defence of lovely sand dunes that stretch away in both directions. That this is one of Britain's most beautiful coastlines is not widely known, and it is hardly surprising that it is classified as an Area of Outstanding Natural Beauty.

On these beaches, and at headlands or wherever nature makes it possible, harbours have appeared. These grew up because of the proximity of the North Sea fishing grounds. Any sandy beach with some degree of shelter was home to a small fleet of boats, and it was out of the need to beach these boats that the coble evolved. There are few deep-water harbours.

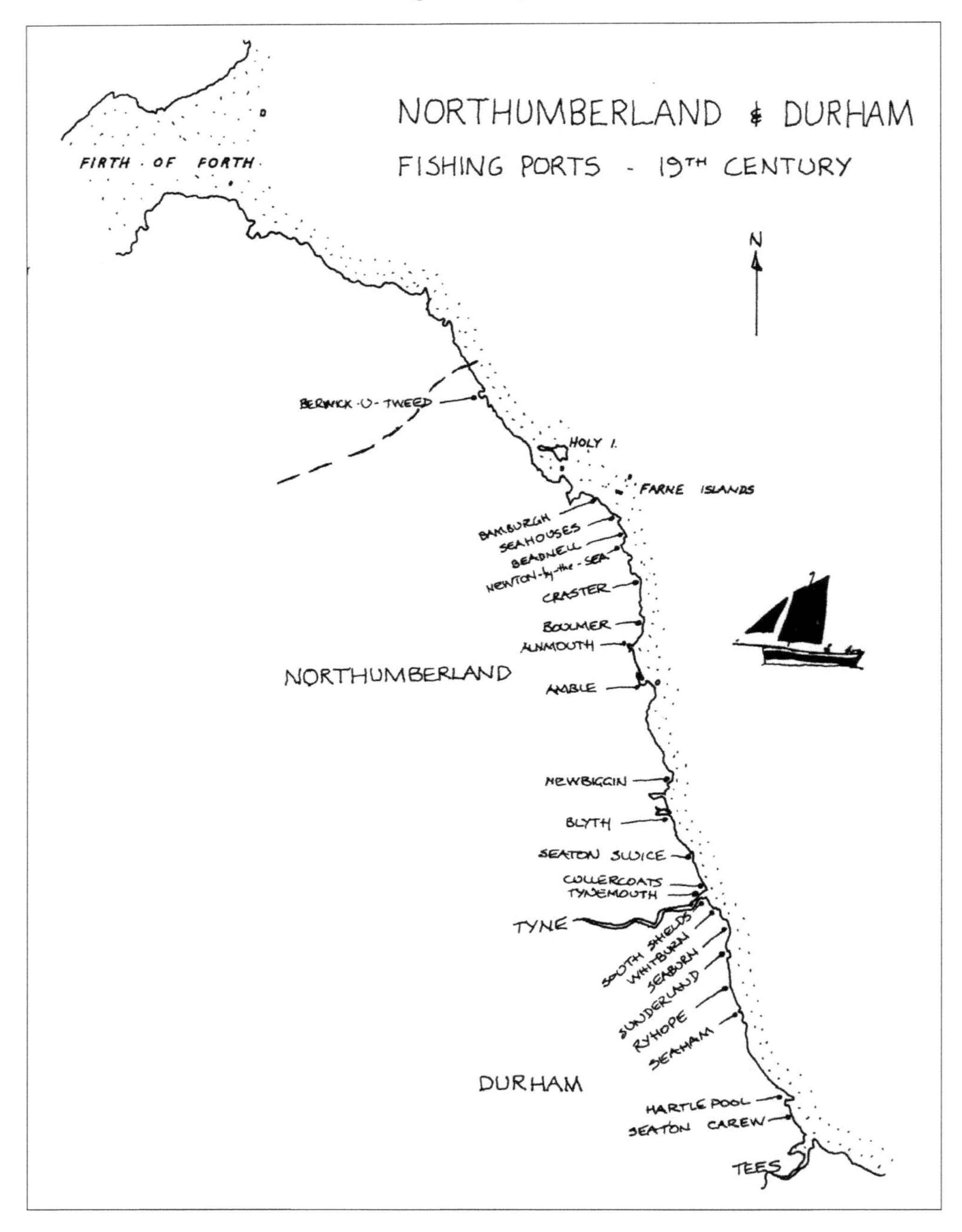

Between Berwick-upon-Tweed and Amble on the River Coquet, there are some eleven harbours or shelters that have had fishing fleets over the last century or more. Perhaps one of the best examples of a fishing community born out of the beach is Low Newton-by-the-Sea. The fishermen's cottages are built around the centre on three sides, facing the sea, and the boats were beached and pulled up into this square to be sheltered from the storms. Today, though, it exists as a holidaymaker's paradise, although many of the places like it do still support one or two boats.

The coble *Gratitude* from Whitby, having just been beached. (*Edgar Readman*)

That the shape of the coble is Viking seems indisputable. Stories of their invasions of these coasts, of the raping and general pillaging, are common knowledge, although probably exaggerated, some claim totally false. But the Vikings certainly left their legacy here, from the River Humber right round the north coast to the Outer Hebrides, Scottish west coast, Isle of Man, Wales and, to a lesser extent, on the south-west and south-east coasts. When the Vikings came after their initial raids, they arrived in their Knorrs – bigger, broader and deeper than their longships – with animals, supplies and families who settled here. These boats had square sails, and a long oar astern that acted as a rudder. Their shape vaguely resembles a coble, especially around the bow section.

Against this background, the coble had developed to meet local needs. On a coast without natural harbours, where the surf can be extremely dangerous, the fishermen needed a boat that could be beached safely. The deep forefoot under the bows of the coble provides a good grip on the water that enables the boat to be swung round and rowed, stern first, through the breaking waves, and onto the shore quickly without being swamped. Once beached, the cobles were hauled out of reach of the surging waves and unloaded. Trailers were used with growing popularity in the nineteenth century; they were pulled into the sea by horse, turned around and the incoming coble rowed onto it and then readily hauled out. Today diesel tractors are to be seen at any beach where modern-day cobles are working. The cobles themselves, though, are hardly any different from their ancestors of a hundred years ago. Their shape remains as it was, and their only advancement in today's world are their engines and the modern fishing gear.

The earliest cobles of which we have definite details date back to the first half of the eighteenth century. They were about 25 feet overall, and 5 feet in breadth. Many of the earlier vessels were said to have had sharp sterns, especially around Holy Island, and a mention of the same is made in Hartlepool in 1816. The sharp-sterned cobles are said to have been easier to handle, yet less buoyant, so that they could only be landed on gently sloping beaches. The heel would dig in on steep landings.

These boats were worked by three men; the fisherman sitting aft would have a pair of oars, while the other two would have one each. Their main occupation was long-lining for cod, halibut and ling. Each of the small communities had twenty to sixty boats. Seahouses – really part of North Sunderland, but so called for being that part with 'the houses by the sea' – and Amble were the bases of two of the main coble builders – Dawsons of Seahouses and Harrisons of Amble. The former is still in existence, and the latter only recently closed down. For the herring season that began in May, larger cobles were used to sail further offshore. These were up to 35 feet overall, and were generally rigged for the journey to the grounds. In between the herring and the winter fishing, the fishermen would set pots for lobsters and crabs.

CONSTRUCTION OF THE COBLE

The clinker construction of the coble is unique among the British boats, and many good books document it in full. In a nutshell, the boat has a ram plank instead of a full-length keel, indeed the boats were measured by the length of this ram plank *(ram* from old Norske meaning strong), which was typically two thirds of the overall length. It was usually of oak, occasionally larch in small cobles, and was laid first with the shorter keel, forefoot and stem being attached by the stomach and apron (deadwood). To this were fitted the planks or strakes as they are called, and then the frames (timbers) were notched over these riveted strakes. The shape of the boat came from the way the strakes were laid – a flat bottom was preferred by the fishermen as it made the boat more stable in strong winds; a rounded bottom perhaps being preferable in lighter winds. The Northumberland cobles were said to have broader strakes and finer lines, while the Yorkshire boats had narrower stakes and a bluffer appearance. Northumberland boats also had lower bows, and many have a canvas cuddy to keep the waves out nowadays. However, as is always the case with wooden boatbuilding, each builder would lay the planks to his own eye, there being no drawings, and these generally took account of local variations such as beach angle, prevailing wind direction, fishing ways, etc. On the bottom of the boat, bilge keels called 'skorvels' (N'land) or 'draughts' (Yorks), of oak, were fitted about two feet apart. Iron shod, they acted as runners for the boat during beaching. They ran 12 inches up the transom, with holes for attaching tow lines when beaching in the days before trailers.

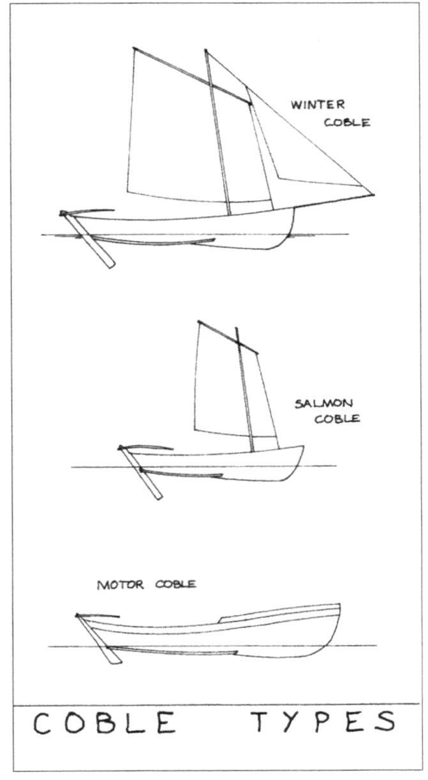
WINTER COBLE
SALMON COBLE
MOTOR COBLE
COBLE TYPES

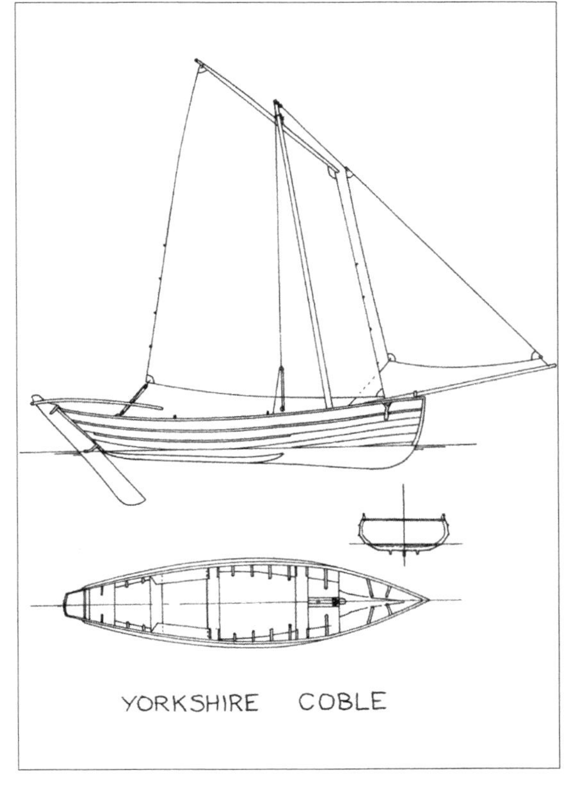
YORKSHIRE COBLE

RIG

The rig was always the fine lugsail on a raking mast. Like many beach boats that have evolved from Norse influence, the angle of the mast is adjustable for sailing into or with the wind. This adjustment was made by using a special rounded wedge that was moved in the 'gantry' that surrounded the mast between the thwarts. The foot always stayed in its housing in the stomach.

The smaller boats relied solely on the lugsail, while the bigger ones set a bowsprit with a jib. Often three jibs were carried, the choice of which was set depending on the weather. Furthermore, this bowsprit was often the smaller winter mast, a larger mast being used in summer.

THE DURHAM COAST

Between the Rivers Tyne and Tees, larger, but fewer, harbours exist. The Tyne itself is home to many great shipyards, or at least was until very recently, and its main fishing harbours spanned its mouth at North and South Shields. The word 'Shields' comes from the Norske *scheles*, which means 'simple fishermen's summer huts along the shore'. N. Shields had a fleet of 489 cobles in 1872, while across the river there were only forty-four.

Sunderland and Seaham grew out of the coalfields of the north-east, and not the fishery, although both had their own small fleets.

The harbour at Staithes with cobles. (*Edgar Redman*)

Hartlepool was, at the turn of the century, the largest dock in the country, and has had an association with the fishing since the thirteenth century, when there were quays where herring, cod, ling and rays were landed and sold to Durham's religious houses. One Hartlepool merchant, Henry Carr, sold 10,000 dried herring to the Convent of Durham in 1490. This same convent owned one herring house in the town. Between the fifteenth and seventeenth centuries, the fishing declined because of the Dutch superiority. A new dock was built in 1880, said to be the best in the whole of the north-east, and Hartlepool's fortunes improved. In 1874, there were 148 fishing boats employing 367 fishermen.

Just south of Hartlepool is Seaton Carew, which was a small fishing station during the Middle Ages, but which became a popular Georgian resort for those merchants and landowners rich enough to be able to take holidays.

Hartlepool has a history of boatbuilding, the company of Cambridge's building generations of cobles. In 1862, they built a coble for Richard Picknett of Redcar for the sum of £14 10*s*. Another coble, the *Seaview*, the last of Redcar's sailing cobles, was built at Pounder's Yard in 1900 for the Boulmer family. In 1909, a Cambridge coble cost £24 10*s*.

YORKSHIRE

Yorkshire, in contrast, has a coastline of high cliffs towering above sometimes sandy, sometimes rocky beaches, often with outlying reefs. Inside these reefs, between reef

The Whitby-registered coble *Lily* sailing up the river with one reef in the sail. (*Edgar Readman*)

and shore, sheltered water could be found, enabling boats to be beached in relative calm. Redcar was one such place. Saltburn-on-the-Sea, a little southwards, was a popular smuggling haunt with a tiny row of fishermen's houses. When its spa was discovered, it rapidly grew into a resort, yet still today it has managed to retain its own fleet of shore-based cobles.

Staithes, as we have seen, nestles at the mouth of the beck and affords good shelter, while Runswick Bay, Sandsend and Robin Hood's Bay all grew up from very small communities on beaches that allowed some protection. Coble fishing out of Robin Hood's Bay has been immortalised in the Bramblewick Trilogy by local writer Leo Walmsley (*Three Fevers*, *Sally Lunn* and *Phantom Lobster* – first published by Collins in 1932).

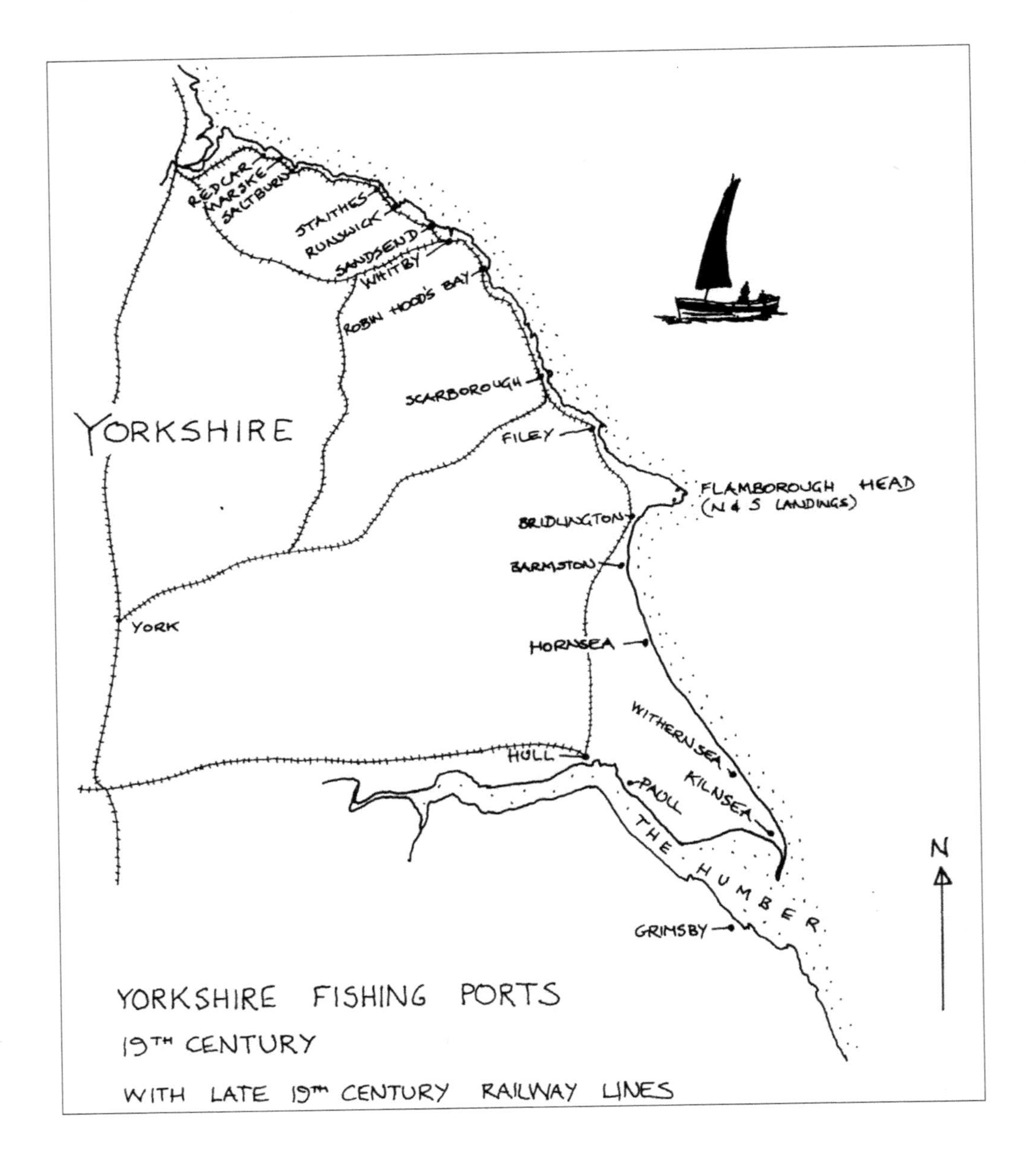

Whitby, on the other hand, has for centuries offered a safe haven along this otherwise inhospitable coast, being the only natural harbour between the Tees and the Humber. Its association with the fishing goes well back – the monastery built in the seventh century traditionally took herrings in; yet the town did not really develop until the seventeenth century when the colliers en route to London called in. By 1850, the railway had opened up what was previously a town completely cut off overland.

Sandwiched between the North Yorkshire Moors and the North Sea, it was only by sea that communication with the rest of the county was possible. As the markets opened up, the North Sea fleets soon realised the new importance of the town. During the herring season, fleets of Scottish and Cornish boats mixed with the local cobles to land herring. These 'keel' boats, as they are still called, were entirely different from the cobles. They were fifies, zulus, Cornish drivers, and in time the Yorkshire fishermen began to copy them, giving birth to the Yorkshire Keel boats.

MULES AND PLOSHERS

Both terms sound rather nostalgic, and seem to be rather vague. A mule is a double-ended coble originating from Viking influence, possibly brought down by Scottish visitors. North of the Scottish border, a mule is a crossbreed of boat – normally a fifie/zulu cross – and it is probable that in Yorkshire the same crossbreed meaning is relevant. Some say that a mule is a cross between a coble and a whaling boat and this

Cobles and a mule on the beach at Filey, *c.* 1880. (*Edgar Readman*)

appears to me quite likely given the importance of whaling on this coast. However, the first mules seem to have appeared about 1875, and were introduced through the need to stay at sea longer so that larger catches could be taken. This meant that the crews needed some form of protection from the elements. The boats also tended to carry more gear. Early mules were around 30 feet overall, half-decked with a tiny forecastle. They retained much from the cobles such as the ram plank and clinker construction, but the coble's transom was replaced with a sharply sloping sternpost, similar to the Scots' boats. The hull shape became fuller, giving better sailing abilities, yet the boats were unsuited to beaching. This, however, was not important at Whitby, from where many sailed.

The rig remained unaltered, the dipping lug staying, and all but the smallest craft still setting a jib on a bowsprit. A few even set a small mizzen sail.

By the turn of the century, these mules were fully decked. Like the Scots, the coble men had seen the advantages of decking-in their boats when fishing the herring. Previously they had deemed it a disadvantage. Within the twenty years or so that the boats had evolved, they had adopted a full-length keel instead of the ram plank, and had in fact become not cobles but keel boats. The planking was narrower, the sternpost became more upright, and then some were carvel-built. They changed so much that in time they were hardly distinguishable from the Scotch and Cornish boats during the herring season. These keelboats are not to be confused with the small beach boats that are to be found at places such as Staithes today, although there are many similarities.

The plosher is also referred to as a herring coble. Although not unlike the mules, the plosher evolved purely for the herring from the five-man boats. Unlike the mules, however, they worked all year round. They were decked forward of the mast, with a forecastle, some were double-enders – although most retained the transom – and were large (up to 42 feet overall), powerful boats with five compartments, only one for the crew, and were capable of carrying 200 cran of herring. They set a mizzen sail when trawling or when racing in the annual regatta. These boats have also been termed 'sploshers' and 'splashers', possibly because they were renowned for being wet boats with one

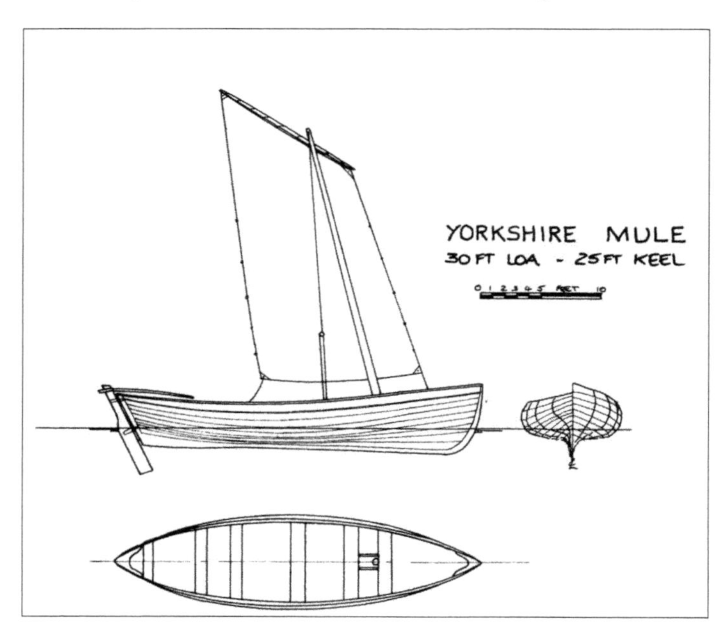

Ploshers at Scarborough. Note the small double-ended coble in the foreground. (*National Maritime Museum*)

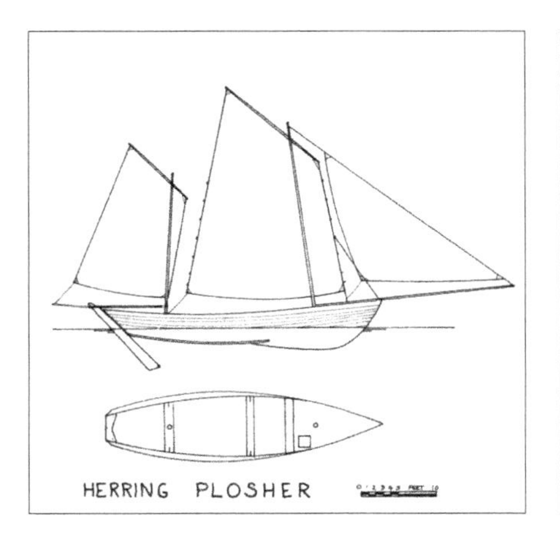

The mule *Dora Ann* at Scarborough *c.* 1900. (*National Maritime Museum*)

crew member having to bale out continuously. In contradistinction to this, 'ploshing' was the act of beating the bows of the coble with a rope at the waterline to drive the herring into the net. It has also been used as the term for the process of packing fresh herring into barrels. I have also heard that 'plosher' comes from the sound that the sea made against their hulls when underway. Whatever the reason, the crews of the ploshers were held in high esteem as they were regarded as fine seamen or 'seadogs' in the true sense of the word. As was common along the coast, different places referred to boats by differing terms: a mule in Whitby was called a herring coble in Flamborough, and was further called a 'hulk' or 'burro' there, while called a 'jinnie' in Bridlington. Designs differed – a Staithes plosher had more 'kilp' (tumble-home) than a Scarborough boat. The same could be said for all around the British coasts, where builders and fishermen alike were terribly insular in their response to things from other areas. Has this changed in the modern world?

SCARBOROUGH, FILEY & BRIDLINGTON

Here the coast is slightly gentler. Scarborough and Bridlington grew in the eighteenth century as resorts, and the harbours were built in the next century. Like Whitby, the arrival of the railway in 1847 opened up the market for fresh herring. Both became major fishing ports. Filey, with no harbour, also retained its long association with fishing, the fishermen sticking to their cobles. In 1866, there were sixty-four cobles, and forty-five in 1920. Scarborough and Bridlington, with their deep-water harbours, were home to fleets of keelboats as well as cobles. Bridlington was so busy in the season with Scots' boats that it was possible to walk across the harbour from the south pier to Crane Wharf without getting one's feet wet. Here, too, the boats were said to be bigger than anywhere else on the coast and were the first to have a cuddy for the fishermen to sleep in.

Flamborough Head, between Filey and Bridlington, is notorious as the graveyard of many an unfortunate sailing vessel. On each side of the point, there is a landing for cobles. Cobles have been beached out here for centuries, and hauled up the steep slipway by chain and drum. Traditionally, owners kept a coble on each side so that they could always get out whatever the weather. Even today, the north landing, which is the main one, is a quiet cove that seems to reflect the peace of country life. This picture would alter radically on a winter's night, with the deck heaving at all angles, and the wind blowing straight from the Arctic. Yet in these conditions the boats still sail out in their quest for a living from the sea. It is easy for us to dwell nostalgically on the great age of the herring, but sometimes it is perhaps best left alone. These boats were workhorses, tried and tested through generations of work, developed specifically for their task, yet many were lost in the sudden gales that sprang up from nowhere. In Bridlington, on 10 February 1871, an unexpected south-easterly gale caused thirty boats to be wrecked and seventy men to be drowned. Many of these vessels were heavily laden colliers, and they sank because of overloading. Samuel Plimsoll used these events to get a law passed making it compulsory for all ships to have a mark on their sides, showing maximum load waterlines, and hence limiting the weight of the

Various boats in Scarborough harbour. Note the two clinker double-enders in the foreground.

The Scarborough plosher *Enid*, SH16, under sail. (*National Maritime Museum*)

Motorised cobles on the beach at Flamborough Head in 1995. (*Author*)

cargo. For more than 120 years since that gale, the nearest Sunday to 10 February has been set aside in Bridlington as Fishermen's Sunday.

A final fact about this coast that is little known: Scarborough was at one time home to the British tuna fishery. Small, open, double-ended boats, 15–20 feet overall, often rowed 20 miles offshore, lining for fish. In 1933, Henry Wray caught an 851 lb tuna (British record) from his small tender *Success*.

MOTORISATION

The introduction of the motor had serious repercussions on the coble fishermen, as it did in all parts of the country. It was an invention that was to change the whole way of fishing, and one that was to lead to serious shortages of the fish through overfishing. Steam had begun this change, but the engine was to compound the problem. Before the First World War, only cobles working from harbours could have engines fitted because of the flat-bottom of the coble, which made it impossible to fit a propeller and beach the boat. Ways were soon found after the war to solve this. Some boats were built with tunnels for the propeller and shaft to run in. Others had a universal joint on a shaft, so that when it was beached, the prop would lift out of the way into a housing in the hull. To begin with, the rig was kept because of the unreliability of the motors. Engines became more powerful, and more reliable. The hull became fuller and beamier as it did not need to have sailing qualities. Within several years some had foredecks, and then wheelhouses. Northumberland boats tend not to have any shelter, although, as mentioned, they do have a canvas cover or cuddy set on a boom for shelter over the front half or more of the boat.

With the demand for greater catches, many Yorkshire fishermen bought in old Scottish motor fifies cheaply in the same way that they had bought in the old fifies and zulus in the 1880s for £50. These larger boats, alongside the larger mules, led to more double-ended keel boats being built locally. Competition was fierce. Today the result of this is apparent. On a recent visit to Staithes, I counted only three cobles out of sixteen open boats. The others were keelboats. The same was so for Scarborough and Bridlington. It seems that only where beaching is still a must do the cobles command preference.

In 1965, the final insult to the tradition of coble building occurred. From a 17-foot coble that had been built by Arg Hopwood of Flamborough came a fibreglass version. Two years later came a 24-foot model. Although one can say that it is a proud moment for the coble to be epitomised in a GRP version, there is altogether something sad when one considers the long tradition of wooden coble building. Nevertheless, I suppose that is progress, but I still think that in our progression we have forgotten how to live. Perhaps the escalating cost is relevant. In 1914, a coble cost £1 per foot to build. In 1951 this figure had risen to £51, by 1975 to £184, and then by 1982 a coble cost a staggering £620 per foot. In an age when cost and profitability are the only important matters, it is surprising that anyone goes to sea at all.

SURVIVING BOATS

Today, although there are lots of motor cobles working all along this coast, and indeed in other parts of the country, there are few examples of sailing cobles afloat.

Gratitude, the last boat built by Hector Handyside at Harrison's Yard in 1976 ,belonged until recently to Dave Wharton of Whitby, and now she is in Scarborough. Dave has taken her all over the place, including to Australia, although I hasten to add that he didn't sail her there. The oldest cobles are *Little Lady*, built 1899; *Venus*, built *c.* 1900 by Cambridge's, and now owned by the Hartlepool Museum; and *Sweet Promise*, built in 1906, again by Cambridge's.

At Bridlington, the visitor can see a coble sitting in the harbour. She is the *Three Brothers*, built by Messrs Baker & Siddall in 1912. Her overall length is 40 feet, and ram length is 27 feet. She has a small forecastle with a stove and supplies. Her cost was £75, excluding sails and ironwork. Sails would have cost another £25, and the ironwork about £15. Today, fully rigged as she was when working, she is owned by the Bridlington Harbour Commissioners and leased to the Bridlington Sailing Coble Preservation Society, who sail her during the summer. Another two Bridlington cobles, the *Kate & Violet* and *Kathleen*, the former built at Flamborough in 1911 by Arg Hopwood for the Hutchinson family, are both now in Sunderland. Various others can be seen all along the coast.

Other sailing cobles in existence but not afloat are *Three Brothers Grant*, HL154, in the Hartlepool Museum, and *Seaview* at the Kirkleatham Museum. There seems to be only one true mule in existence, *Blossom,* built in Berwick-upon-Tweed for a Burnmouth fisherman in 1887. She worked in Seahouses until 1969, and is now in the hands of the Tyne & Wear Museums at Newcastle awaiting refurbishment. But perhaps the most famous coble of all is the 21.5-foot boat that Grace Darling made her famous rescue in 1838, now on show in the museum of her name at Redcar. This was a normal fishing coble that was used on the lighthouse until it was sold in 1857, after which it fished out of Seahouses until 1873. It was then sold to Col. John Joicey, whose wife presented it to the museum. This, then, must surely, be the oldest of a class of boat that has evolved alongside man and nature for generations; a type unique among the fishing boats of Europe and a type that, while many designs disappear as fast as the fish they chase, has continued to flourish as a workboat. And too, because of the way it has grown out of the beaches upon which it has served, is likely to continue to do so for generations to come.

CHAPTER 18

The New Age of Steam and Motor Power

These newfangled things will never catch on.
—Fisherman's observation *c.* 1910

Steam power first entered the fishing industry during the 1850s, when smacks and luggers were towed into and out of harbours by steam-powered paddle tugs. A decade later, smacks were towed by tugs around the fishing grounds with their trawl gear down. However when, in 1877, one tug owner, William Purdy of North Shields, decided to use his tug as a trawler without the need for the smack, the incredulous fishermen looked on in astonishment, believing that this 'absurdly new-fangled and impractical' idea was a waste of money. Purdy's tug *Messenger* produced surprisingly good results on his first trip, and, after his second successfully showed a handsome profit, the idea was immediately copied by up to fifty tug owners.

The smack owners felt relatively unconcerned, deeming these experiments to be unreliable and unseaworthy, and no threat to their livelihood. But they were shocked when, in 1881, the first purpose-built steam trawler, the *Zodiac*, was launched by the Great Grimsby Steam Trawling Company. Its immediate success was apparent from its catches, which were four times bigger than those of the smacks. The *Zodiac* was capable of steaming at nine knots, burning 4 tons of coal a day, yet still her profits were greater than those of the smacks. This produced instant changeover within the fleets.

Like floodgates opening, there was this sudden rush to build wooden steamers. The result was that within a couple of years the numbers of trawlers had risen sharply, and the catches increased as rapidly. Although no one recognised it at the time, this signalled the beginning of the end for the herring fishing. The necessary investment of £3,500 for a 100-foot steamer represented the equivalent of about 2½ first-class smacks, yet owners were eager to join in with the 'fashionable' steamers. Within ten years of Purdy's experiments, 1,000 steam trawlers were built in Britain.

North of the border the story was the same. The first steam trawler built appears to have been the *Rob Roy*, LH92, launched at Leith in 1882. Although, at 56 feet overall, she retained the fifie look with the rig, and was built specifically for long-lining; she had a two-stage expansion engine that gave her a speed of eight knots. The

WATERWITCH LH961
EARLY SCOTTISH STEAM DRIFTER

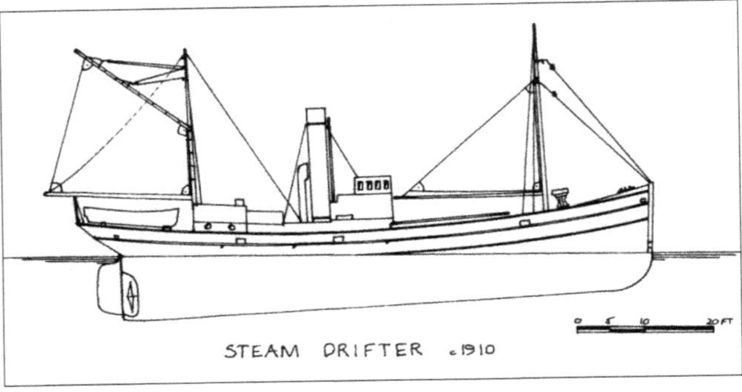
STEAM DRIFTER c1910

second trawler came later that year, again at Leith, when the 87-foot *Hawk* was launched. Lowestoft, Great Yarmouth, Ramsgate and Brixham were slower to adopt steam, yet soon they had their own fleets. On the west coast, the steamers arrived in Fleetwood and Milford Haven, while a few worked out of Swansea and Liverpool. By the turn of the century, these vessels were about 120 feet overall and 20 feet in the beam. Each had a triple-expansion engine of some 50–60 hp that was situated well aft. This is turn meant that the centre of gravity was, too, well aft and thus made life in the forward accommodation at times unbearable. In a rough sea, the bows travelled a long vertical distance because of this great pivot length, and anything not fixed down tended to remain in the air as the great bow plunged downwards after its ascent up a wave. Then, as the bow lifted again, the objects which were left in mid-air and which had only just begun their own freefall descent, met the rising boat with a loud, bone-crunching thump. To supplement life in this living hell, water poured in nearly continuously from every direction. Yet, by 1909, there were 1,336 trawlers in England and Wales, 514 of which were based at Grimsby and 449 at Hull. Scotland had another 278, although fishermen north of the border tended to be more sceptical about trawling, and retained the belief that it destroyed more than it benefited.

THE PIPE STALKIES

Steam was delayed in its effect upon the drifting fleets, the transition not beginning until the building of the first drifter in Lowestoft in 1897. Although experiments had been done upon Scottish drifters previously, the *Consolation*, LT718, represented the first real change in attitudes. She was built in the yard of Chambers & Colby, and appeared as any oilier sailing drifter did, the only exception being the tall, thin funnel. Like the trawlers, the drifters retained the rig because of the unreliability of the engines. Boats

The steam drifter *Ugie Brae*, PD 231, entering Peterhead. (*Michael Craine collection*)

were also wooden until the beginning of the twentieth century, when riveted iron vessels were built, and later to give way to steel. Within a couple of years of the launching of *Consolation*, there were 100 such vessels working out of East Anglia. By 1913 there were more than 1,800 steam drifters, all about 90 feet overall and 19 feet in the beam, and costing about £2,000. When riding the nets they set the mizzen as in the days of sail, yet the foremast stayed down nearly all the time, being raised only when in harbour as it housed the lifting derrick for hauling the catch ashore, or when the engine played up.

Unlike the trawlers, the accommodation for the ten crew was aft; otherwise, there were distinct similarities between the two types. The drifters only had some three feet of freeboard, with low counter sterns and high bows. The engine casing protected the engine room, with the galley at its after end. Forward of this was the tall funnel from which these boats got their Scotch nickname of 'pipe stalkie'. Further forward again was the 'telephone kiosk' wheelhouse at the front edge. Below decks the fish hold and net rooms were forward.

The year 1913 was the most successful on record for the autumnal herring fishers. 1,359,213 cran of herring was landed at Yarmouth and Lowestoft that year, a cran officially

being about 1,000 fish, although in reality it was often nearer 1,300. In fourteen weeks, Yarmouth received 854 million fish and Lowestoft 436 million. Some 6,000 Scots herring lassies gutted and packed a catch that was worth in excess of £1.5 million. There were 1,776 boats working from the two ports, 1,163 of which were Scottish – 854 steamers, 100 motor and 209 sail. Average earnings of these Scots boats were, respectively, £794, £365 and £235 per type. Sailboats did manage to survive in Scotland because they were so much cheaper to run, but numbers did fall. From about 2,000 luggers in 1905, numbers declined to about 1,500 in 1910, but within a few years these were fitted with petrol/paraffin engines, which ensured their survival for another decade. When hostilities arrived in 1914, some 1,502 drifters and 1,467 trawlers were requisitioned by the Admiralty for mine-sweeping and patrol duties. 49 per cent of fishermen joined the navy aboard these boats, while others joined the army. Out of all the communities, fishermen contributed more to the war effort in terms of percentage. During the war, 394 fishing vessels were lost on naval service, of which 246 were trawlers, 130 were drifters and eighteen were Admiralty trawlers. Of the 2,058 men lost on active duty, most were fishermen.

Yet boats did continue to fish, and 439 fishermen were killed while working with the 675 fishing boats that were lost, 249 of these being smacks, 156 trawlers and 270 drifters. By the end of the war, numbers were seriously down, yet fishing began again in earnest. The decline, however, had already set in, and the price of fish had rocketed. With the coming of the Second World War in 1939, there were still 277 drifters working, of which the Admiralty took 200 into what became known as the 'kipper patrol'. Of the 816 trawlers that were requisitioned, 146 were lost, but many of these were fitted with engines. By the end of the war few steamers had survived, and today only the *Lydia Eva*, YH89, built by the King's Lynn Slipway Company in 1930, remains in her original style. Happily she survives under the ownership of the Lydia Eva and Mincarlo Trust and can normally be seen in Yarmouth or Lowestoft, where she is often open to the public.

More recently the 1912, Aberdeen-built drifter *Feasible* has come to light and now awaits restoration.

The picture of these fine boats belching forth black smoke into the sky and jostling for space alongside the unloading quay is one that most of us never saw. But when fishermen, scarred by years working at the nets, themselves reminisce about these craft, it surely says something about the vessels themselves. Their story is an evocation of that great silver age that sadly seems to have gone awry in today's bustling existence.

Above: Fishing boats alongside the pier at Bridlington.

Right: The steam herring drifter *Lydia Eva*. Owned by a trust, she is normally moored in Great Yarmouth and is open to the public. (*Lydia Eva & Mincarlo Charitable Trust*)

THE ADVENT OF THE MOTOR

The idea of fitting motors into fishing boats seems to have originated in Denmark where in 1895 a fishing boat had a motor installed that ran on paraffin. Five years later, an Esbjerg smack's small boat had an engine from Mollerup of Esbjerg fitted. But the idea never developed in the closing stages of the nineteenth century because steam vessels showed greater advantages over the primitive motors.

However, the first British vessel to be powered with an internal combustion engine was suitably called *Pioneer*, LT368. She was built by Henry Reynolds at his yard at Oulton Broad, Lowestoft, in 1901 at a cost of £1,600. Although built on smack lines with a normal rig, she had a 38 hp four-cylinder Globe Marine Gasoline engine from Philadelphia, USA, that itself cost £680. Within a couple of years, *Pioneer* produced some startling results, and in 1905 she earned £788.

Another *Pioneer* appeared in 1905 from Scotland, this one registered as ML30. She had been built at Anstruther on typical fifie lines, and measured 72 feet overall, 21 feet beam and 8 feet draught aft. At the recommendation of the Scottish Fisheries Board, she had a 25 hp single-cylinder four-stroke Dan motor fitted. The vessel cost £700, the engine installation £446 and a later steam capstan £114. Although the boat was generally considered to be underpowered, results from her first year's fishing were deemed a success.

At the same time, other vessels were being built in Lowestoft with motors fitted. One such open beach yawl was the 35-foot Sheringham whelk boat *Reaper*, YH34, which had a Gardner paraffin motor fitted. Her speed amazed watching fishermen and tug owners alike as she managed to out-pace a tug during her first trials. But over the next two years only a handful of new boats were built with motors, and only another handful of existing sailing craft adopted motors.

Ring-netters alongside the pier at Carradale *c.* 1935. (*Naomi Mitchison*)

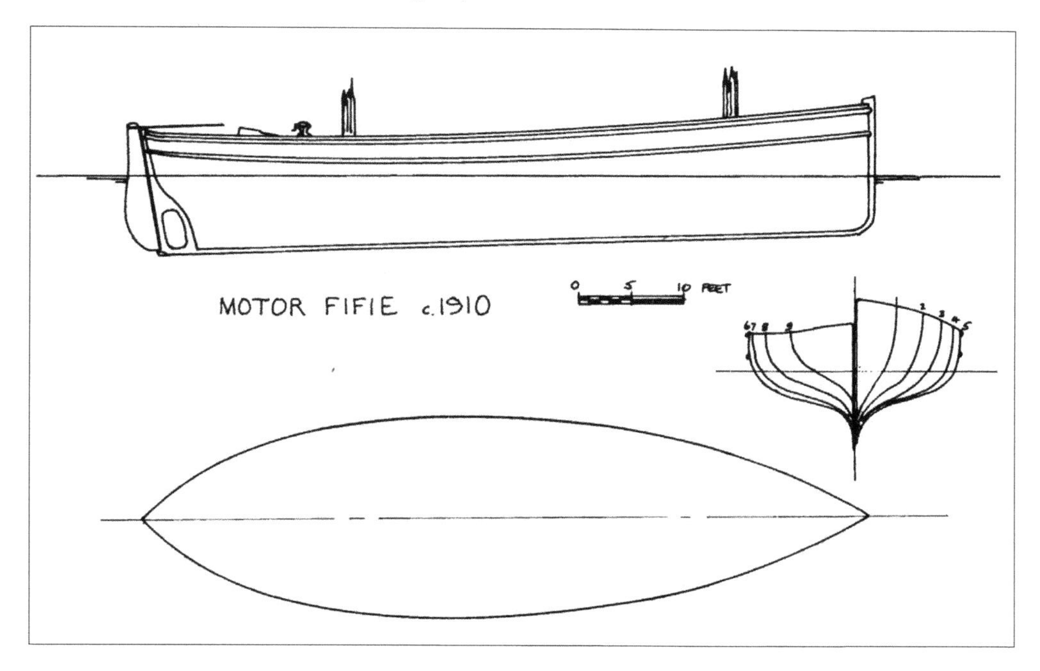

The first really successful Scottish conversion was in 1907 with the fitting of a 55 hp Gardner 3KM engine into the 1901-built fifie *Maggie Jane*, BK146, from Eyemouth. She proved to be a good sea boat, and was financially worthwhile. The Gardner brothers had made their first vertical oil engine at their premises at Cornbrook, Manchester, in 1894, but it wasn't until 1902–03 that they fitted a unit into a fishing boat. Between 1907 and 1911 however, some sixty-four British fishing boats received Gardner engines.

By 1902, four more local fifies followed the example of *Maggie Jane*. As well as fifies being converted, zulus too were having engines fitted, although with their raking sternposts they were never as successful. In 1909, the 48-ton Arbroath fifie *Ebenezer*, AH46, had a 60 hp Blackstone heavy-oil engine fitted.

Meanwhile in Lowestoft, Henry Reynolds built twenty-one herring drifters in 1970, twenty of which were steam powered, and the last one *Thankful,* LT1035, although built on the same lines as the others, had a 60 hp Brauer & Betts engine that was built locally.

In Ireland, the first motor-driven fishing boat was by Tyrrell & Sons of Arklow, who built *Ovoca* in 1908. Built for the Fisheries Department of the Board of Agriculture for Ireland, she had a 20 hp Dan engine that gave her a mean speed of 7 knots over a measured mile on her first trial in rough sea conditions. One unique point about *Ovoca* was that she had a canoe-stern, the first fishing boat, it is believed, to have one in Britain.

The first large boat motorised on the Moray Firth was in 1909, when the 42-ton zulu *Mother's Joy*, BF892, built in 1902, had a 60 hp Fairbanks-Remington engine installed. In Fraserburgh it was the 50-ton fifie *Vineyard* in 1908.

An Orkney Yole in Stromness as an example of an early motor conversion. (*Orkney Library Archive*)

On the west coast of Scotland, the first boat to have a unit fitted was in 1907 when, as we've already seen, the *Brothers*, CN97, owned by Robert Robertson, had a Kelvin 7.9 hp engine fitted. Later that year the *Lady Carrick Buchanan*, CN38, another similar Lochfyne skiff, had a 7.5 hp Thornycroft paraffin motor fitted, and by the end of the year there were ten boats on the west coast with motors. Because of the configuration of the sternpost of these skiffs, the engines were fitted on the starboard side, well away from the ring-net that was always shot on the port side.

And so the conversions and new-built vessels were eventually nearly all powered by motors. Although these motorised boats worked alongside the steam drifters and trawlers, it soon became apparent that hull shapes would alter, mainly because by the 1920s the rigs had disappeared now that the engines were more reliable. The figures tell the story: in 1919 there were 8,124 fishing vessels in Scotland, 324 of which were steam trawlers, 872 steam drifters, 1,844 motor boats and the remaining 4,058 were sailing craft, mainly of the small variety.

The first two of the new breed of boats were, as we've seen earlier, *Falcon* and *Frigate Bird* of Campbeltown, which were both fitted with Gleniffer 18–22 hp paraffin engines. These canoe-sterned types were regarded with suspicion to begin with, and three new decked versions of the Lochfyne skiffs that were added to the Campbeltown fleet in 1926 proved that the fishermen were utterly sceptical about Robertson's claims. However, within a couple of years, new canoe-sterned vessels built by Millers of St Monans entered the fleet, and local opposition finally crumbled before the 'new ways'.

Typical Scottish motorised fifes and early cruiser-sterned fishing boats at Portree, Skye.

On the east coast of Scotland, the new 'motor fifie' gained recognition as a perfected class of fishing boat in its own right. These vessels retained a sloping sternpost, albeit of a different angle, and they were used extensively for long-lining, shellfish catching and, of course, the herring. Later on, *Marigold*, a 50-foot fifie built by William Wood & Sons of Lossiemouth, was the first motor boat built specifically for seine-netting.

We've already seen how the sloping sternpost of the zulu was deemed unsuitable for motor conversion, yet, ironically, a range of smaller zulus – some will say not *true* zulus – was built in the 1920s and 1930s on the east coast. These were identical to the earlier zulus in all but length and the angle of the sternpost; they were generally about 40–50 feet overall, and had a 30 degree rake aft. Many came from Nobles yard in Fraserburgh, and various examples remain today, such as *True Love* and *Hirta*.

Nobles also developed a range of smaller motorized yoles, based on earlier fifie and zulu types. These yoles were all about 25 feet with a 9-foot 10-inch beam. They had a large dipping lugsail set on a mast of the same length as the hull, and were all fitted with motors – mostly Kelvins. They proved adaptable and up to fifty worked from the harbour at any one time. Both J&G Forbes and Thomas Summers built several, with the last clinker hull being built in Gardenstown about 1942. Today a handful remain in different stages of well-being.

The final stage in the development of the Scottish motor fishing vessel (MFV) appears to have come with the arrival of the J&G Forbes-built *Cutty Sark* in 1928 with its cruiser stern, although it can be said that Robertson had already started the change. Influence seems to have come from Denmark, where the cruiser-stern had

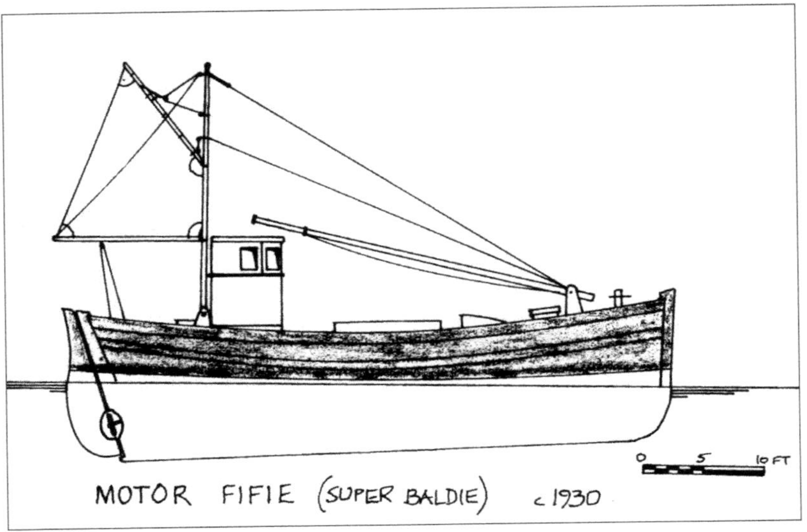

Above: The *Falcon* at Campbeltown soon after this innovative canoe-sterned vessel had arrived on the Clyde. (*Angus Martin*)

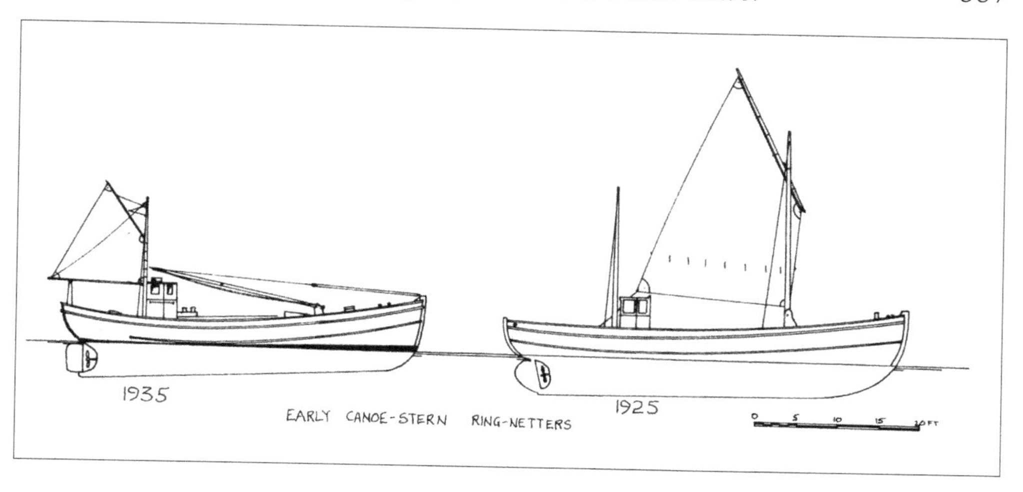

EARLY CANOE-STERN RING-NETTERS

'VERONICA' IN THE KILBRANNAN SOUND C.1936

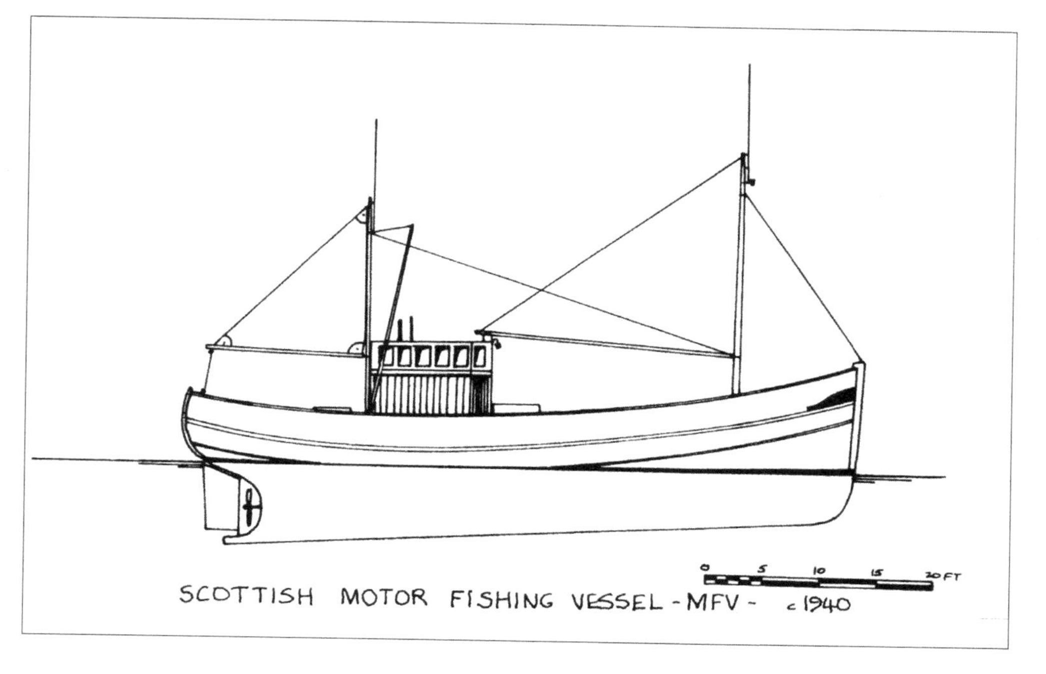

SCOTTISH MOTOR FISHING VESSEL - MFV - c1940

Motorised keelboats at Scarborough in 1995. Although retaining something of a mule's shape, they have a continuous keel instead of a ramplank.

been incorporated into fishing boats for centuries. Indeed, some say that Robertson was so impressed with Swedish double-enders that he was inspired to build *Falcon* and *Frigate Bird*. However, *Cutty Sark* was an all-round fishing boat, being able to ring-net, drift-net, seine, or longline by adapting her gear. By the Second World War, boats were having the new diesel engines fitted, and vessels as we know them today were fast replacing the older ones. Although differences remained between boats built from different yards, the boats retained similarity through the necessities of their work. West coast ring-netters retained the canoe-stern and were normally beautifully varnished n the same way that they had earlier varnished their Lochfyne skiffs, while the east coast boats were gaily painted. The one thing in common was that they were all kept in fine condition, as they had been for centuries. Sizes advanced, and eventually steel boats began replacing wooden ones, and shapes changed, transoms arrived, electronic aerials decorated the upper decks, and the fish disappeared. Today's boats will never tug the heart as those of yesterday and neither are today's fishermen comparable. Hunting skills have been exchanged for TV screens, listening and watching have given way to electronic gear, sails and oars have been replaced by perfected motors. Fishing simply is not what it used to be! Although there are many advantages to be had from this, the loss of Britain's fishing heritage through the enforced destruction of fishing boats is indeed sad. One thing is certain: they are lost forever.

Right: An early motorised canoe-sterned ring-netter with a Lochfyne skiff alongside at Tarbert *c.* 1930. (*National Maritime Museum*)

Below: Fred's Last, a typical beach boat working from Dunwich in 1996.

Providence, BH70, at Amble. Known as the 'Banana Boat' because of her shape, this coble was built by Victor Henderson for his own use in the 1950s. He built the keel boat *Endurance* from Whitby and at least three cobles and one steel trawler. *Providence* fished for salmon with 'T' nets. In the background are three cobles: two built by Harrisons (LHS & RHS) and between them is *Sweet Promise*, built by Cambridge in 1902 and owned by Hector Handyside. (N.B. 'BH' stands for Blyth.)

Appendix 1

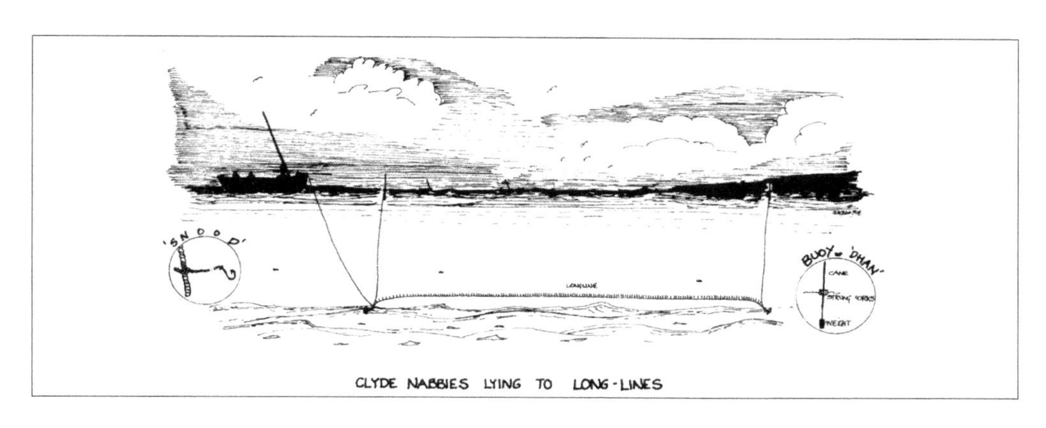

CLYDE NABBIES LYING TO LONG-LINES

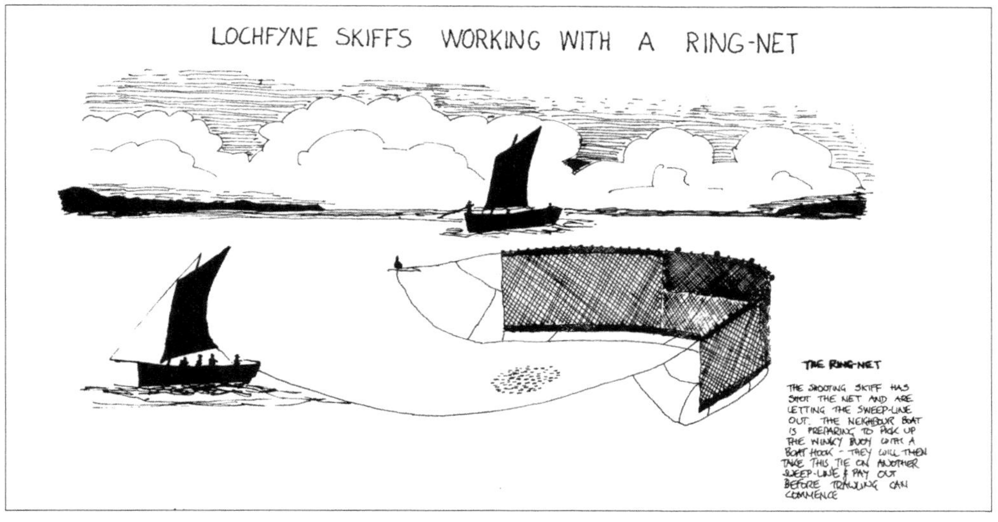

LOCHFYNE SKIFFS WORKING WITH A RING-NET

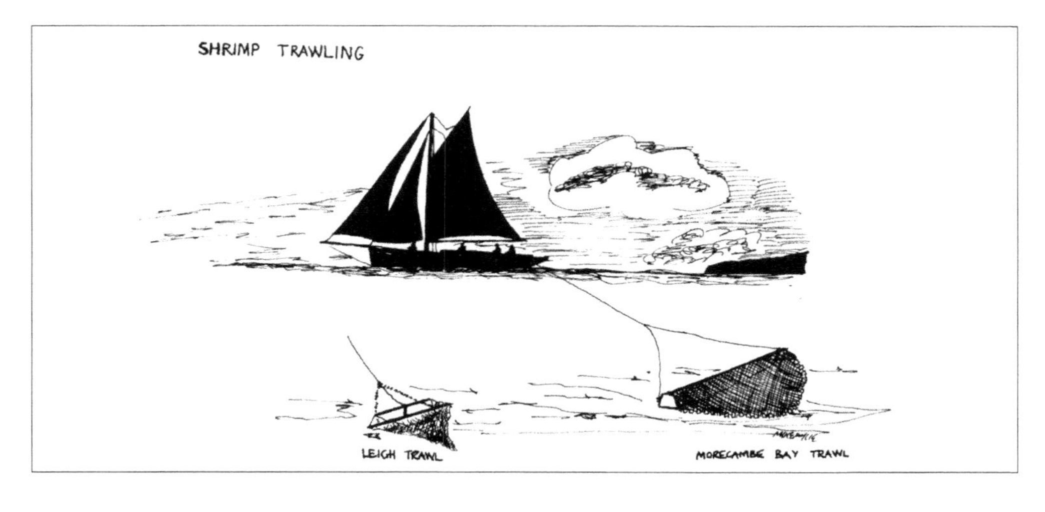

SHRIMP TRAWLING

OYSTER DREDGING

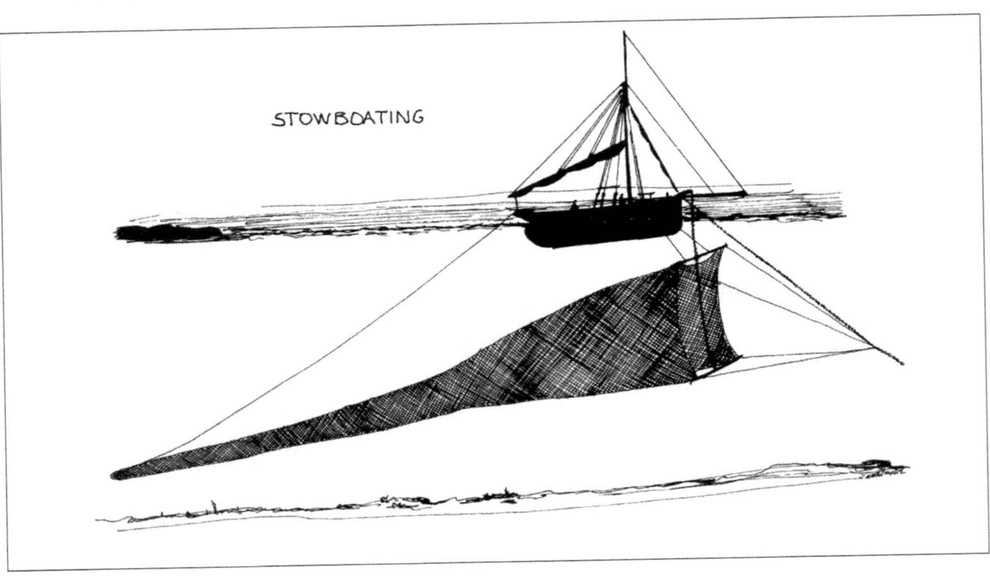
STOWBOATING

DRAGGING A TRAWL-NET

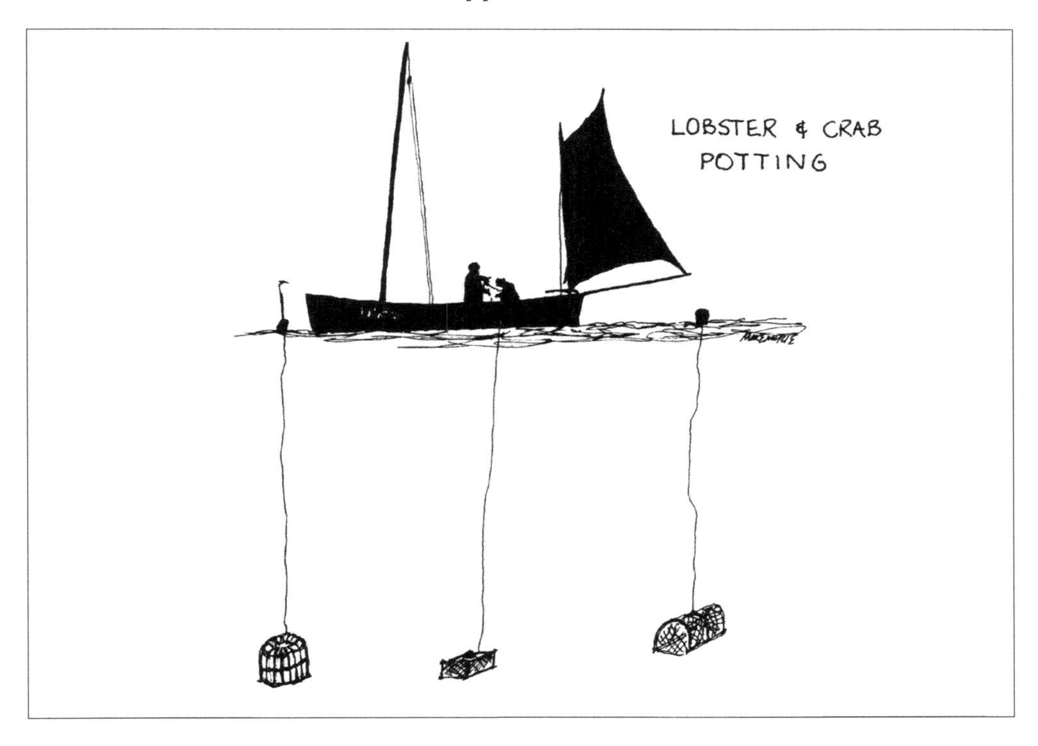
LOBSTER & CRAB
POTTING

SEINING FROM THE SHORE

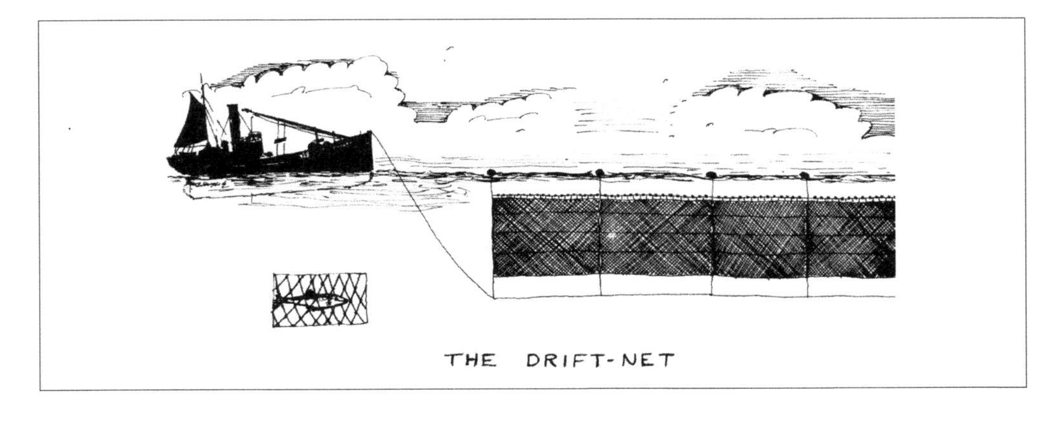
THE DRIFT-NET

Appendix 2
Numbers of Fishing Boats and Fishing Harbours

ENGLAND AND WALES 1872

Port	*Port Letters*	*First class*	*Second class*	*Third class*
Carlisle	CL	0	18	1
Maryport	MT	4	48	2
Workington	WO	0	14	0
Whitehaven	WA	17	15	34
Lancaster	LR	5	122	101
Fleetwood	FD	53	27	0
Preston	PN	3	46	0
Liverpool	LL	35	137	10
Runcorn	RN	3	35	17
Beaumaris	BS	2	58	68
Caernarfon	CO	14	115	36
Aberystwyth	AB	2	6	75
Cardigan	CA	0	54	35
Milford Haven	M	13	159	112
Llanelli	LA	2	0	0
Swansea	SA	3	184	0
Cardiff	CF	0	1	0
Bristol	BL	1	3	42
Bridgwater	BR	0	69	25
Barnstaple	BE	1	76	35
Bideford	BD	8	117	11
Padstow	PW	0	83	36
Hayle	HE or SS	56	162	91
Penzance	PZ	41	466	73
Falmouth	FH	8	280	165
Truro	TO	0	28	111
Fowey	FY	16	324	148
Plymouth	PH	66	290	16
Dartmouth	DH	136	270	93
Teignmouth	TH	1	108	172
Exeter	E	0	210	103
Lyme Regis	LE	1	37	119
Bridport	BT	0	21	51
Weymouth	WH	3	83	253
Jersey Guernsey	J GU }	72	583	147
Poole	PE	0	61	16
Southampton	SU	17	141	102
Cowes	CS	5	182	104
Portsmouth	P	18	247	21
Littlehampton	LI	2	156	36
Shoreham	SM	18	174	132
Newhaven	NN	8	52	97
Rye	RX	28	97	73
Folkestone	FE	9	89	72
Dover	DR	21	36	60
Deal	DL	8	79	27
Ramsgate	R or RE	139	43	0

Faversham	F or FM	28	201	30
Rochester	RR	9	124	1
London	LO	149	56	0
Maldon	MN	16	261	127
Colchester	CK	132	250	40
Harwich	HH	17	60	49
Ipswich	IH	6	36	1
Woodbridge	WE	15	99	51
Lowestoft	LT	269	258	15
Yarmouth	YH	493	462	47
Wells	WS	14	66	5
King's Lynn	LN	11	100	0
Wisbech	WI	0	16	0
Boston	BN	5	89	7
Grimsby	GY	330	16	25
Goole	GE	6	1	0
Hull	H	313	340	1
Scarborough	SH	109	153	0
Whitby	WY	20	277	4
Middlesbrough	MH	0	55	0
Hartlepool	HL	1	172	6
Sunderland	SD	5	103	0
Newcastle-upon-Tyne	NE	0	0	1
South Shields	SSS	0	44	2
North Shields	SN	2	489	6
Berwick	BK	60	599	6
TOTAL		*2,849*	*9,933*	*3,346*

TOTAL NUMBERS OF BOATS 16,128
TOTAL TONNAGE 100,3223

SCOTLAND 1872

Port	*Port Letters*	*First class*	*Second class*	*Third class*
Leith	LH	137	407	1
Borrowstones	BO	0	38	2
Alloa	AA	1	0	0
Perth	PEH	0	9	0
Kirkcaldy	KY	135	585	3
Dundee	DE	40	135	26
Arbroath	AH	7	146	4
Montrose	ME	46	291	19
Aberdeen	A or AN	36	463	2
Peterhead	PD	529	277	348
Banff	BF	541	607	9
Inverness	I or INS	90	1,923	355
Wick	WK	313	1,180	24
Kirkwall	K or KL	58	1,187	102
Lerwick	LK	66	1,613	701
Stornoway	SY	28	983	55
Campbeltown	CN	44	277	1
Glasgow	GW	10	42	27
Port Glasgow	PGW	0	27	17
Greenock	GK	22	1,657	337
Ardrossan	AD	7	178	1
Troon	TN	1	50	4
Ayr	AR	3	160	10
Stranraer SR		6	163	9
Wigton	WN	0	19	73
Dumfries	DS	0	93	5
TOTAL		*2,120*	*12,510*	*2,135*

TOTAL NUMBER OF BOATS 16,765

ISLE OF MAN 1872

Port	*Port Letters*	*First class*	*Second class*	*Third class*
Douglas	DOS of DO	227	82	66
Peel	PL			
Ramsay	RY			

TOTAL NUMBER OF BOATS 375

IRELAND 1872

Port	*Port Letters*	*First class*	*Second class*	*Third class*
Dublin	D	156	368	122
Wexford	WD	25	228	58
New Ross	NS	1	70	25
Waterford	W	19	179	39
Youghal	Y	3	47	33
Cork	C	61	361	398
Skibbereen	S	11	247	507
Tralee	T	11	183	434
Limerick	L	0	0	79
Galway	G	10	635	880
Westport	WT	2	34	881
Ballina	BA	0	1	191
Sligo	SO	1	191	396
Londonderry	LY	7	444	183
Coleraine	CE	0	120	40
Belfast	B	43	198	45
Newry	N	35	194	148
Dundalk	DK	0	43	4
Drogheda	DA	1	46	12
TOTAL		*386*	*3,589*	*4,475*

TOTAL NUMBER OF BOATS 8,450

N.B. Different figures to those of the Register of General Shipping are given by the Coastguard: first class 372, second class 3,091, third class 5,551, total 7,914 (i.e. 536 fewer).

NUMBER OF FISHING BOATS

Year	*England & Wales*	*Scotland*
1872	16,503	16,765
1886	8,826	14,391
1900	7,190	N/A
1920	9,235	8,177
1930	7,269	5,969
1938	6,614	5,217
1950	7,192	5,222
1970 (over 40 feet)	1,063	1,079

Bibliography

A complete list of all books that touch on the subject of traditional fishing and the boats used by the coastal fishermen would fill pages, but the following list includes those that give more than just a localised picture. Each geographical area, however, does have one or more books written by local historians, which most certainly cover the local aspects of fishing in the past.

Anson, Peter, *Fishing Boats & Fisher Folk on the East Coast of Scotland* (J. M. Dent & Sons, 1930).

Anson, Peter, *Fishermen & Fishing Ways* (George Harrop & Co., 1932).

Bagshawe, J. R., *Wooden Ships of Whitby* (Horne & Son, 1933).

Benham, Hervey, *Once Upon a Tide* (Harrap & Co., 1955).

Benham, Hervey, *The Stowboaters* (Essex County Newspapers, 1977).

Benham, Hervey, *The Codbangers* (Essex County Newspapers, 1979).

Benham, Hervey, *Essex Gold* (Essex Record Office, 1993).

Butcher, David, *The Driftermen* (Tops'l Books, 1979).

Butcher, David, *Living from the Sea* (Tops'l Books, 1982).

Cameron, A. D., *Go Listen to the Crofters* (Acair, 1986).

Castleton, Frank, *Fisher's End* (F. Castleton, 1988).

Catching Stories: Voices from the Brighton Fishing Community (QueenSpark Books, 1996).

Combe, Derek, *The Bawleymen* (Pennat Books, 1979).Coull, James, *Sea Fisheries of Scotland* (John Donald, 1996).

Combe, David, *Fishermen from the Kentish Shore* (Meresborough Books, 1989).

Dade, Ernest, *Sail and Oar* (J. M. Dent & Sons, 1933).

Davies, Alun, *History of the Falmouth Working Boats* (A. Davies, 1989).

Dickinson, M. G., ed., *A Living from the Sea: Devon's Fishing Industry and Its Fishermen* (Devon Books, 1987).

Dunlop, Jean, *British Fisheries Society 1786–1893* (John Donald, 1978).

Dyson, John, *Business in Great Waters* (Angus & Robertson, 1977).

Finch, Roger, and Hervey Benham, *Sailing Craft of East Anglia* (Terence Dalton, 1987).

Goodlad, G. A., *Shetland Fishing Saga* (Shetland Times, 1971).

Gray, Malcolm, *The Fishing Industries of Scotland 1790–1914* (Oxford University Press, 1978).

Halcrow, A., *Sail Fishermen of Shetland* (T. & J. Manson, 1950).

Harris, Ken, *Hevva!* (Dyllansow, 1983).

Hawthorne, W. R. and D. J., *Sailing Ships of Mourne* (The Mourne Observer, 1971).

Hill, H. Oliver, *The English Coble*, monograph no. 30 (National Maritime Museum, 1978).

Jenkins, J. Geraint, *Inshore Fishermen of Wales* (Cardiff: University of Wales Press, 1991).

Kennerley, Elija, *The Old Fishing Community of Poulton-le-Sands*, monograph (Lancaster Museum).

Leather, John, *Smacks and Bawleys* (Terence Dalton, 1991).

Lloyd, Len, *The Lancashire Nobby* (North West Model Shipwrights, 1994).

MacCullagh, Richard, *The Irish Curragh Folk* (Wolfhound Press, 1992).

MacPholin, Donal, *The Drontheim* (D. MacPholin, 1992).
Manx Sea Fishing 1600–1990s (Manx Heritage Foundation, 1992).
Martin, Angus, *The Ring-Net Fishermen* (John Donald, 1981).
Martin, Angus, *Fishing & Whaling* (National Museums of Scotland, 1995).
Mathieson, Colin, *Wales and the Sea Fisheries* (National Museum of Wales & the Press Board of the University of Wales, 1929).
Noall, Cyril, *Cornish Seines and Seiners* (Bradford Barton, 1972).
Pain, E. C., *The Last of Our Luggers* (T. F. Pain & Sons, 1929).
Peak, Steve, *Fishermen of Hastings* (NewsBooks, 1985).
Robinson, Robb, *Trawling* (University of Essex Press, 1996).
Robinson, Robb, *History ofthe Yorkshire Coast Fishing Industry 1780–1914* (Hull University Press, 1987).
Sandison, Charles, *The Sixareen & Her Racing Descendants* (T. & J. Manson, 1954).
Scott, Richard, *The Galway Hooker* (Ward River Press, 1983).
Smylie, Mike, *Fishing Boats of Cornwall* (The History Press, 2009).
The Story of the Lydia Eva (The Maritime Trust, 1975).
Sutherland, Iain, *From Herring to Seine Net Fishing* (Wick Society).
Sutherland, Iain, *Wick Harbour & the Herring Fishing* (Wick Society).
Wilson, Gloria, *Scottish Fishing Craft* (Fishing News Books, 1965).

General books on the subject:

De Caux, J. W., *The Herring and the Herring Fishery* (Hamilton, Adams & Co., 1881).
Finch, Roger, *Sailing Craft of the British Isles* (Collins, 1976).
Glasspool, John, *Boats of the Longshoremen* (Nautical Publishing Co., 1977).
Greenhill, Basil, *The Archaeology of Boats & Ships* (Conway Maritime Press, 1995).
Hawkins, L. W., *Early Motor Fishing Boats* (L. W. Hawkins, 1984).
Holdsworth, E. W. H., *Sea Fisheries of Great Britain & Ireland* (Edward Stanford, 1883).
Holdsworth, E. W. H., *Deep Sea Fishing & Fishing Boats* (E. Stanford, 1975).
Leather, John, *Gaff Rig* (Adlard Coles, 1970).
Leather, John, *Spritsails and Lugsails* (Adlard Coles, 1979).
March, Edgar, *Sailing Drifters* (Percival Marshall & Co., 1952).
March, Edgar, *Sailing Trawlers* (Percival Marshall & Co., 1953).
March, Edgar, *Inshore Craft of Britain*, vols 1 and 2 (David & Charles, 1970).
McKee, Eric, *Working Boats of Britain* (Conway Maritime Press, 1983).
Simper, Robert, *British Sail* (David & Charles, 1977).
Simper, Robert, *Beach Boats of Britain* (Boydell Press, 1984).
Smylie, Mike, *The Herring Fishers of Wales* (Gwasg Carreg Gwalch, 1998).
Smylie, Mike, *Herring: A History of the Silver Darlings* (Tempus, 2004).
Thurston Hopkins, R., *Small Sailing Craft* (Philip Allan, 1931).
Warrington Smyth, *H., Mast & Sail in Europe and Asia* (W. Blackwood, 1929).
White, E. W., *British Fishing-Boats & Coastal Craft* (Science Museum, 1950).

Government papers and other sources:

Buckland, Frank, and Spencer Walpole, 'Report on the Sea Fisheries of England & Wales' (1878).
Buckland, Frank, Spencer Walpole, and Archibald Young, 'Report on the Herring Fisheries of Scotland' (1878).
Mariner's Mirror, various extracts.
The Old & New Statistical Accounts for Scotland.
Washington, Capt., 'Report on the Loss of Life, and on the Damage caused to Fishing Boats on the East Coast of Scotland, in the Gale of the 19th August 1848' (1849).

Index

Place Names

Boat Types and Names

People